Renewable Economies in the Arctic

This book offers multidisciplinary perspectives on renewable economies in the Arctic and how these are being supported scientifically, economically, socially, and politically by the Arctic states.

The economic development of the Arctic region is witnessing new, innovative trends which hold promise for the sustainable development of the region. This book discusses the emerging forms of renewable economies to understand where intellectual and technological innovations are being made. It draws on the expertise of scholars from across the Arctic and provides the reader with a foundation of knowledge to identify the unique challenges of the region and explore opportunities to unlock the immense potential of renewable resources to boost the region's economy. This book offers a holistic Arctic perspective against the backdrop of prevailing social, economic, and climatic challenges.

With critical insights on the economic state of play and the role of renewable resources in the development of the Arctic region, this book will be a vital point of reference for Arctic scholars, communities, and policy makers.

David C. Natcher is a Professor in the Department of Agricultural and Resource Economics at the University of Saskatchewan, Canada. Trained as a cultural anthropologist, he conducts research in the areas of Indigenous social-economies and sustainable development in the Arctic.

Timo Koivurova is a Research Professor at the Arctic Centre, University of Lapland, and has a multidisciplinary specialisation in Arctic law and governance but has also conducted broader research on multi-level governance.

Routledge Research in Polar Regions
Series Editor: Timothy Heleniak,
Nordregio International Research Centre, Sweden

The Routledge series in Polar Regions seeks to include research and policy debates about trends and events taking place in two important world regions: the Arctic and Antarctic. Previously neglected periphery regions, with climate change, resource development and shifting geopolitics, these regions are becoming increasingly crucial to happenings outside these regions. At the same time, the economies, societies and natural environments of the Arctic are undergoing rapid change. This series seeks to draw upon fieldwork, satellite observations, archival studies and other research methods which inform about crucial developments in the Polar regions. It is interdisciplinary, drawing on the work from the social sciences and humanities, bringing together cutting-edge research in the Polar regions with the policy implications.

Collaborative Research Methods in the Arctic
Experiences from Greenland
Edited by Anne Merrild Hansen and Carina Ren

Greenland's Economy and Labour Markets
Edited by Laust Høgedahl

Young People, Wellbeing and Placemaking in the Arctic
Edited by Florian Stammler and Reetta Toivanen

Renewable Economies in the Arctic
Edited by David C. Natcher and Timo Koivurova

Indigenous Peoples, Natural Resources and Governance:
Agencies and Interactions
Edited by Monica Tennberg, Else Grete Broderstad, Hans- Kristian Hernes

For more information about this series, please visit: www.routledge.com/Routledge-Research-in-Polar-Regions/book-series/RRPS

Renewable Economies
in the Arctic

Edited by
David C. Natcher
Timo Koivurova

Routledge
Taylor & Francis Group
LONDON AND NEW YORK

First published 2022
by Routledge
2 Park Square, Milton Park, Abingdon, Oxon OX14 4RN

and by Routledge
605 Third Avenue, New York, NY 10158

Routledge is an imprint of the Taylor & Francis Group, an informa business

British Library Cataloguing-in-Publication Data
A catalogue record for this book is available from the British Library

Library of Congress Cataloging-in-Publication Data
A catalog record has been requested for this book

ISBN: 978-1-032-00030-5 (hbk)
ISBN: 978-1-032-00034-3 (pbk)
ISBN: 978-1-003-17240-6 (ebk)

DOI: 10.4324/9781003172406

Typeset in Times New Roman
by KnowledgeWorks Global Ltd.

Contents

Figures

Tables

Contributors

Jón Árnason
Matis
Iceland

Bradley Barr
University Centre of the Westfjords
Iceland

Ruth Beer
Emily Carr University of Art and
    Design
Canada

Paul Bowles
University of Northern British
    Columbia
Canada

Karin Buhmann
Copenhagen Business School
Denmark

Dorothée Cambou
University of Helsinki
Finland

Catherine Chambers
University Centre of the Westfjords
Iceland

Ken Coates
University of Saskatchewan
Canada

David Cook
University of Iceland
Iceland

Glen Coutts
University of Lapland
Finland

Sigridur Dalmannsdottir
Norwegian Institute of Bioeconomy
    Research
Norway

Herminia Din
University of Alaska Anchorage
USA

Níels Einarsson
Stefansson Arctic Institute
Iceland

Silje Elde
Norwegian Institute of Food,
    Fisheries and Aquaculture
    Research
Norway

Florent Govaerts
Norwegian Institute of Food,
    Fisheries and Aquaculture
    Research
Norway

Hilde Halland
Norwegian Institute of Bioeconomy
 Research
Norway

Rakel Halldórsdóttir
Matis
Iceland

Kristin Hansen
Norwegian Institute of Food,
 Fisheries and Aquaculture
 Research
Norway

Morten Heide
Norwegian Institute of Food,
 Fisheries and Aquaculture
 Research
Norway

Theresa Henke
University of Iceland Research
 Center of the Westfjords
Iceland

Óli Þór Hilmarsson
Matis
Iceland

Carin Holroyd
University of Saskatchewan
Canada

Maria Huhmarniemi
University of Lapland
Finland

Shawn Ingram
University of Saskatchewan
Canada

Timo Jokela
University of Lapland
Finland

Brooks Kaiser
University of Southern Denmark
Denmark

Ögmundur Knútsson
Directorate of Fisheries
Iceland

Timo Koivurova
University of Lapland
Finland

Matthias Kokorsch
University Centre of the Westfjords/
 Stefansson Arctic Institute
Norway

Olaf Kuhlke
University of Minnesota Duluth
USA

Ingrid Kvalvik
Norwegian Institute of Food,
 Fisheries and Aquaculture
 Research
Norway

Patrick T. Maher
Nipissing University
Canada

Dieter K. Müller
Umeå University
Sweden

David Natcher
University of Saskatchewan
Canada

Bjørg Helen Nøstvold
Norwegian Institute of Food,
 Fisheries and Aquaculture
 Research
Norway

Martin Mohr Olsen
Danish Technical University and
  University of the Faroe Islands
Faroe Islands

Páll Gunnar Pálsson
Matis
Iceland

Timothy J. Pasch
University of North Dakota
USA

Barry Costa Pierce
University of New England
USA

Greg Poelzer
University of Saskatchewan
Canada

Outi Rantala
University of Lapland
Finland

Ólafur Reykdal
Matis
Iceland

Rune Rødbotten
Norwegian Institute of Food,
  Fisheries and Aquaculture
  Research
Norway

Norma Shorty
Yukon University
Canada

Anna-Sofie Hurup Skjervedal
Allaffeqarfimmi pisortaq - Head of
  Secretariat
Greenland

Chris Southcott
Lakehead University
Canada

Mark Stoddart
Memorial University of Newfoundland
Canada

Trent Sutton
University of Alaska Fairbanks
USA

Gunnar Thór Jóhannesson
University of Iceland
Iceland

Gunnar Þórðarson
Matis
Iceland

Eivind Uleberg
Norwegian Institute of Bioeconomy
  Research
Norway

Svetlana Usenyuk-Kravchuc
Tomsk State University
Russia

Þóra Valsdóttir
Matis
Iceland

1 Introduction

Renewable economies in the Arctic

David C. Natcher and Timo Koivurova

Introduction

Arctic economic development has long been synonymous with resource extraction. For centuries, the Arctic has been exploited for its vast minerals, fisheries, marine mammals, water, oil, gas, and timber resources. While non-renewable resource extraction has created considerable wealth for some, the extraction of non-renewable resources has also left a wake of devastation in Arctic ecosystems and has threatened the wellbeing of Indigenous and other Arctic peoples who are left to bear the costs of past developments. In Canada, Justice Thomas Berger (1977, p. 123) forewarned nearly a half-century ago that "[i]t is a self-deception to believe that large-scale industrial development [will] end unemployment and under-employment of people in the North. We have never fully recognized that industrial development has, in itself, contributed to social, economic, and geographical dislocation." Despite these warnings, non-renewable resource extraction continues to hold a prominent role in the development strategies of Arctic states and continues to be promoted by those of influence as the most expedient route to improving the socio-economic conditions of Arctic communities.

While non-renewable resource extraction remains the economic linchpin of most Arctic states, local communities across the Arctic are making important strides in diversifying their economies through new and innovative ways, all of which hold great promise for the sustainable development of Arctic regions. There are a number of drivers that justify the transition from extractive to renewable Arctic economies. The ethical and environmental values are of extreme importance and are at the core of the United Nations' Sustainable Development Goals (SDGs). Adding additional urgency are the findings of the Intergovernmental Panel on Climate Change Special Report 4 that calls for action to limit global warming to no more than 1.5°C by 2030. In the Arctic, where the impacts of climate change are projected to be most extreme, local action is not only warranted but critical. In response to these global challenges, Arctic communities are making novel technological advancements in digital technologies, renewable energy capabilities,

DOI: 10.4324/9781003172406-1

sustainable food systems, and other social economy enterprises. These advances are creating a paradigm shift in Arctic development where new technical and entrepreneurial skills are emerging. This is particularly apparent in the many small- and medium-sized enterprises that have developed around the circular economy and industrial biotechnologies.

Notwithstanding the important advances being made in renewable economies, many parts of the Arctic continue to be challenged by under-developed and badly depreciated infrastructure. Arctic communities have also established tenuous relationships and ineffective interaction with research and development organisations. Some Arctic regions are also being affected by declining and limited skilled labor forces, inadequate education and training opportunities, and ineffective legislative and policy support. These and other constraints have made it difficult to diversify local economies and lesson their dependence on non-renewable resource extraction. While much of the transitionary burden will be on communities whose responses will be place-based and challenge-led, there will remain an important role for governments and international organisations to provide the necessary legislation and policy support. This will be particularly necessary to overcome the infrastructure and labor constraints noted above. Yet, there is also a need for governments to disseminate practical information that can inform local actions and decision making. The sharing of knowledge and business development experiences can inform the efforts of others in ways that may lead to scalable outcomes across the Arctic.

This commitment is well reflected in Arctic Council's Sustainable Development Working Group (SDWG) Strategic Framework (2017), where it has reaffirmed its commitment to supporting self-sufficient, resilient, and healthy Arctic communities. This commitment includes protecting the Arctic environment and to ensure the sustainable development of local livelihoods and the preservation of cultural traditions. These commitments are premised on the harmonisation of three core elements of Arctic sustainable development: social equity, economic development, and environmental protection. To facilitate this harmonisation, the SDWG has set out to compile practical knowledge that can be used by Arctic communities as they transition to more sustainable forms of development. This volume is an outcome of that commitment.

This volume was led by the SDWG's Social, Economic, and Cultural Expert Group (SECEG) in their capacity to provide practical knowledge that can be used to advance the social, economic, and cultural well-being of Arctic peoples through sustainable and integrated approaches to renewable economic development. This is the first in a planned series of publications that will address various themes of Arctic sustainable development. In this inaugural volume, we have drawn on the expertise of scholars from across the Arctic, asking them to explore the challenges and unique opportunities that exist for renewable economies in Arctic regions. This volume offers various perspectives on Renewable Economies in the Arctic and how these forms of economy are being supported scientifically, economically, socially, and politically. This volume is designed to provide the reader with a broad understanding of the current status and contribution of renewable resources to the Arctic economy and to create a foundation of knowledge on which to build policy, practice, and future research. We believe this can be an important contribution to scholarship, policy, and future economic development in the Arctic.

Our intention was to provide a holistic Arctic perspective, against the backdrop of prevailing social, economic and climatic related challenges. However, a challenge in achieving this holistic perspective is to adequately capture the enormous geographical, cultural, and economic complexity that defines the Arctic. The diversity of the Arctic means the different regions will have their own assets and challenges, which require appropriate place-based responses that elude generalisation. Given the enormity of this challenge, we make no claims of absolute representation. Rather, we intend only to provide a glimpse, albeit well-informed by a group of international experts in their respective fields, of the breadth of advances being made in the Arctic's renewable economic sectors. Undoubtedly some important examples have been excluded, and we encourage others to take up the challenge of bringing them to light for others to learn from. We hope the 14 chapters presented here serve that purpose.

Chapter outlines

Following this introductory chapter, Tim Pasche and Olaf Kuhlke explore digital creative entrepreneurship as it is impacted by data connectivity and communication infrastructure in remote communities of the North American Arctic. In addition to summarising details related to access, data speeds, and bandwidth in specific regions of the North, this chapter looks at values-based Arctic digital entrepreneurial curricular development, collaborative possibilities between Nunavut and Alaska, and cites opportunities and challenges for the Arctic's Indigenous creative economy. Similarities and differences between the United States and the Canadian Arctic in terms of opportunity and networking based on digital connectivity and cost of access are also explored. The chapter offers specific examples related to opportunities and barriers for Arctic small business development given variances in digital access. The chapter concludes with a number of important policy recommendations for government and industry.

In Chapter 3, Ken Coates and Carin Holroyd show that Arctic regions have a great deal to gain and, equally, much to lose from the twenty-first century-onslaught of new technologies. Led globally by such large firms as Samsung, Apple, Nokia, Panasonic, Alphabet/Google, Microsoft, Sun Systems, Cisco, TenCent, and Huawei, the recent technological and economic transformation has had profound effects on the global economy. New technologies like food factories, small modular nuclear reactors, autonomous vehicles, and drones are already disrupting existing industries, while high-speed Internet and satellite systems provide the foundation for radical changes in health care, education, governance, and business. Advocates of Artificial Intelligence (AI) argue that AI, along with 5G wireless capabilities, will have profound economic and social effects globally. Yet, the impact of these technologies on the Arctic is rarely mentioned in international discussions, and it remains unclear if the Arctic will see net benefits, dramatic losses, or mixed results from them. To some extent, the impact will depend on how Arctic states and communities approach the challenges and opportunities these new technologies represent.

In Chapter 4, Timo Jokela and his co-authors examine the impact of creative industries on renewable economies in the Arctic. Until recently, understanding of the frameworks of the creative industry and renewable economy has remained vague, especially in the field of art and design. In this chapter, the potential of art and design in promoting renewable economies is explored, using the concepts of ecosystem services, particularly cultural ecosystem services, and place-making as our theoretical and practical framework. This framework allows us to rethink the ways that creative entrepreneurs, businesses, and communities may collaborate, through art and design, in place-based development in the rapidly changing Arctic. By presenting case studies drawn from Alaska (United States), Canada, Finland, and Russia, the authors not only share experiences and findings but also suggest future lines of enquiry. The takeaway finding from this chapter is that creative, renewable economies in the fields of art and design can play an important role in the future of sustainable development in peripheral and remote areas in the Arctic.

Patrick Maher and his colleagues (Chapter 5) continue this discussion in their examination of Arctic tourism. Viewed through the lens of the "destination," this chapter explores the various ways tourism has developed, and continues to develop, in the Arctic. Many Arctic actors assume that the publicity of a specific place or region will lead to increased number of tourists and investors. But this has not poroven to be the case. Rather, an important success criterion for the tourism industry is to provide the right experience to the right visitor. For this to happen, the image of the Arctic alongside the realities of small communities must be addressed.

In Chapter 6, Chris Southcott reviews the role of the social economy in Nunavut, Canada. The social economy is made up of organisations in the not-for-profit sector that seek to enhance the social, cultural, health, economic, and environmental conditions of communities. These organisations continue to be an important part Nunavut's effort to resist an overdependence on extractive resource development in the region. While extractive resource industries will continue to be an important part of Nunavut economy, this chapter offers direction for how communities can leverage those resource revenues to hasten the transition to renewable economic development opportunities.

In Chapter 7, Martin Olsen argues that smaller institutions of higher learning within the Arctic must play a significant role in tackling the issues facing the region in a more *practical* sense. Olsen proposes that educational institutions should work with geographically embedded knowledge in a real work setting and focus on solutions relevant to the area and its stakeholders. However, for this to become a reality, changes to how many small Arctic universities currently operate must be made. To that end, an outline of an operational framework is offered that universities in the Arctic can consider as they strive to minimise reliance on input resources in order to maximise their sustainability output. Most instructively, this chapter describes a basic framework, and provides a point-by-point analysis of each step in the process, outlining a theoretical basis and practical considerations.

Norma Shorty (Chapter 8) extends the knowledge-economy discussion through an exploration of Indigenous-led research that embraces Indigenous knowledges, philosophies, methods, and healing. In order for Indigenous knowledge to be sustainable, Indigenous peoples must return to their philosophies, methods, and heritage in order to fully embrace what their ancestors left behind for them to decipher and put to use. This chapter makes a call for long-term Indigenous-led research, for the purpose of articulating, defining, and implementing Indigenous Knowledge as a critical and the most historically relevant renewable resource in the Arctic.

In Chapter 9, Karin Buhmann and her colleagues examine how climate change has spurred projects in Arctic countries to shift to low-carbon renewable energy sources. Several of these projects have been met by protests by local communities including Indigenous groups concerned with environmental and social impacts. These tensions underscore the need for stronger and meaningful involvement of communities and Indigenous peoples in impact assessments and consultation processes in order to identify and address concerns from the local perspective. Based on cases from Sápmi, Greenland, and Canada, this chapter shows that in some cases renewable energy projects can have perversely negative impacts on community health and safety as well as the traditions and income-generating activities of Arctic Indigenous groups. The authors argue that the need for energy justice highlights the importance of approaching climate change responses and renewable energy transitions in ways that adequately address local concerns, needs, and rights in a manner that is meaningful to those who may be adversely affected.

Dorothée Campou and Greg Poelzer (Chapter 10) carry this discussion forward by exploring the extent to which Indigenous communities participate in the transition to renewable energy in the Arctic region. Using the concept of energy justice, this chapter provides legal and empirical arguments to demonstrate the need to consider energy justice in order to ensure that the transition to renewable energy in the Arctic region addresses the rights of Indigenous peoples. In so doing, the authors outline the importance of renewable energy as a means to achieving sustainable development and to fulfilling human rights in accordance with the international commitments of Arctic states adopted under the auspices of the 2030 Agenda and the SDGs in 2015. Second, the chapter outlines the broader contexts and corresponding patterns of renewable energy development in the Arctic. Against this backdrop, the chapter examines the actual state of play of the energy transition and its impact on Indigenous peoples in the Arctic based on illustrative examples. For this purpose, the chapter includes examples from Canada, Alaska, and Russia and in the Nordic countries of Norway and Sweden. Based on this appraisal, the authors offer important recommendations for policymakers and business leaders to achieve greater justice for Arctic Indigenous peoples during this current period of the global energy transition.

In Chapter 11, Bjørg Helen Nøstvold, Ingrid Kvalvik, and Morten Heide share the results from a study they completed on "Arctic Origin" as a marketing opportunity for food producers in Arctic Norway. The assumption was that it is possible to achieve added value based on Arctic origin in strategic marketing, but to do this, it is vital to know what consumers perceive as Arctic qualities. The chapter shows that consumers associate reindeer, seafood, and game as Arctic species, and associate food from the Arctic as natural, pure, healthy, tasty, and traditional. The perception is quite similar in the north and south of Norway. This means that producers generally can use the same branding, unless they have a strong focus on a local food image. Furthermore, many of these characteristics are in line with current international food trends related to health and environmental sustainability.

In Chapter 12, Catherine Chamber and her colleagues examine fisheries that center on both capture fisheries and its related industries (e.g., fish processing, gear manufacturing, harbor operations, etc.) and subsistence fisheries that contribute to local mixed economies. The Arctic marine socio-ecological ecosystem is experiencing a continuous, rapid change including shifts in the range of fisheries, decreasing sea ice coverage, increased risk of pollution, and varying forms of economic development and governance changes that can have both positive and negative impacts. The specific objective of this chapter is to use the best available data to contribute to scholarship, policy, and future development by identifying opportunities and threats for current and fisheries and aquaculture activities in the Arctic. The chapter concludes by identifying key considerations for Arctic communities and decision-makers interested in renewable economies that include fisheries and aquaculture.

In Chapter 13, David Natcher and his co-authors present the results of their recently completed study on the Arctic's food producing potential. The aim of the project was to assess the potential for increased production and added value of foods originating in the Arctic, with the overarching aim of improving northern food security, and enhancing the social and economic conditions of Arctic communities. The results of the project affirmed that the Arctic region is a considerable producer of commercial foods. Food industries are producing large volumes of food commodities that are culturally compatible with Indigenous/local food preferences and also have high export value. However, the research also found that the Arctic foods value chain is challenged by a host of social, economic, logistical, and political obstacles. While these challenges are experienced unevenly across the Arctic regions, Arctic food industries: (1) tend to be fragmented; (2) have tenuous professional connections; and (3) have limited communication streams. In this chapter, the authors make a call for a cluster-based approach to food innovation that can draw together Arctic food producers with governments, Arctic Indigenous communities, universities, research centers, vocational training providers, and industry associations. A cluster-based approach to food innovation would be guided by the combined efforts to respond to regional challenges in food security and renewable economic development.

In the concluding chapter (Chapter 14), Natcher and Ingram present the results a regional study that examined the nexus between water, energy, and food systems in northern Canada. In 2017, the Arctic Council, under the Finnish Chair, adopted the United Nations' SDGs to inform its strategic policy direction; noting that the SDGs are global in scope but are amendable to the sustainable development of Arctic regions. In that same year, the Arctic Council's SDWG made a commitment to use SDG targets as guideposts for advancing its sustainable development agenda. However, before those guideposts could be determined, the SDWG emphasised the need to better understand the nexus—or the connections and interactions—that occur between SDG targets. The SDWG cautioned that failing to consider the nexus between SDG targets could result in ill-informed and unintended policy outcomes, whereas an accurate accounting of the synergies and trade-offs between SDG targets could inform more sustainable policy solutions. With this direction, Natcher and Ingram examined the nexus between SDG 2—Ending hunger and achieving food security for all, SDG 6—Ensuring the availability and sustainable management of water and sanitation for all, and SDG 7—Ensuring access to affordable, reliable, sustainable, and modern energy for all. Their focus on WEF-related SDGs is particularly warranted in northern Canada, given the high rates of WEF insecurities experienced by Indigenous communities. By assessing the positive and negative interactions between the WEF-SDGs, Natcher and Ingram concluded that 87 percent of interventions to alleviate WEF insecurities would be synergistic at some magnitude, meaning that efforts to address insecurity in one WEF sector will have positive spillover effects toward the others. With synergies significantly outweighing trade-offs, this chapter demonstrates that important opportunities exist to simultaneously address WEF insecurities through mutually beneficial actions that capitalise on and promote synergetic policies.

Summary

A theme that runs throughout all chapters in this volume is the notion that Arctic economic development has historically been weighted by "southern-based" investments and influenced by a development discourse originating in the global south. While parallel experiences of colonisation, environmental stress, remote access, resource constraints, and food insecurity can be found, opportunities to amplify Arctic perspectives and resilient economic strategies warrant deliberate attention. The lessons learned from the chapters in this volume represent our intentional efforts to shape research and introduce a development discourse that is unique to the people and conditions of the Arctic. To that end, we call upon the Arctic Council and Arctic States to help facilitate local and international relationships, investments in research capacity, and the utilisation of research results to guide future Arctic economic policy. Through such a development platform, emerging and well-established renewable economies can have enduring benefits and can be a critical component to promoting the future prosperity and wellbeing of all Arctic communities.

References

Berger, T. R. 1977. Northern Frontier, Northern Homeland. The Report of the Mackenzie Valley Pipeline Inquiry. Vol. 1. Ottawa: Supply and Services Canada.
Sustainable Development Working Group. 2017. The Human Face of the Arctic: Strategic Framework 2017. Arctic Council Secretariat. Available at: https://oaarchive.arctic-council.org/bitstream/handle/11374/1940/SDWG-Framework-2017-Final-Print-version.pdf?sequence=1&isAllowed=y

2 Arctic broadband connectivity and the creative economy

Access, challenges, and opportunities in Nunavut and Alaska

Timothy J. Pasch and Olaf Kuhlke

Introduction

The North American Arctic geographic region is increasingly attracting economic and political interest (Keil & Knecht, 2017; Smith, 2010). Along with climate change and geostrategic concerns, new business opportunities are emerging, and Arctic stakeholders are playing a key role in facilitating and creating favorable conditions for boosting Arctic economic activity (Petrov, 2017). There is a wide spectrum of Arctic business activities underway, of which oil and gas, mining, and shipping are perhaps among the most well-known.

Beyond extractive and transport-related industries, there is growing interest in the North American Arctic region's human economic potential in areas such as eco-tourism, innovation and entrepreneurship. Indigenous-owned, digitally connected culturally related businesses show increasing potential to address specific sustainable opportunities and needs of the region (Dana & Anderson, 2007; Patvardhan, 1990; Shadian, 2018).

In parallel to these economic trends, the North American Arctic is becoming more digitally connected (Christensen, 2003; Hudson, 2011; Warwick, 2019). Some researchers have argued that in order to succeed, new business development in the North American Arctic must be rooted in its people(s) and grounded in traditional knowledge toward the goal of innovative development, in order that the Arctic's unique resources and human capital become competitive on a global scale (Pasch, 2015; Rodon & Lévesque, 2015). Rural Arctic communities are under pressure, as younger generations increasingly desire the services, opportunities, and education offered in greater metropolitan areas (Seyfrit, Hamilton, Duncan, & Grimes, 1998). Arctic business development is, therefore, an existential issue.

This chapter argues that creating economic opportunities of a sufficient size and critical mass, if properly balanced with community informatics and values, has the potential to offer rural and remote Arctic residents aspects of the quality of life they seek while celebrating and leveraging

DOI: 10.4324/9781003172406-2

traditional knowledge, languages, and lifestyle. New technologies are revising the concept of "remote" and are rendering previously inaccessible projects economically feasible (Caspary & O'Connor, 2003). Moreover, new communication technologies especially are disrupting traditional business models and forcing business to innovate and to reinvent themselves (Graham, Hjorth, & Lehdonvirta, 2017). With this said, challenges related to impacts of ICT on traditional knowledge and stability/integrity of Northern communities are significant and must be addressed (Young, 2017, Young, 2019).

A number of regional, national, and international reports have carefully documented the rise of the creative economy (Bakhshi, Freeman, & Higgs, 2012; Duisenberg, 2010; Harris, Collins, & Cheek, 2013; Restrepo & Marquez, 2013). As 2013 UNESCO Creative Economy Report points out, "culture is now a *driver* [emphasis in the original document] of economic development, led by the growth of the creative economy in general and the cultural and creative industries in particular, recognized not only for their economic value but also increasingly for their role in producing new creative ideas or technologies, and their non-monetized social benefits" (p. 9).

As a consequence, scholars and policy makers alike have paid close attention to scalable, specific strategies and policy instruments that boost both public and private investment in cultural activities and creative occupations (Florida, 2002; Hagoort, 2003; Kooyman, 2011; Oakley, 2004; Sorin & Sessions, 2015). Furthermore, the question of educating a digital creative workforce has resulted in considerable dialogue concerning how to develop culturally specific or culturally sensitive entrepreneurship training curricula for the creative economy and to make these accessible, for broad community audiences, online (Naudin, 2017; Röschenthaler & Schulz, 2015).

In this chapter, as a component of the section related to renewable Arctic economic opportunity, we examine the potential of developing such culturally specific or culturally guided entrepreneurship training curricula, and the access and/or barriers faced by aspiring entrepreneurs in remote Arctic communities, as they seek to utilize new communication technologies and digital tools (photography, videography, code, etc.) to generate sustainable job opportunities. Specifically, we seek to address the following questions:

1 What access do remote Arctic communities have to broadband technology and digital tools to participate in the creative economy?
2 What concerns and barriers do remote Arctic communities face in developing viable business opportunities in the global creative economy?
3 What opportunities already exist in remote Arctic communities for participation in the global creative economy? What conversations about jobs outside of the resource extraction economy and traditional subsistence are being had?

Background

Currents trends in Arctic environments and economies

The North American Arctic is undergoing measurable, visible and significant environmental, economic, and social changes. Observable warming trends in the Arctic present significant and sustained challenges including, but not limited to ecological shifts such as observable differences in animal migratory patterns, hunting range, ice thickness, and numerous other threats to traditional practices and ways of life (Stroeve, 2017; Wang, 2017). Such developments represent a significant challenge to traditional economies, along with associated sociocultural, environmental, traditional knowledge and linguistic/communicative concerns (Emmerson & Lahn, 2012; Giles, 2003; Romero Manrique, Corral, & Guimarães Pereira, 2018; Stephen, 2018).

The warming/heating trends, despite the significant challenges mentioned above, may also paradoxically present some economic opportunities in the North American Arctic, through enhanced shipping and transport, mineral exploration, ecotourism, and infrastructure (Drewniak, Dalaklis, Kitada, Ölçer, & Ballini, 2018; Tol, 2009). This is evidenced in-part by enhanced global interest in Arctic shipping routes, increased activity in Arctic-based diplomatic and negotiating activity and geopolitical interests and stakes proposed by an increased number of (thirteen as of this writing) Arctic observer nations such as China (Arctic Council,n.d.)

These trends in the North American Arctic have raised concerns related to sovereignty over Arctic waters. Canada is particularly focused upon Arctic sovereignty and definitions (by the United States and other nations) of the Northwest Passage as either internal or international waters, issues that will impact oversight of environmental conditions and resources, reduce the ability to control shipping traffic and collect duties, and influence access (Geddert, 2019; Lajeunesse & Huebert, 2019; Lalonde, 2018; Stein, 2018). Initiatives such as China's Polar Silk Road and guidebook for Chinese ships navigating the Northwest Passage are examples of international attention to and attempts to begin using the passage at large scale (Byers & Lodge, 2019; Koivurova, 2018; Liu, 2019). Such national sovereignty concerns directly impact economic opportunities and are linked to the sovereignty of the Indigenous populations of the Arctic.

While much of the North American Arctic economy is currently focused on commodity extraction and exports, resource-dependent communities prone to boom-and-bust cycles are looking toward digital and creative economic models as tools to diversify their economic base (Alvarez, Yumashev, & Whiteman, 2019; Avango, Nilsson, & Roberts, 2013; Bennett, 2016; Johnston, Dawson, & Stewart, 2019; Nong, Countryman, & Warziniack, 2018; Poppel, Flaegteborg, Siegstad, & Snyder, 2015). Developments such as the Arctic Investment Protocol of the World Economic Forum and early adopters such as Guggenheim Partners indicate a global interest in how Arctic development proceeds (World Economic Forum, 2015).

The rise of the creative economy and its economic and social impacts have been widely documented and one tool for enhancing economic development of this type is through the creation of curricula in accelerators and incubators (Connell, 2013, Felton, 2010, Flew, 2011, Heinsius and Lehikoinen, 2013, Oakley, 2004; Sorin & Sessions, 2015). A key goal of these programs has traditionally been to prepare entrepreneurs to start businesses in the creative and cultural sectors, and some of these programs have been specifically designed to address underserved minority populations (Leung, 2019; Salemink, Strijker, & Bosworth, 2017; Sum & Jessop, 2013).

As of this writing, a large number of these efforts have focused on major urban centers despite the fact that it is especially in the highly remote and underdeveloped regions of the planet where the digital/creative economy can be the most effective for economic diversification and poverty reduction (Florida 2006, Markusen and Gadwa, 2010; Duisenberg, 2010; Philip & Williams, 2019; Restrepo & Marquez, 2013).

Digital entrepreneurship education: From North American tribal colleges to community efforts

Tribal Colleges and other Native American Tribal groups in the United States are engaging with culturally centered digital/creative incubators designed around economic development (Dana & Anderson, 2007; Dixon, 2019; Wuttunee, 2004). And in a most recent example, an Australian not-for-profit accelerator, Barayamal, is now offering a variety of entrepreneurship education programs for tech startups, specific sensitivity and inclusion of Kamilaroi language elements ("Barayamal – Indigenous Entrepreneurship Australia," 2019.)

In the Pueblo Nation, an accelerator was conducted entirely in the Zuni language, enabling youth to participate in the Indigenous Comic-Com focused on Native Superheroes and creating new series focused on Native Realities: Superheroes of Past, Present, and Future (Murphy, 2018; Simón, 2016).

Shoshone-Bannock Tribal Member and Indigenous Journalist Mark Trahant has written extensively on the importance of digital content creation by Indigenous Peoples. As co-author and presenter with Pasch, Bjerklie, & Trahant (2016) at the ArcticNet Annual Scientific Meeting, he wrote:

> So what can we in the Academy do about this? How do we make certain that the opportunity for a digital future is as real for an Indigenous Arctic as it for other global citizens? As we develop digital media we must not allow ourselves to repeat mistakes from the past, and especially the missing voices from Indigenous communities. I am fond of the work of the 1947 Hutchins Commission, formally the Commission on Freedom of the Press (Hutchins, 1947). That Commission said underrepresented

"constituent groups" are too easily misrepresented in any democracy. "We cannot assume (that) the mere increase in quantity and variety of mass communications will increase mutual understanding,' Hutchins wrote. "It may give wider currency to reports which intensify prejudice and hatred" (Hutchins, 1947, p. 35). So, again, what can we in the Academy do about this? What can we do to make sure that in the Arctic indigenous voices are developing stories, content in digital-speak, and more important, being heard?

One of the challenges for Indigenous communities, including the Arctic region, is that we need better data about Internet or cell phone service. If you look at Facebook, Instagram, Snapchat, Twitter, and other social media, it's clear that Native young people use social media in similar ways to other young people. It's also important to remember that the Indigenous population skews younger than the general population, and in many communities the under-18 age group constitutes the largest population segment of typically fast-growing communities (Census Canada, 2017).

Indigenous communities already have a data gap. Many of our statistics, ranging from unemployment to health metrics, are unreliable and out of date. We need better, faster data collection in Indigenous communities, in North America and globally (Kukutai & Taylor, 2016). The National Congress of American Indians says when it comes to data Native Americans are too often the "asterisk nation" because the information we see is scant and presented as a footnote ("Data Disaggregation NCAI," 2019). http://www.ncai.org/policy-research-center/research-data/data

A second opportunity here is the lifting of geography as a barrier. In the digital world, location does not matter. A business can operate successfully anywhere there's a good connection to the Internet. The retail site Etsy is a good example. According to Mary Meeker's research (2018), 35% of Etsy sellers started a business without much capital (compared to 21% of all small business owners). It's the perfect space for authentic Indian art.

The important thing is that we are at the beginning of the digital transformation. American Indians and Alaska Natives have a long history of adapting to new technology. This is just a new and exciting chapter. What's interesting to us is that we need more of this across the Arctic. The talent is already there. We just need to open up more opportunities for its use and an exposure to a broader audience.

Our incubator and accelerator approach establishes a visible physical presence for sustainable economic development through the creation of Indigenous-owned businesses, assisting with training including but not limited to digital content production, expanding conceptual and practical possibilities for data sharing and collaboration in Arctic communities, and broader impacts (Mark Trahant, in Pasch, Bjerklie, & Trahant, 2016).

One major barrier to successful entrepreneurship education and training in the Arctic relates to historical and present-day postcolonial and neo-colonial ramifications. Some previous efforts to adapt entrepreneurial education to the specific needs of Indigenous communities have inadequately addressed significant aspects of cultural imperialism/hegemony (Pinto & Blue, 2016). The concept of the institutionalized othering of Indigenous peoples in their own lands, and aspects related to glorifying "saviors" of a highly idealized/ romanticized Arctic have been explored in the literature (Huggan & Jensen, 2016; Jensen, 2015; Pasch, 2015). Some existing Western entrepreneurship approaches have primarily utilized traditional models of capitalistic entrepreneurship training and merely "translate" or localize them into a different cultural and linguistic context, without questioning the fundamental concept of entrepreneurship itself—or its fit for Indigenous communities. Programs of this type may not be aligned with local community values and the various types of remote-community entrepreneurship and ownership (from sole proprietorship to tribally owned cooperatives) that are possible, and which of these might fit the community and its individual's best.

This chapter argues that attempts to "train" Arctic residents in curricula created elsewhere may impart little result or benefit, or may arguably even be highly detrimental to remote communities in the Arctic and other regions—unless the curriculum is the result of a collaborative effort including culturally specific ways to integrate Indigenous knowledge directly into entrepreneurship training.

With expanding digital connectivity in the Arctic, the question of how future economic development in the North might proceed has become a more pressing question, as seen by increasing research into Arctic entrepreneurship, youth engagement, small business development through resilience and other programs in the circumpolar Arctic. A number of new programs have emerged across the Canadian and Alaskan North related to these goals.

Inspire Nunavut is one example of current Canadian-funded teams working on social entrepreneurship training in the Canadian North, focused on incorporating Indigenous values into curricular design and partnering with local participants and leaders. Merging the driving principles of traditional business with the mechanics of social entrepreneurship, while also integrating Inuit culture and values, "Inspire Nunavut provides youth in Nunavut with an opportunity to create new solutions in their communities via entrepreneurship" ("Inspire Nunavut,"n.d.). Through integrating Inuit traditional knowledge principles with entrepreneurship training this program has partnered with the Government of Nunavut and Employment and Social Development Canada, and has already shown success in the creation of new businesses in remote communities in Nunavut ("Inspire Nunavut,"n.d. ; Cornik, 2016)

EntrepreNORTH is a program based out of Yellowknife, Northwest Territories with its mission focused toward empowering Indigenous and community-based entrepreneurs to build sustainable businesses and livelihoods across Northern Canada. EntrepreNORTH provides a Northern Entrepreneur Support Program in addition to powerful thematic approaches (for example, an On-the-Land Tourism Experiences and Services program) that has resulted in numerous Indigenous-owned companies. As indicated in the Impact Stories of the site, "Across Northern Canada, Indigenous entrepreneurs are catalysts of prosperity and drivers of social change within their communities. Their success has a far-reaching ripple effect that leads to greater Northern self-determination and sustainability" ("EntrepreNorth," n.d.).

TakingItGlobal has developed the ConnectedNorth program offering enhanced educational experiences for remote Indigenous communities via partnerships with Cisco and other organisations. ConnectedNorth's principles build upon empowerment through Indigenous role models and principles incorporating a diversity of voices and making strong use of technologies to share educational experiences at a distance.

In Alaska, the Tanana Chiefs Conference (TCC), and its corporation, Doyon Ltd. is also focused on economic development. This tribal consortium consists of 6 sub-regions and 42 member tribes of the (primarily Athabaskan) Alaskan Interior. Their territory covers an area of 235,000 square miles, an area equal to about 37%of the entire state, and just slightly smaller than the state of Texas.

In addition to leading initiatives in areas such as health, housing and environment, wellness, sustainability and energy; rural economic small-business development has become an increasing priority for the Conference. The TCC is working toward multiple economic initiatives including ecotourism as well as investing in small businesses viewed as viable and based on/ aligned with community plans ("Tanana Chiefs Conference," 2019). The Alaskan fieldworks informing this chapter took place in collaboration with the TCC and most particularly in the communities of Hughes and Huslia in the Yukon Tanana and Yukon Koyukuk subregions (see Figure 2.1).

While the Tanana Chiefs Conference represents a large part of the Alaskan interior and focuses on economic development and job training programming, currently they have not yet developed an incubator or accelerator program. The creation of these opportunities, especially as informed by currently existing initiatives in Canada and other parts of the Arctic we believe, is fortuitous given increasing connectivity among and with remote Alaskan communities. At a recent *Navigating the North Summit* in August 2019, the theme focused on "Telecommunications and Technology to Arctic Economic Development and Public-Private Partnerships: Corporate, Government and Industry Leaders Focusing on the Rising Potential of Alaska" ("Navigating the North – Innovation Summit 2019," n.d.). One panel in particular discussed future investment opportunities in Native communities, and the necessity of broadband expansion. This connectivity could enable not only in-person incubator and accelerator programs, but also allow for online and hybrid programming to be supported by broadband communication technology such as conferencing, webinars, and other forms of online learning.

Traditional urban accelerator programs, focused on the tech industry, who invest in the companies they help launch, also already exist in Alaska.

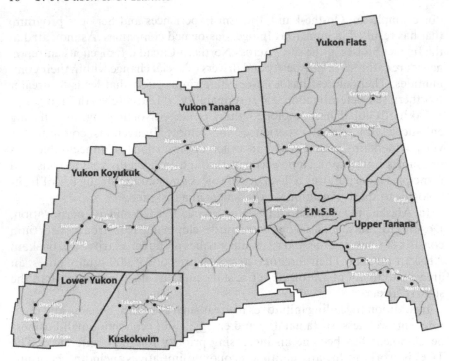

Figure 2.1 Territory and subregions of the Tanana Chiefs Conference.

Take for example *Launch Alaska,* an Anchorage based tech start-up accelerator with a variety of companies in their portfolio ("Launch Alaska," 2019). Launch Alaska provides the typical model for start-up companies: they guide promising new companies through an educational program—acceleration—and then invest in a selection of them, receiving part of the company's equity in return for the continued technical assistance and networking. In contrast to this urban program in Alaska's largest city, few other accelerator or comprehensive start-up entrepreneurship education programs exist across the state, especially in more remote communities.

In summary, there are currently a variety of efforts on the way to integrate Canadian First Nation knowledge and curriculum with existing entrepreneurship training approaches. In Alaska, that process is very much in its infancy, with a more traditional urban technology accelerator in place. Yet, all of these programs have successfully helped launch a number of (Indigenous) businesses that either serve their communities with modern services or sell Indigenous goods and services to a larger market. One area where both the Canadian and Alaskan Arctic continue to struggle in remote communities, relates to the challenge of digital access and bandwidth.

The current State of Arctic digital access: Community informatics, digital media, and bandwidth

North American Arctic Bandwidth has the potential for enhancements in the near future, which (if successful) may significantly impact Arctic economic development in remote regions. In 2017, the Quintillion Corporation succeeded in finalizing subsea connections linking the Alaskan community of Nome with Prudhoe Bay (and communities between), after which point terrestrial cable connects with Anchorage and Seattle. Phase Two of the project proposes submarine fiber optic cable between Alaska and Tokyo.

For Alaska and the Canadian Arctic, the proposed eastern extension of the network in Phase Three extends through the Northwest Passage in Canada and from there onwards to the United Kingdom. In a September 2019 telephone discussion with Zach Naramore of Quintillion it was described that the exact route East through the Northwest passage has not yet been finalized. Ultimately, the route will depend on building a business case that would allow us to successfully build the infrastructure, and then have customers benefit from … being in that location (based on) the geography.

Madeleine Redfern, the Mayor of Iqaluit Nunavut, indicated to the Special Senate Committee on the Arctic that Canada had missed out on connecting Baffin Island to fiber-optic lines from Nuuk, Greenland to Newfoundland. Redfern, 2018) stated that "you effectively cannot put a branching unit in after it has been built. It is as expensive as a new build." She called for the construction of a fiber link from "Iqaluit to Nuuk at a cost of approximately $80 million; a cost that would have been reduced by half had a branching unit been installed in the original design phase" ("Senate of Canada- Northern Lights: A Wake-Up Call for the Future of Canada," n.d.).

Given that client needs will impact the final route selection of the Quintillion Network proposed in 2021/2022 through the Northwest Passage, it may potentially be advantageous for Internet connectivity in Nunavut and Nunavik (and other Canadian Arctic regions) if successful negotiation between policymakers in the Canadian North could link remote Canadian Arctic communities during the third phase of this major North American Arctic infrastructure development.

Beyond Alaska, other Arctic regions including Cinia (Finland) and MegaFon (Russia) have created memoranda of understanding related to sub-sea data cable across the Arctic Sea. These connections also link Russia and China with the network, which raises a number of potential security concerns for the North American Arctic.

As one example of Arctic data security challenges, Huawei Corporation (China) indicated that it would partner with Ice Wireless and Iristel to help them connect (by 2025) rural communities in the Arctic as well as remote areas of north-eastern Quebec and Newfoundland and Labrador. Huawei added that some 25 communities in the largely Inuit areas of the Nunavut

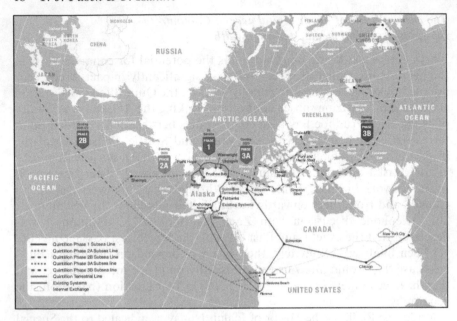

Figure 2.2 Quintillion Arctic broadband development strategy.

Source: https://www.qexpressnet.com/system/#FAQ

territory would also benefit from the deployment. One concern related to Huawai's presence as a communication leader in the Canadian Arctic mentioned that this "does create a vulnerability in the event of an escalated Canada-China dispute, where Huawei could potentially be ordered to shut down those services, thus cutting Canadian Arctic communities off the Internet" (Byers & Lodge, 2019; Levinson-King, 2019).

Beyond the risk of services being shut down in the Arctic, questions of security in terms of United States/Chinese data sovereignty challenges have been frequent in the media as of this writing, and data for the Canadian North managed by Huawei could have further contraindications on United States/Alaskan collaboration and data management.

While some specific Arctic centers have achieved significantly fast(er) data speeds than previously due to 4G LTE connectivity, there remain numerous Arctic communities where data is either completely lacking, or provided at a significant premium resulting in higher costs and major limitations in what is possible for North American Arctic entrepreneurship and small business development. Eric Anoee, an Inuit member of the board of the Nunavut Broadband Development Corporation expressed his view with the authors when asked if Internet speeds in the Arctic are an issue: "Yes, it is definitely an issue but we make do with it ... ideally it should be on par with the rest of Canada. They (Internet speeds) should be seen as a Basic Human Right to have the same level of service with other Canadians."

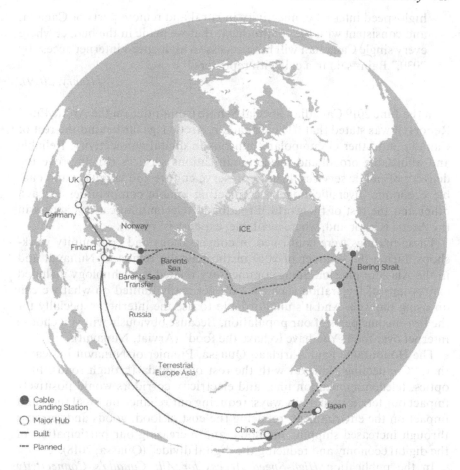

Figure 2.3 Cinia/MegaFon proposed Arctic sea digital communication strategy (Knaapila, 2019).

Some data services to Nunavut are currently delivered via Telesat satellites running through gateways in each of the 25 Nunavut communities. Despite the 4G upgrades, wireless internet is a far cry from broadband connectivity enjoyed by most of southern Canada. This may change due to significant investment from the Government of Canada into R&D for Telesat's broadband satellite constellation that may enable connectivity for the most remote Arctic communities.

Minister of Innovation, Science and Economic Development Navdeep Bains said July 24 that the Canadian government views Telesat's future low Earth orbit broadband constellation as the only means to connect the country's most remote citizens. "This is going to provide us with privileged access to this constellation, which will enable us to get

high-speed internet connectivity in rural and remote parts of Canada,
and consistent with the commitment that we made in the budget where
every single Canadian will have access to high-speed internet access by
2030," Bains said in a call with reporters"

(Henry, 2019).

In the June 2019 Canadian Special Senate Committee on the Arctic Final
Report, it was stated that "The Canadian Arctic lags far behind the rest of
Canada and other circumpolar countries in digital connectivity. Reliable
and affordable broadband telecommunications services can improve the
delivery of public services, help to preserve culture and language and enable economic diversification by connecting remote communities to each
other and the rest of the world. Broadband telecommunications access in
the Arctic is poor and, where available, expensive" (p. 45).

These findings were confirmed in conversations and community workshops during the summer of 2019 in the Kivalliq region of Nunavut and
in the Alaskan interior. Some commentary related to technology included
expressions of frustration such as: "We get really limited on what we can
do using the web. And it's unaffordable to use the internet, especially for
the low-income parts of our population. Because obviously will you choose
internet over food? You have to have the food" (Arviat, Nunavut).

The Honourable Paul Aarulaaq Quassa, Premier of Nunavut indicated
that: "Connecting Nunavut with the rest of Canada through roads, fiber
optics, telecommunication lines and electricity corridors would positively
impact our territory in many ways: reducing our reliance on fossil fuels and
impact on the environment; reducing the cost of food, goods and services
through increased shipping methods; and increasing our participation in
the digital economy and reducing the digital divide" (Quassa, 2018).

In the publication *High-Speed Access for All: Canada's Connectivity
Strategy* produced by Innovation, Science and Economic Development
Canada (ISED) in 2019, the Minister of Rural Economic Development
describes that in 2019 "we made a bold commitment to connect all
Canadians to reliable high-speed Internet" (p. 2). The document continues that "Canada faced a national connectivity gap" with rural communities facing "the daily challenge of slower, less reliable Internet access than
those in urban centres" (Innovation, Science and Economic Development
Canada, 2019, p. 4). The report additionally confirms that "limited Internet
has proven to be a real and significant challenge for Indigenous communities" with focus on affordability and connectivity made available within
communities rather than requiring relocation (Innovation, Science and
Economic Development Canada, 2019, p. 13). "The Rural and Northern
Stream of the Investing in Canada Infrastructure Program provides up to
CAN $2 billion to support various infrastructure projects that improve the
quality of life in rural and northern communities" (Innovation, Science and
Economic Development Canada, 2019, p.18).

Infrastructure of this type, while transformative for Arctic economies in the Canadian North, at present are "still very fragile: a single event can cause a mass outage where people can't send or receive emails, people can't get money out of the bank machine, people can't buy gasoline or groceries at the store," stated Madeleine Redfern (Levinson-King, 2019).

Beyond land-based Internet possibilities a number of other space-based Internet options may potentially render undersea and terrestrial cable redundant. In particular for the Arctic, even if undersea connections are made, laying fiber on the tundra brings with it a host of other challenges related to permafrost, extreme weather conditions and the challenges of access to such remote regions.

OneWeb is a company proposing future space-based blanket coverage for the Arctic, in addition to Amazon's Project Kuiper, SpaceX's Starlink, Facebook's Athena and Google's Loon (among other companies). While Iridium phones are presently commonplace in the Arctic, until recent times connection speeds for data at 2.4kbps have been largely insufficient to enable large-scale remote-region digital communication and data transfer. The new Iridium NEXT satellite constellation array has the potential to blanket the North with high-speed communication technology. While in coming years there may indeed be a host of new devices connecting with new networks enabling enhanced digital communication in the Arctic, at present the reality is that when away from major communities data can be extremely sparse indeed.

Community informatics: Beyond data speed and focusing on community values

The importance of training in information technology, as well as job creation and participation in markets for digital products is widely recognized as essential for economic growth in modern economies. Key to the implementation of creative economic initiatives in rural and underserved regions is local access to and familiarity with information technology and digital media (Robert and Townsend, 2016). The importance of training in information technology, as well as job creation and participation in markets for digital creative products is widely recognized as essential for economic growth in modern economies (Comunian, Gilmore, & Jacobi, 2015). And yet a digital divide still remains a limiting factor in many regions, including much of the circumpolar Arctic.

Even when high speed data is available (for example, the 4G LTE service available in Arviat via the qiniq system), the high cost of data coupled with immediate data caps when limits are exceeded means that monthly data limits can be reached extraordinarily quickly, effectively limiting the ability of Nunavummiut to upload films, transmit datasets, conduct live video meetings, and conversations, and participate in training sessions, as only some examples.

Regardless of when broadband arrives in Nunavut, it is not merely speed of connectivity that can create value for a community. This chapter argues that economic and social value for a community is not solely based on the rapidity of digital connectivity. A number of researchers have proposed that it is not sufficient to merely overcome this digital divide by providing faster access, but rather by focusing on the real needs of users. "A preoccupation with the digital divide as a problem of technical connectedness more often than not serves the commercial interests of Internet service providers (ISPs), without necessarily empowering or addressing the critical needs of those one is striving to connect" (Clement, Gurstein, Longford, Moll, & Shade, 2012, p. 16).

Some research like Landzelius (2006) has shown that increased passive consumption of digital commercial content (especially by youth) may be strongly at odds with cultural values and traditional knowledge. Community Informatics Theory argues that communities collaboratively accomplishing meaningful self-determined goals utilizing technology/data provide a far more valid measure than data speeds alone, of enhancing economic opportunity while overcoming digital disparity (Kuersten, 2018; McMahon, Gurstein, Beaton, O'Donnell, & Whiteduck, 2014). Jason C. Young (2019) focuses on the impact of information and communication technologies on communities and especially on Inuit qaujimajatuqangit and knowledge transmission. In this piece, the author (2019) discusses how digital technology is negatively impacting the amount of time spent on the land and outdoors, degrading interactions between children and elders, and creating concerns not only by serving as a distracting element while on the land, but also replacing actual skill (such as navigation and situational awareness) with digital tools functioning as ersatz replacement for traditionally learned abilities. Both in terms of social interactions and learning, technology carries with it numerous aspects of the post-colonial and real dangers in terms of cultural appropriation, and the balance that the author discusses between "good" and "bad" uses of ICT are highly applicable to digital entrepreneurial curricular economic development as it is being developed in the North American Arctic. Community Informatics Theory and the thoughtful, value-driven creation of economic initiatives may assist in addressing some of the challenges researched in Young's research.

In summary, there is tremendous potential to expand the fast-growing global creative economy to Arctic regions. A key advantage of this sector is the fact that many products and services developed under this rubric are digital in nature, such as (but not limited to) audio and video production, podcasting, still photography, augmented and virtual reality games and applications, coding, telepresence and numerous others. With broadband access, the remoteness of the Arctic can be overcome, and the resource extraction economy could be supplemented and even replaced in the long run by a digital creative economy that reduces the typically high transportation costs and creates location-independent jobs. Once proper, equal, reliable and affordable broadband options are in place, value-based entrepreneurship programs (incubators/accelerators) can offer opportunities for remote populations to participate in opportunities the creative digital economy has to offer. There are challenges related to such efforts, which are described in detail in the following chapter.

While this research argues that a culturally aligned and community-focused digital creative business incubator can assist in the development of new companies and jobs in remote regions, the implementation of any entrepreneurship incubator/accelerator requires careful study of the values that communities themselves see in such a program, especially listening in terms of self-determining strengths, challenges and opportunities. Toward this goal of listening we conducted fieldwork in May and June 2019, as described in the following sections.

Methodology and research approach

The fieldwork informing this chapter was funded by the National Science Foundation grant: *Developing and Testing an Incubator for Digital Entrepreneurship in Remote Communities*. The goals of the project include conducting listening sessions toward informing the co-creation of a culturally centered curriculum on digital entrepreneurship for the purpose of enhancing economic and social opportunities in remote North American Arctic regions.

Conceptually, the project is based on the belief that any curricular design focused on sustainable digital economic development in the Arctic will be ultimately unsuccessful without fully incorporating (from the earliest phases of development) traditional knowledge, respect of each Arctic community's communicative and linguistic preferences, and the foundational incorporation of traditions and community-based consensus of value and needs. We argue that digital technologies and communication infrastructure and curriculum intelligently leveraged toward creative and culturally situated small business opportunities can assist in diversifying the economic development of the Arctic's remote and resource-dependent communities.

To assess the likelihood of successfully creating a curriculum for digital entrepreneurship training in the Arctic, we first examined existing access to the telecommunication infrastructure in the North American Arctic. This was done through an extensive literature review as well as interviews with telecommunications providers and specialists in our host communities. Second, we conducted interviews with community members, workshop participants and key stakeholders in order to assess the personal, cultural, economic, and social barriers to both telecommunication access as well as business development, including entrepreneurship. Third, we examined the many existing opportunities for telecommunication access as well as business development, and we discuss our findings below.

Our initiative additionally proposed to connect regions of Arctic Canada with regions/communities in Alaska, thus enabling the cross-pollination of ideas and sharing of digital infrastructure and communicative networks/ databases that may assist in elevating the scale and scope of small busi- nesses to ultimately become Indigenous-owned Arctic digital enterprises, potentially with linkages connecting Canada and the United States.

In May and June 2019, we traveled to Nunavut and to the Alaskan Interior for initial discussion and workshops with Inuit and Alaska Native tribal mem- bers. We conducted research in the form of workshops, interviews, survey collection (in person and online), and in-person focus groups in the Hamlet of Arviat, Nunavut, Canada, and with the Tanana Chiefs Conference in Alaska, specifically in the Alaskan interior communities of Hughes, and Huslia as well as Fairbanks. We interviewed local tribal leaders, small business owners, and residents of the areas and conducted participatory workshops with mem- bers of the respective communities (ranging from one to five days) to learn more regarding shared traditions, values, small business opportunities, and economic challenges related to living and working in the North American Arctic. We additionally brainstormed regarding possible economic collabo- rations between Alaska and Nunavut, as well as about similarities and dif- ferences between Alaskan and Nunavummiut perceptions of challenges and possibilities for renewably growing the creative economy in both regions. While in Alaska we traveled together with an Inuk colleague working toward facilitating in-depth discussions between Inuit and Alaska Native peoples in order to further enhance understandings of collaborative economic potential between Indigenous peoples of the United States and Canadian Arctic.

This initiative takes the perspective that any work with Arctic Indigenous communities on economic development, including that made possible through digital media and technology must respect and work with traditional value sys- tems, and should result in a co-designed program, strategy or curriculum that ends up being owned by and led by Indigenous people (Castleden, Morgan, & Lamb, 2012; Ermine, Sinclair, & Jeffery, 2004; Louis, 2007; Nicholls, 2009; Weber-Pillwax, 2004). Thus, our workshops with the community sought to establish a conversation about the value system that guides community-based economic thinking and identifies perceived strengths and challenges; and based on that, develops ideas for the most needed businesses and the digital components that help promote them. The Technology of Participation (ToP) approach assisted in organizing the process of inquiry designed around a highly inclusive, listening focused approach (Nelson, 2017).

According to this phenomenological approach to inquiry, we began with a focused conversation topic, and asked participants to identify, without prompting or interruption, various characteristics of their communities: strengths and challenges, and the traditional as well as modern values they perceive as guiding them. Furthermore, participants discussed digital tools and aspects of data access that were viewed as economic opportunities for the community (and region), while also calling out aspects of Arctic digital infrastructure that were viewed as significant barriers to overall communi- cation and successful small business development.

As facilitators, we stepped back, observed and did not interrupt or guide this process. This brainstorming was followed by a group consensus workshop, the second part of a typical ToP session. Here, participants begin to discuss, sort, categorize and prioritize various ideas, and come to a collective group consensus on their key values, strengths and challenges. Participatory strategic and action planning—the next steps of ToP—followed, and community participants identified which business ideas and future economic development would address the strengths and challenges, while being true to community values. Technology of Participation techniques have been widely used in a number of contexts and communities (including Indigenous communities) as small and large group interaction methods to facilitate difficult conversations about major community change efforts and challenges, and to develop action plans for the opportunities that exist (Bryson & Anderson, 2000; Nelson, 2017; Spencer, 1989).

The means by which ToP is framed is critical (Lee, McQuarrie, & Walker, 2015). "Technologies of participation refers to arrangements of practices, metrics, discourses, and actors that perform community self-determination in ways that are designed to realize specific goals ... deliberation can be sold as a commodity or it can help create group solidarity and identity ... it can enable democratisation or elite rule. It derives its significance from how it is situated" (Lee, McQuarrie, & Walker, 2015, p. 83). For this reason, the synergy of Indigenous-centered research methodologies and community informatics theory with ToP was critically important for developing a conceptual framework designed toward enhancing benefits to communities while reducing negative influences of technology, economic development and digital in-roads.

As part of the workshops, many discussions took place outside hamlets/villages and out on the land. The conceptual framework guiding these discussions was in part driven by the perspective that even being indoors inside fixed-style houses was a colonial construct, and that being outside in traditional ways could create new opportunities for freedom of expression and heightened communication. For this reason on day 3 of the workshop in Arviat the group traveled overland via snowmobile and discussed at length in an environment much more conducive to communication. Many conversations took place in Inuktitut. In Alaska, similarly the group left the tribal meeting halls to meet by the riverbanks and discuss, in addition to boat travel upriver and forest walks/ATV travel.

Developing an Arctic digital creative economy of the future: From values to challenges and strengths

Values

During our fieldwork in both Arviat, NU and Hughes, AK, we began our community workshop with a lengthy discussion of values. We asked the community members to identify the values that guide their work, and their life in

Arviat, NU Values			
Love	Family	Unity	Listening
Teaching	Respect (elders)	Loyalty	Mutual Help/Cooperation
Respect	Love	Appreciation	Wisdom
Serenity	Preparing for the Future	Trust	Commitment
Honesty	Patience	Cultural Awareness/Steadfastness	Integrity
Hughes, AK Values			
Knowledge of the land	Unity	Knowledge of Culture and Tradition	Willingness to learn
Subsistence	Community Unity	Hospitality	Caring for each other
Seasonal Changes	Cultural unity	Cohesion: Everyone pulls together in time of need	Love and Care for Children in Community
Working together to keep drugs out of community	Progressive attitude	Respect for Elders	Friendliness

Figure 2.4 Graphical summary of the most frequently used terms in discussions in Arviat and Nunavut.

the community. All participants were asked to identify key values that guide their work and their actions in the community, and to place these on the large interactive wall ("sticky wall") used in Technology of participation sessions.

Each participant then ranked their top two (2) priorities in each category. The priorities were then ranked in descending order with the most frequently ranked priority at the top of the column. A graphic summary of the most frequently used terms during discussions in Arviat and Nunavut follows in Figure 2.4.

Some elements that rapidly emerged in discussions included

a   The necessity of co-creation in all business endeavors. This means that in small communities, no business should operate without prior consultation with the community, and with a clear mission to serve the community. In other words, there must be value alignment between individual business owners and the community.
b   The focus on incorporating strong Indigenous socio-linguistic, traditional, elder-guided curricular aspects into all enterprises. New business development should strengthen the cultural fabric of the community, not weaken or dilute it.
c   The development of trust and resilience. The communities we worked with expressed a strong need for emphasizing historical continuity and

the value of tradition. It was expressed repeatedly that elders working with youth would be a critical piece of identifying new business opportunities, since the elders highlighted the need for continuing traditions, whereas youth was interested more in modern technology and communication. Negotiating that divide, and identifying ways to bring the elders and youth together in business and in the community is critical to build sustainable businesses that are embraced by the entire community.

d   The importance of having community-led data sovereignty mechanisms strongly in place. Shared knowledge and clear acknowledgement of historical past wrongs (and current pitfalls), along with the importance of leadership held by both Indigenous and non-Indigenous researchers working in a joint partnership were described. Individual and social/societal values were described as being a critical part of any economic effort.

There was a great deal of discussion related to Inuit traditional knowledge or Inuit Qaujimajatuqangit (Inuit IQ) and many of the concepts discussed during the workshop were drawn from these values and described in Inuktitut. The anonymized results of these conversations on community values were shared with Alaskan communities in order to compare and contrast values informing economic development between Nunavut/Alaska. Similar discussions were held in Hughes, Alaska where it became rapidly apparent that there are strongly shared guiding values across the North American Arctic in terms of economic and business development.

In summary, our discussions included a strong sense that any small business development should emerge first and foremost from community values, which they should be value-based and should not lose sense of their values, no matter how successful they became. In Arviat, for example, we were told about Hinaani, a local textile design/fashion company that firmly operates in the modern world (by selling apparel across North America) but that also based all of their work and designs on Inuit art, culture, and stories.

One of the first key insights we gained from our work with our host communities was that this discussion of values as part of a business and entrepreneurship development workshop was highly unusual. We were repeatedly told that while there are other job training and entrepreneurship workshops offered in both communities, we were the first in their experience to start with an in-depth conversation about values, and how values should guide business. Following a discussion of values, we then guided our workshops through an exploration of perceived strengths, utilizing the same methodology as described above.

Strengths

It became very evident throughout the weeks spent together in Nunavut and Alaska that culture and tradition were deeply important for the group, and regarded as their greatest strength. Respecting Elders and keeping

Indigenous languages vibrant and respected were topics frequently mentioned. A mutually shared sense of volunteerism, providing community help and support to others, and continuing the critically important traditions of hunting, being on the land, and keeping the environment healthy were all mentioned as foundational priorities.

A number of individuals mentioned that so many aspects of their communities were so beautiful and attractive. "Similar to fall colors down south, up here we have very small changes in colors, in the moss and lichen, but so beautiful. The tea is so delicious. The colors on the (Hudson) bay. The Northern Lights have a deep meaning for our ancestors and for us." Discussions of sustainable, Indigenous-guided ecotourism opportunities emerged from these conversations, with a focus on what specific community-related events could be of economic potential, without overtly disrupting the community or causing negative impacts to the communities.

In terms of education, business and entrepreneurship, a focus on practical training for needed industries was another strong focus area. The importance of more Indigenous medical staff and members of the police force were included, as were numerous aspects of traditional skills and talents already present in the communities that could become the base for renewable/sustainable small businesses. A traditional seamstress commented in Nunavut that:

> "Now we have access to everything (patterns) on computer/via the internet. We can design our own clothes and trade patterns. With amouti (traditional Inuit parka for carrying a baby) for example, we combine traditional patterns with new designs. Our designs can help people around the world. Amouti bonds the mother and child more. For the first 3-4 months some people never even see the baby as they are nestled in the Amouti. Feeding is so much easier. Skin to skin, the mother knows what the child needs so much better. The child can "go" by simply "going" on the ground while moving and traveling. There is no waste."

Numerous ideas and concepts were raised in terms of new industries and opportunities that the digital economy could make available across Nunavut and Alaska, and these are discussed in the opportunities section below.

Remoteness challenges

In order to assess what kind of entrepreneurship training the community would co-develop with us in the future, it was incumbent to address the many challenges—historical and contemporary—that remote Indigenous communities are facing.

Hughes, AK Strengths			
Recreational Activities	**Subsistence**	**Local Services**	**Environment and Nature**
ATV riding/transportation	Gardening	Community Hall	Trees/Forest
4-Wheeling	Hunting	Clinic	Animals
Dog sledding	Berry Picking	Post Office	River (access)
Seasonal Recreational Events: Community events such as summer solstice activity and dog sled races	Fishing	Washeteria/Laundry	Hills
Walking	Animal Skins	Water and Sewer	Natural Beauty
Boating	Bonding time	Strong local government/city office	Remoteness
	Jam/Syrup	Excellent Tribal School	
	Canning meat	Bifelt Store	
	Trapping (fur for clothing)	Easy access to community services	
		Daily flights to other communities/Fairbanks	

Arviat, NU Strengths				
Practical Education	**Talents and Skills**	**Culture and Tradition**	**Wildlife and Environment**	**Volunteerism**
Graduation Expansion (growing population)	Homemade school Clothing, arts and crafts	Respecting Elders	Rich wildlife, community, land and animals	Search and Rescue, life saving
Heavy Equipment Operators	Traditional hunting	Inuktitut language speakers	Polar bears	Lots of musicians
Training workshops and courses	Seamstresses	Young Hunters Association	Keeping the environment clean	Music festivals
Travelling out of country (college etc)		Food Bank and Food Sharing		Fundraising and donations for the needy
People participating in church groups		Keeping our culture and family		
Good work performance and ethic		Experienced, expert hunters		
Increasing number of small businesses		Culture of mutual help and support		
Council, board, and committees. MLA.		Friendliness		

Figure 2.5 Hughes, AK and Arviat, NU strengths.

The challenges expressed had many areas of intersection between Arviat, Nunavut and Hughes, Alaska.

a Cost of travel to and from the Arctic, both for personal and professional travel, and for shipping of needed goods and services, is clearly a factor of significant import.

b High cost of food and necessary supplies for remote communities is also a constant consideration.

c The need to travel significant distances for certain health care is a clear concern.

d In terms of communication, Internet speeds and telecommunication were also at the top of most participants' minds. Limited bandwidth and the high cost of data were described in detail in both Nunavut and Alaska.

e While telephone (landline and cellular) connectivity was strong in Arviat, in Hughes especially the lack of cell towers coupled with landline connections prone to going out for extended periods of time caused connectivity issues that can directly impact the ability to communicate outside the community (both for personal and business purposes).

Technology challenges

The need for enhanced training and education in digital tools was another area of focus, in particular digital imaging/photographic skills were mentioned in both communities as sources of income for photographing artistic and creative works for sale, and in using for social media. Screen time was a concern in both communities with concerns expressed related to the culturally negative impact of some video games, movies and television not in-tune with traditional values. As a result, we had a number of discussions that emerged related to how digital tools were leading to a number of community

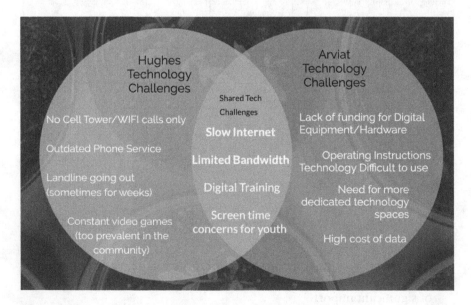

Figure 2.6 Venn diagram of hughes, AK and arviat, NU challenges.

members never leaving their homes, with this impacting negatively on time spent with family, on the land, engaging in traditional activities, and in some cases causing loss of jobs or reducing motivation to engage socially.

Housing challenges

Additional challenges described in the United States and Canadian Arctic communities we visited also showed similarities across borders. A housing shortage is a critical factor across the North American Arctic and was described in detail by many individuals with whom the authors had the opportunity to speak.

"Housing is a major issue. Looking at the increasing number of children in town, where will they live?."

"A dire need is housing, overcrowding, societal problems. Overcrowding causes domestic violence, often there is not enough food for everyone in the same house, etc."

"The issue is still there. Why is this still happening when the rest of Canada and the US is so different?"

Housing has been expressed in multiple TCC community reports as one of the most pressing issues in rural Alaska. With recent budget cuts in the State there is concern regarding reduced housing construction and homelessness support (Kesslen, 2019; Restino, 2018).

Drug and alcohol challenges

Alcohol abuse was repeatedly described in discussions in both Nunavut and Alaska as a challenge so serious as to be a cross-generational crisis. In some communities in Alaska, as one resident stated, "the (corporations) had to separate the dividend out, divvy it out (in smaller sizes), because there would be too much crime like drug use and alcohol abuse when people received too much at once."

Bootleggers can make significant profits by bringing alcohol into dry communities in the Arctic. A fifth of alcohol (750ml) can sell for many hundreds of dollars. Bootleggers can allegedly do this regularly and often, in conversation we heard anecdotes of bootleggers who had regular businesses of bringing and shipping bottles into communities, resulting in significant negative social impacts caused to the communities. Links between alcohol (and other drugs) and suicide were also described as major issues in the North American Arctic that impacted the entire region significantly. Significant research well beyond the scope of this article investigates alcoholism and suicide in the Arctic (Brown, Dickerson, & D'Amico, 2016; Ogilvie, 2018; Seyfrit et al., 1998; Skewes & Blume, 2015; Wexler et al., 2015).

Food insecurity challenges

Food insecurity was described as a major issue. This can take many forms including the extremely high costs of living in rural communities due to the greatly enhanced costs of shipping and transport, including the temptation of rural residents to purchase cheaper yet much less healthy options resulting in health concerns. One Arviat resident stated:

> "Obviously we do not have highways, up here, everything has to be flown up here, and it's really expensive, and we do have sealift, but it only once a year, early summer, This is by water. What about fresh produce? Milk, or fruits and vegetables, you cannot ship them by sealift for the year, because obviously they have a short shelf life."

> "We have no choice but to use air to bring in things like food, we need more road systems to assist with import(ation), however we are less than 1% of the voting members of Canada, we do not make enough of a difference to the big politicians."

> "Within the federal level we do not have good representation from our cultural group. We need new members of parliament, senators. Again it goes back to the fact that we are too small, it's hard but that doesn't mean that we do nothing. One Inuit group on Facebook talking about food insecurity posted pictures of the high cost of food, and started a dialogue. They organized protests, and that kind of thing is good, and needed, and powerful. Even a small group on social media can have a significant impact in remote areas."

In Alaska, food insecurity is also an ongoing and serious issue that impacts quality of life and economic development equally. For Inuit and Alaska Natives traditional hunting and fishing for *country food* is a major source of needed nutrition that is often unavailable in store-bought goods (or is simply unaffordable). Global heating is observably impacting traditional hunting ranges and ability of Arctic residents to reach herds due to costs of gasoline and other necessary equipment to undertake long-distance expeditions (Baskin, 2016; Chapin & Brinkman, 2016; Descamps et al., 2017; Koptseva & Kirko, 2015; Poppel et al., 2015; Tse, Weiler, & Kovesi, 2016).

The challenges of discrimination, nepotism,
and the "Arctic welfare trap"

During conversations in communities in Alaska and Nunavut, the topic of economic nepotism and favoritism emerged regularly. Because communities can often be so small, issues of equitability and equal opportunities can quickly come to the forefront. In scholarly research issues of nepotism have been described in other circumpolar Arctic research (Gremaud, 2017; Russel, 2015; Sejersen, 2015; Wang, Degeorges, & Forsvarsakademiet, 2014). During one workshop discussion the following was mentioned:

"If you are part of the family you get the job. Some companies pre-hire before they even do interviews. They know that want this person, but they go through the procedure. That is really common in a lot of workplaces and companies. You are related to me, you are going to get hired."

As some businesses in Arctic Indigenous communities may be owned by non-natives, there was a sense in workshop discussions that native people in their own communities are often being passed by in favor of individuals living outside the Arctic who are flown in (at great expense) to do a job at very high contractor rates rather than training and growing local talent and expertise in order to support local sustainable economic development.

"They (non-natives) would rather hire their own families, from down south, than someone capable from town like an (Inuk, or Alaska Native) mechanic. Even if you have the qualification, even if you have the certificate, they want them because they are white and they think they know more. Even if you have the qualifications, there is a preconceived idea of this person as better compared to a local."

Inequities in access to health services were described including descriptions of time-delays in accessing healthcare for Indigenous peoples (as opposed to non-native people who were perceived as being seen immediately). Delays in law enforcement attention for natives as opposed to non-natives were also described. Some quotes related to these aspects included:

"We have a story when the mother (who is Inuk) had to wait many months to get health care, but the qallunaat (non-inuk) got help immediately".

"Native people or people of color, if they are missing, you do not hear about them."

One aspect that arose during discussions related to concerns that Indigenous residents of the Arctic held as regards travel outside the Arctic, or into larger towns or cities where more non- Indigenous residents were present. These elements of discrimination and even fear were noted in both Nunavut and Alaska:

"It is scary. People look at you (outside of the Arctic), and they're like: 'she doesn't look white, she must be Native, Dene, Inuit.'"

"It is dangerous to be down south with brown skin. Even when you are walking down the street, there are people who do not know your nationality, but they are like '(expletive) Indians, or just like 'savages'".

"It is especially bad, because you hear of these white people going missing, they are on the news 10 minutes later. But native people, you do not hear about it for 6 weeks, 2 months later, or maybe never. A white person, 10 minutes later they are on the news, amber alert".

Governmental welfare support in the Arctic was a topic that emerged on several occasions during workshop discussions. In some communities there was a sense expressed of the "welfare trap" or "welfare spiral" consisting of welfare resulting in a contraindication for willingness to start a small business or even commence new employment.

For some, the risk of starting a job, or creating a personally owned small business could result in the potential loss of welfare and an increase in the cost of housing that was far too great a risk to take. For this reason welfare was described by some as preferable to employment due to the regular income and reduced cost of housing. While future research into this phenomenon is certainly required it would seem that further governmental incentives for those interested in remote community small business development could be beneficial, so that a system designed to assist those in need would not be revoked for those who are taking the potentially significant risk of attempting to build their personal, local, and regional economies.

All of the issues described above are only a small fraction of the total challenges described. This section is not meant to be a comprehensive or exhaustive synopsis of challenges impacting Arctic sustainable entrepreneurship, but rather to serve as a brief and highly limited example of the scale and scope of issues that those living in the North American Arctic are encountering on a daily basis. If a resilient and sustainable North American economy beyond extractive resource economies is to be developed, it is incumbent that social issues be focused upon in tandem with culturally aligned entrepreneurial curricular and training elements in order to ensure healthy communities wherein enhanced business development may truly flourish with resilience.

Unique opportunities

Digital entrepreneurial nomadic economies

As a final step, in both our workshops and in individual interview in Arviat and Hughes, we asked participants about the unique opportunities they see for a digital creative economy in the Arctic.

One incredibly powerful motivator and mindset that was expressed by Indigenous residents of both Nunavut and Alaska during this fieldwork was that of the desirability of self-reliance, living off the land, with freedom to leave and return to town at will while sharing resources in community: a lifestyle that was indicated as being highly attractive to many among whom the authors discussed. This is also an entrepreneurial mindset: to use the resources available, and find solutions to problems while thriving in the harshest environments. Entrepreneurial definitions today look at the utilisation of scarce resources for the benefit of the community.

When the opportunity arose for participants in the Arviat workshop to travel onto the tundra to hunt and fish and spend time on the land, it was perceived by all with great anticipation. A number of participants indicated that this was the first time that they had had the opportunity to travel away from town in years. For others (Inuit residents born in the community), it was the very first time that they had ever gone on a hunting expedition. When a discussion began surrounding this issue, some comments included that:

"Access to travel on the land has become a luxury for the rich"

"We do not have access to our culture on the land because the cost of gasoline is so expensive".

"An ATV is over 10k, Snowmobiles up to 15k. I can't afford it and I never leave town."

Connections between the traditional Indigenous nomadism of the past and the concept of digital nomadism of today were discussed in a hunting cabin near Maguse Lake, Nunavut. The concept discussed was that of possibilities for Inuit and other remote-region Arctic entrepreneurs living as digital nomads going back out on the land, using solar power for electrical power and leveraging (forthcoming) broadband digital connectivity outside of towns. This concept could incorporate digital currency transfer enabling payment to remote communities (including payments for services rendered on the land), and could enable digital Arctic nomadic entrepreneurs the possibility of conducting economic activity (possibilities mentioned included photography, drone filming/tracking of wildlife, videography, patterns for clothes, ecotourism, guiding, hunting, games and apps based on traditional knowledge, etc.).

The concept of digital nomadism exists in the research literature however often in terms of project-based. "Western professionals using a range of information systems (IS) and information technology (IT) tools to work digitally over the Internet while travelling perpetually" (Schlagwein, 2018). A number of articles describe the concept in terms of dissatisfied young urban digital workers constantly searching for more desirable, exotic locales within which to work (Jarrahi, 2019; Müller, 2016; Nash, 2018). The concept as it is being described in this writing however does not relate to attempting to wander in search of another locale to replace one's own, but rather to enable the full immersion in one's own born environment, in a sense to assist in reconnecting with the land/disconnecting from the colonial past which necessitated the loss of nomadic heritage due to residential schools, forced relocation, and other factors (Bombay, Matheson, & Anisman, 2014; Dombrowski et al., 2016; Elias et al., 2012; Evans-Campbell, Walters, Pearson, & Campbell, 2012; Felt, Procter, & Natcher, 2012; Tester & Kulchyski, 2011).

Presently the resources necessary to enable culturally focused Arctic digital nomadic economic development are still lacking. Transportation (ATV/ Snowmobiles) still requires fossil-fuels, costs of which can be extremely high (and transport of which is extremely costly). Renewable power sources with reliable, strong batteries that can run vehicles (and function well in extremely cold temperatures) would be game-changers in the North. Power for small electronics is improving in terms of solar charging (GoalZero[1] for example) however even beyond power sources, the key component of digital connectivity for cellular and data is greatly lacking outside of towns/hamlets. Iridium technology offers excellent solutions for connectivity (calling) and limited data (currently improving under the IridiumNEXT development) however the possibility of true satellite or terrestrial broadband could enable connectivity even in regions now considered disconnected.

Economic activities that could take place on the land are far-reaching in terms of possibilities. Eco-tourism is one area of great potential, providing sustainable experiences to tourists from across the world with a focus on environmental sustainability, stewardship and culturally guided/ values-driven locally owned initiatives. One of TakingItGlobal/Connected North's Digital Media Kits provided to some Arctic program participants consists of a MacBook Pro, a DJI Drone, a GoPro, a 360° camera, Microphone, and iPhone. This kind of technology can generate an entire broad spectrum of potential economic activity able to be leveraged with adequate power, connectivity, and training

Culturally enhanced AirBnB experiences have generated interest in a number of communities in the North American Arctic in terms of ecotourism. Renee Linton of the Tanana Chiefs Conference has been spearheading initiatives related to a rural AirBnB Pilot Project designed to encourage entrepreneurism in interior Alaska TCC Ecotourism readiness surveys and village planning and development programs are designed to assess the suitability of ecotourism initiatives and alignment with community plans and tribal values (TCC, 2020). Some discussion during workshops centered around what would be needed in order to incentivise tourist travel to remote regions, and what would be viewed as "minimum necessities" in order to make ecotourism more attractive. Some of these necessities described included running water, electricity, communications (telephone and especially Internet connectivity), temperature control (heating and, increasingly cooling in summer months) and insect barriers among other factors. The potential possibility of allowing tourists to observe or even participate in some cultural events (only if approved by the communities) was mentioned as a possible incentive for increased ecotouristic activities. "Combining the AirBnB experience with cultural experiences, with beading, with country food, with participation in creating crafts" (Participant in Hughes, AK).

It should be mentioned that AirBnB is not meant here to serve as a utopian expression of ecotourism, but rather as one potential option for encouraging sustainable ecotourism in the Arctic. A number of research articles have described successful (and unsuccessful) attempts at leveraging this and other initiatives toward encouraging rural entrepreneurship, describing potential benefits but also challenges and concerns related to significant increases of tourism in rural circumpolar Arctic and other Indigenous regions (E. Henry, Newth, & Spiller, 2017; Jæger, 2019; Koninx, 2018; Leeming, 2016; Müller, 2019; Outi et al., 2019; Sisneros-Kidd, Monz, Hausner, Schmidt, & Clark, 2019; Spangler, 2018; Veijola & Strauss-Mazzullo, 2019).

The interest in including art and traditional crafts into the ecotourism experience was expressed during workshop discussions: "When one does not know a lot about small businesses, it boils down to what we already know, culturally, it is our passion, what we like to do. For example, our art, jewelry making, or traditional art, if we market it well especially online, coupled with social media...before where you were so far from the market-place, now that gap is now bridgeable due to the web."

In Hughes specifically, local residents who had made beautiful works of art had very little confidence in marketing these productions even though internet access was available. Taking high-quality images of the work was extremely helpful in terms of providing the best visibility and sale price possible for residents. Such training can be accomplished quite rapidly and can have measurable impacts in terms of financial takeaway.

Conclusions and recommendations for further research

The North American Arctic is entering into a unique yet tumultuous period where changes in environmental conditions coupled with extraordinarily increases in primarily extractive industries, are resulting in the exacerbation of significant social issues in Indigenous and other isolated communities in Alaska and Nunavut. Developing a value-based knowledge economy and providing an avenue for creative digital expression not simply for its own sake, but as a source of locally owned and operated economic development driving investment and national/international support, is potentially trans-formative economic activity for the future of the North American Arctic economy. Linkages between Nunavut and other Canadian Arctic territories with Alaska in terms of economic development hold potential for dynamic international circumpolar economic collaboration.

There is an extraordinarily untapped economic and social opportunity in facilitating North American Arctic Indigenous communities' efforts toward the development and application of their unique values, cultural traditional knowledge and expertise for the specific purpose of stimulating the knowl-edge economy in the North. Value-driven, community-centered digital cre-ative industries focused on the unique environment of the North American Arctic, with collaborative hubs linking Alaska with Nunavut, Nunatsiavut, Nunavik and other regions, could result in the alleviation of a number of critical social issues currently hindering human and economic development in the Arctic.

In order to fully enable resilient, sustainable economic development in the North American Arctic, it is incumbent upon Canada and the United States to comprehensively address the major social and infrastructural issues currently present in the North. Such investment can, this chapter argues, assist in the development of robust, sustainable, resilient and highly unique Arctic economies that will far outweigh the infrastructural and social commitments required.

Acknowledgment

This material is based upon work supported by the National Science Foundation under Grant No. 1758781 and Grant No. 1758814. Any opinions, findings, and conclusions or recommendations expressed in this material are those of the author(s) and do not necessarily reflect the views of the National Science Foundation.

Note

1. https://www.goalzero.com/story/

References

Alvarez, J., Yumashev, D., & Whiteman, G. (2019). A framework for assessing the economic impacts of Arctic change. *Ambio.* doi: https://doi.org/10.1007/s13280-019-01211-z.

Arctic Council. (n.d.). *Observers. Arctic Council.* https://arctic-council.org/index.php/en/about-us/arctic-council/observers

Avango, D., Nilsson, A. E., & Roberts, P. (2013). Assessing Arctic futures: Voices, resources and governance. *The Polar Journal, 3*(2), 431–446. doi: https://doi.org/10.1080/2154896X.2013.790197.

Bakhshi, H., Freeman, A., & Higgs, P. L. (2012). *A dynamic mapping of the UK's creative industries.* Retrieved from https://eprints.qut.edu.au/57251/

Barayamal (2016). *Barayamal – Indigenous Entrepreneurship Australia.* Barayamal. https://barayamal.com.au/

Baskin, M. (2016). *Climate Change Related Impacts on Food Insecurity and Governance in the United States and Canadian Arctic* (PhD Thesis). The George Washington University.

Bennett, M. M. (2016). Discursive, material, vertical, and extensive dimensions of post-Cold War Arctic resource extraction. *Polar Geography, 39*(4), 258–273. doi: https://doi.org/10.1080/1088937X.2016.1234517.

Bombay, A., Matheson, K., & Anisman, H. (2014). The intergenerational effects of Indian Residential Schools: Implications for the concept of historical trauma. *Transcultural Psychiatry, 51*(3), 320–338.

Brown, R. A., Dickerson, D. L., & D'Amico, E. J. (2016). Cultural identity among urban American Indian/Alaska native youth: Implications for alcohol and drug use. *Prevention Science, 17*(7), 852–861. doi: https://doi.org/10.1007/s11121-016-0680-1.

Bryson, J. M., & Anderson, S. R. (2000). Applying large-group interaction methods in the planning and implementation of major change efforts. *Public Administration Review, 60*(2), 143–162. doi: https://doi.org/10.1111/0033-3352.00073.

Byers, M., & Lodge, E. (2019). China and the northwest passage. *Chinese Journal of International Law, 18*(1), 57–90. doi: https://doi.org/10.1093/chinesejil/jmz001.

Caspary, G., & O'Connor, D. (2003). *Providing low-cost information technology access to rural communities in developing countries*. https://www.oecd-ilibrary.org/content/paper/675385036304

Castleden, H., Morgan, V. S., & Lamb, C. (2012). "I spent the first year drinking tea": Exploring Canadian university researchers' perspectives on community-based participatory research involving Indigenous peoples. *The Canadian Geographer/ Le Géographe Canadien, 56*(2), 160–179.

Census Canada (2017). *Population size and growth in Canada: Key results from the 2016 Census*. (11), 14.

Chapin, T., & Brinkman, T. J. (2016). Human adaptation responses to a rapidly changing Arctic: A research context for building system resilience. *AGU Fall Meeting Abstracts*.

Christensen, N. B. (2003). *Inuit in cyberspace: Embedding offline, identities online*. https://books.google.com/books?id=9LfHN8xLILsC

Clement, A., Gurstein, M., Longford, G., Moll, M., & Shade, L. R. (2012). *Connecting Canadians: Investigations in community informatics*. Athabasca University Press.

Comunian, R., Gilmore, A., & Jacobi, S. (2015). Higher education and the creative economy: Creative graduates, knowledge transfer and regional impact debates. *Geography Compass, 9*(7), 371–383.

Cornik, K. (2016, October 31). *An Ottawa millennial is changing lives and inspiring new businesses for Nunavut's youth*. National Observer. https://www.nationalobserver. com/2016/10/31/news/ottawa-millennial-changing-lives-and-inspring-new-businesses-nunavuts-youth

Dana, L. P., & Anderson, R. B. (2007). *International handbook of research on indigenous entrepreneurship*. http://www.loc.gov/catdir/toc/fy0905/2006934134.html

National Congress of American Indians (NCAI). (2019). *Data Disaggregation*. NCAI. http://www.ncai.org/policy-research-center/research-data/data

Descamps, S., Aars, J., Fuglei, E., Kovacs, K. M., Lydersen, C., Pavlova, O., ... Strøm, H. (2017). Climate change impacts on wildlife in a high Arctic archipelago – Svalbard, Norway. *Global Change Biology, 23*(2), 490–502.

Dixon, T. (2019). *The role of social entrepreneurship and education in addressing native American social and economic challenges: A transformative study* (Publication No. 13899185) [Doctoral Dissertation, University of Seattle]. ProQuest Dissertations Publishing. https://search.proquest.com/openview/2d2ba6f6f378dab8d13b58bd-58f1f64b/1?pq-origsite=gscholar&cbl=18750&diss=y

Dombrowski, K., Habecker, P., Gauthier, G. R., Khan, B., Moses, J., Arzyutov, D. V., & Ween, G. (2016). Relocation redux: Labrador Inuit population movements and inequalities in the land claims era. *Current Anthropology, 57*(6).

Drewniak, M., Dalaklis, D., Kitada, M., Ölçer, A., & Ballini, F. (2018). Geopolitics of Arctic shipping: The state of icebreakers and future needs. *Polar Geography, 41*(2), 107–125. doi: https://doi.org/10.1080/1088937X.2018.1455756.

Duisenberg, E. D. S. (Ed.). (2010). *Creative economy report 2010: Creative economy: A feasible development option*. Geneva: United Nations Conference on Trade and Development.

Elias, B., Mignone, J., Hall, M., Hong, S. P., Hart, L., & Sareen, J. (2012). Trauma and suicide behaviour histories among a Canadian indigenous population: An empirical exploration of the potential role of Canada's residential school system. *Social Science & Medicine, 74*(10), 1560–1569.

Emmerson, C., & Lahn, G. (2012, May 30). Arctic opening: Opportunity and risk in the high north [Monograph]. http://library.arcticportal.org/1671/

EntrepreNorth. (n.d.). Impact. *EntrepreNorth.* https://www.entreprenorth.ca/impact.html

Ermine, W., Sinclair, R., & Jeffery, B. (2004). *The ethics of research involving Indigenous peoples.* Saskatchewan: Indigenous Peoples' Health Research Centre Saskatoon.

Evans-Campbell, T., Walters, K. L., Pearson, C. R., & Campbell, C. D. (2012). Indian boarding school experience, substance use, and mental health among urban two-spirit American Indian/Alaska natives. *The American Journal of Drug and Alcohol Abuse, 38*(5), 421–427.

Felt, L., Procter, A. H., & Natcher, D. C. (2012). *Settlement, subsistence, and change among the Labrador Inuit: The nunatsiavummiut experience* (Vol. 2). Univ. of Manitoba Press.

Flew, T. (2011). *The Creative Industries: Culture and Policy.* Sage.

Florida, R. (2002). *The rise of the creative class: And how it's transforming work, leisure, community and everyday life.* Basic Books.

Florida, R. (2006). The flight of the creative class: The new global competition for talent. *Liberal Education, 92*(3), 22–29.

Geddert, J. S. (2019). Right of (northwest) passage: Toward a responsible Canadian Arctic sovereignty. *Canadian Journal of Political Science/Revue Canadienne de Science Politique,* 1–18. https://doi.org/10.1017/S0008423919000052

Giles, B. D. (2003). The Earth is faster now: Indigenous observations of Arctic environment change, edited by Igor Krupnik and Dyanna Jolly. Arctic Research Consortium of the United States, Fairbanks, Alaska, 2002. No. of pages: xxviii + 356. ISBN 0-9 720 449-0-6. *International Journal of Climatology, 23*(10), 1265–1265. https://doi.org/10.1002/joc.936

Graham, M., Hjorth, I., & Lehdonvirta, V. (2017). Digital labour and development: Impacts of global digital labour platforms and the gig economy on worker livelihoods. *Transfer: European Review of Labour and Research, 23*(2), 135–162. doi: https://doi.org/10.1177/1024258916687250.

Gremaud, A.-S. N. (2017). Icelandic futures: Arctic dreams and geographies of crisis. In L.-A. Körber, S. MacKenzie, & A. Westerståhl Stenport (Eds.), *Arctic environmental modernities: From the age of polar exploration to the era of the anthropocene* (pp. 197–213). https://doi.org/10.1007/978-3-319-39116-8_12

Hagoort, G. (2003). *Art management: Entrepreneurial style.* https://books.google.com/books?id=qgAq1vxxYgQC

Harris, C., Collins, M., & Cheek, D. (2013). *America's creative economy: A study of recent conceptions, definitions, and approaches to measurements across the USA.* Oklahoma City: National Creativity Network.

Heinsius, J., & Lehikoinen, K. (2013). Training Artists for Innovation: Competencies for New Contexts.

Henry, C. (2019). Canadian government pledges $521 million for Telesat LEO constellation. *SpaceNews.* https://spacenews.com/canadian-government-pledges-521-million-for-telesat-leo-constellation/

Henry, E., Newth, J., & Spiller, C. (2017). Emancipatory indigenous social innovation: Shifting power through culture and technology. *Journal of Management & Organization, 23*(6), 786–802. doi: https://doi.org/10.1017/jmo.2017.64.

Hudson, H. E. (2011). Digital diversity: Broadband and indigenous populations in Alaska. *Journal of Information Policy, 1,* 378–393. https://doi.org/10.5325/jinfopoli.1.2011.0378

Huggan, G., & Jensen, L. (2016). *Postcolonial perspectives on the European high North: Unscrambling the Arctic.* Springer.

Hutchins, R. M. (1947). *By the commission on freedom of the press. A free and responsible press: A general report on mass communication: Newspapers, radio, motion pictures, magazines and books...* [Foreword by Robert M. Hutchins.]. University of Chicago Press.

Innovation, Science and Economic Development Canada. (2019). *High-Speed Access for All: Canada's Connectivity Strategy.* https://www.ic.gc.ca/eic/site/139.nsf/eng/h_00002.html

Inspire Nunavut. (n.d.). *Inspire Nunavut. Inspire Nunavut.* https://www.inspirenunavut.com

Jæger, K. (2019). *Tourists and Communities in Rural Festival Encounters: A mutually beneficial relationship?.* https://uis.brage.unit.no/uis-xmlui/handle/11250/2593180

Jarrahi, M. H. (2019). Personalization of knowledge, personal knowledge ecology, and digital nomadism. *Journal of the Association for Information Science and Technology, 70*(4), 313–324.

Jensen, L. (2015). Greenland, Arctic orientalism and the search for definitions of a contemporary postcolonial geography. *KULT. Postkolonial Temaserie, 12,* 139–153.

Johnston, M., Dawson, J., & Stewart, E. (2019). Marine tourism in Nunavut: Issues and opportunities for economic development in Arctic Canada. In R. L. Koster & D. A. Carson (Eds.), *Perspectives on rural tourism geographies: Case studies from developed nations on the exotic, the fringe and the boring bits in between* (pp. 115–136). https://doi.org/10.1007/978-3-030-11950-8_7

Keil, K., & Knecht, S. (2017). *Governing Arctic change. Global perspectives.* Palgrave Macmillan.

Kesslen, B. (2019). Homeless services in Alaska face uncertain future as state cuts back. *NBC News.* https://www.nbcnews.com/news/us-news/death-sentence-homeless-services-alaska-face-uncertain-future-state-cuts-n1057021

Knaapila, A.-J. (2019). Arctic Telecom. *Cinia.* https://www.cinia.fi/en/archive/arctic-telecom-cable-initiative-takes-major-step-forward.html

Koivurova, T. (2018). *China and the Development of International Law on Arctic Shipping: Challenges and Opportunities* (SSRN Scholarly Paper No. ID 3115830). Social Science Research Network. https://papers.ssrn.com/abstract=3115830

Koninx, F. (2018). Ecotourism and rewilding: The case of Swedish lapland. *Journal of Ecotourism, 0*(0), 1–16. doi: https://doi.org/10.1080/14724049.2018.1538227.

Kooyman, R. (Ed.). (2011). *The entrepreneurial dimension of the cultural and creative industries.* Hogeschool vor den Kunsten (HKU).

Koptseva, N. P., & Kirko, V. I. (2015). The impact of global transformations on the processes of regional and ethnic identity of indigenous peoples Siberian Arctic. *Mediterranean Journal of Social Sciences, 6*(3 S5), 217.

Kuersten, A. (2018). The Arctic digital divide. In B. O'Donnell, M. Gruenig, & A. Riedel (Eds.), *Arctic summer college yearbook: An interdisciplinary look into Arctic sustainable development* (pp. 93–105). https://doi.org/10.1007/978-3-319-66459-0_8

Kukutai, T., & Taylor, J. (Series Ed.). (2016). *Indigenous Data Sovereignty* (Vol. 38). http://www.jstor.org/stable/j.cttlqlcrgf

Lajeunesse, A., & Huebert, R. (2019). Preparing for the next Arctic sovereignty crisis: The northwest passage in the age of Donald Trump. *International Journal, 74*(2), 225–239. doi: https://doi.org/10.1177/0020702019849641.

Lalonde, S. (2018). *Canada's Influence on the Law of the Sea.* (7), 26.

Landzelius, K. (Ed.). (2006). *Native on the net: Indigenous and diasporic peoples in the virtual age* (1 edition). Routledge.

Launch Alaska. (2019). Fall 2019. *Launch Alaska.* http://www.launchalaska.com/fall-19

Lee, C. W., McQuarrie, M., & Walker, E. T. (2015). *Democratizing inequalities: Dilemmas of the new public participation.* NYU Press.

Leeming, J. (2016). *Addressing Cultural Vulnerabilities in Arctic Tourism: Kindness as "Third Space."* https://uwspace.uwaterloo.ca/handle/10012/10878

Leung, W.-F. (2019). Cool, creative, but not so equal. In W.-F. Leung (Ed.), *Digital entrepreneurship, gender and intersectionality: An East Asian perspective* (pp. 161–196). https://doi.org/10.1007/978-3-319-97523-8_5

Levinson-King, R. (2019, September 9). The superpower fight for internet near the Arctic. *BBC News.* https://www.bbc.com/news/world-us-canada-49415867

Liu, N. (2019). China's Arctic policy and belt and road initiative: Synergy or conflict? *The Yearbook of Polar Law Online, 10*(1), 431–434. doi: https://doi.org/10.1163/22116427_010010020.

Louis, R. P. (2007). Can you hear us now? Voices from the margin: Using indigenous methodologies in geographic research. *Geographical Research, 45*(2), 130–139.

Markusen, A., & Gadwa, A. (2010). Arts and culture in urban or regional planning: A review and research agenda. *Journal of Planning Education and Research, 29*(3), 379–391.

McMahon, R., Gurstein, M., Beaton, B., O'Donnell, S., & Whiteduck, T. (2014). Making information technologies work at the end of the road. *Journal of Information Policy, 4*, 250–269. https://doi.org/10.5325/jinfopoli.4.2014.0250

Meeker, M., & Wu, L. (2018). *Internet trends 2018.* Kleiner Perkins.

Müller, A. (2016). The digital nomad: Buzzword or research category? *Transnational Social Review, 6*(3), 344–348.

Müller, D. K. (2019). *A research agenda for tourism geographies.* Edward Elgar Publishing.

Nash, C. (2018). Digital nomads beyond the buzzword: Defining digital nomadic work and use of digital technologies. *International Conference on Information* (pp. 207–217). Springer.

Murphy, A. (2018, December 6). Native Heroes Take Center Stage at Indigenous Comic Con. *PBS.* https://www.pbs.org/native-america/blogs/native-voices/native-heroes-take-center-stage-at-indigenous-comic-con/

Naudin, A. (2017). *Cultural Entrepreneurship: The Cultural Worker's Experience of Entrepreneurship.* https://books.google.com/books?id=ujQ8DwAAQBAJ

Navigating the North – Innovation Summit 2019. (n.d.). Retrieved September 30, 2019, from http://www.navnorthsummit.com/

Nelson, J. (2017). *Getting to the Bottom of Top: Foundations of the Methodologies of the Technology of Participation.* https://books.google.com/books?id=CuNFDwAAQBAJ

Nicholls, R. (2009). Research and indigenous participation: Critical reflexive methods. *International Journal of Social Research Methodology, 12*(2), 117–126.

Nong, D., Countryman, A. M., & Warziniack, T. (2018). Potential impacts of expanded Arctic Alaska energy Resource extraction on US energy sectors. *Energy Policy, 119*, 574–584. https://doi.org/10.1016/j.enpol.2018.05.003

Oakley, K. (2004). Not so cool britannia the role of the creative industries in economic development. *International Journal of Cultural Studies, 7*(1), 67–77.

Ogilvie, K. A. (2018). Unintended consequences of local alcohol restrictions in rural Alaska. *Journal of Ethnicity in Substance Abuse, 17*(1), 16–31.

Outi, R., de la B., Suzanne, Brynhild, G., Þór, J., Gunnar, K, M., Dieter, Jarkko, S., … Maaria, N. (2019). *Arctic tourism in times of change: Seasonality*. Nordic Council of Ministers.

Pasch, T., Bjerklie, D., & Trahant, M. (2016). *Arctic IDEA: Nunavut Broadband and digital entrepreneurialism*. Presented at the ArcticNet Annual Scientific Meeting, Winnipeg, MB, Canada.

Pasch, T. J. (2015). *Towards the enhancement of Arctic digital industries: "Translating" cultural content to new media platforms*.

Patvardhan, V. S. (1990). *Growth of indigenous entrepreneurship*. Bombay: Popular Prakashan. (4848136).

Petrov, A. N. (2017). Human capital and sustainable development in the Arctic: Towards intellectual and empirical framing. In G. Fondahl & G. N. Wilson (Eds.), *Northern Sustainabilities: Understanding and addressing change in the circumpolar world* (pp. 203–220). https://doi.org/10.1007/978-3-319-46150-2_16

Philip, L., & Williams, F. (2019). Remote rural home based businesses and digital inequalities: Understanding needs and expectations in a digitally underserved community. *Journal of Rural Studies, 68*, 306–318. https://doi.org/10.1016/j.jrurstud.2018.09.011

Pinto, L. E., & Blue, L. E. (2016). Pushing the entrepreneurial prodigy: Canadian Aboriginal entrepreneurship education initiatives. *Critical Studies in Education, 57*(3), 358–375. doi: https://doi.org/10.1080/17508487.2015.1096291.

Poppel, B., Flaegteborg, M., Siegstad, M. O., & Snyder, H. T. (2015). The Arctic as a "Hotspot" for natural Resource extraction and global warming. *The Economy of the North*, 129–135.

Quassa, P. A. (2018, February). Proceedings of the Special Senate Committee on the Arctic. *Proceedings of the Special Senate Committee on the Arctic*. Ottawa. https://sencanada.ca/en/Content/Sen/Committee/421/ARCT/03ev-53837-e

Restino, C. (2018). Housing shortage in rural Alaska must be met with comprehensive solution. *The Arctic Sounder*. http://www.thearcticsounder.com/article/1835housing_shortage_in_rural_alaska_must_be_met

Restrepo, F. B., & Marquez, I. D. (2013). *The Orange economy: An infinite opportunity*. Washington, D.C: Inter-American Development Bank.

Roberts, E., & Townsend, L. (2016). The contribution of the creative economy to the resilience of rural communities: Exploring cultural and digital capital. *Sociologia Ruralis, 56*(2), 197–219. https://doi.org/10.1111/soru.12075

Rodon, T., & Lévesque, F. (2015). Understanding the social and economic impacts of mining development in inuit communities: Experiences with past and present mines in inuit nunangat. *Northern Review*, (41), 13–39.

Romero Manrique, D., Corral, S., & Guimarães Pereira, Â. (2018). Climate-related displacements of coastal communities in the Arctic: Engaging traditional knowledge in adaptation strategies and policies. *Environmental Science & Policy, 85*, 90–100. https://doi.org/10.1016/j.envsci.2018.04.007

Röschenthaler, U., & Schulz, D. (2015). *Cultural Entrepreneurship in Africa*. https://books.google.com/books?id=z4_4CgAAQBAJ

Russel, G. (2015). *Canadian Arctic Policy and Program Development and Inuit Recognition: A Neoliberal Governmentality Analysis of Canada's Northern Strategy and "The Missing Piece"* [Doctoral Thesis, University of Toronto]. University of Toronto TSpace. https://tspace.library.utoronto.ca/handle/1807/69754

Salemink, K., Strijker, D., & Bosworth, G. (2017). Rural development in the digital age: A systematic literature review on unequal ICT availability, adoption, and use in rural areas. *Journal of Rural Studies, 54,* 360–371. https://doi.org/10.1016/j.jrurstud.2015.09.001

Schlagwein, D. (2018). "Escaping the rat race": Justifications in digital nomadism. *ECIS,* 31.

Sejersen, F. (2015). *Rethinking Greenland and the Arctic in the era of climate change: New northern horizons.* Routledge.

Canada, S. of. (2019, June 10). Senate of Canada - Northern Lights: A wake-up call for the future of Canada. Senate of Canada. https://sencanada.ca/en/info-page/parl-42-1/arct-northern-lights/

Seyfrit, C. L., Hamilton, L. C., Duncan, C. M., & Grimes, J. (1998). Ethnic identity and aspirations among rural Alaska youth. *Sociological Perspectives, 41*(2), 343–365. doi: https://doi.org/10.2307/1389481.

Shadian, J. D. (2018). The Emerging Economy of the North American Arctic. *Arctic360,* 9.

Simón, Y. (2016, July 22). *The First-Ever Indigenous Comic Con Puts Native Artists in the Spotlight.* Remezcla. https://remezcla.com/features/culture/this-comic-convention-aims-to-put-indigenous-art-and-pop-culture-in-dialogue/

Sisneros-Kidd, A. M., Monz, C., Hausner, V., Schmidt, J., & Clark, D. (2019). Nature-based tourism, resource dependence, and resilience of Arctic communities: Framing complex issues in a changing environment. *Journal of Sustainable Tourism, 27*(8), 1259–1276. doi: https://doi.org/10.1080/09669582.2019.1612905.

Skewes, M. C., & Blume, A. W. (2015). Ethnic identity, drinking motives, and alcohol consequences among Alaska native and non-native college students. *Journal of Ethnicity in Substance Abuse, 14*(1), 12–28.

Smith, L. C. (2010). *The world in 2050: Four forces shaping civilization's northern future.* Dutton.

Sorin, G. S., & Sessions, L. A. (2015). *Case studies in cultural entrepreneurship: How to create relevant and sustainable Institutions.* Lanham: Rowman & Littlefield.

Spangler, I. (2018). "ONE MORE WAY TO SELL NEW ORLEANS": Airbnb and the commodification of authenticity through local emotional labor. *Theses and Dissertations–Geography.* https://doi.org/10.13023/etd.2018.241

Spencer, L. J. (1989). *Winning through participation: Meeting the challenge of corporate change with the technology of participation.* Kendall/Hunt Pub. Co. (3874708).

Stein, D. (2018). Protecting the Arctic environment from northwest passage shipping in the era of climate change drinking water issue: Center for environmental law prize winning comment. *Tulane Environmental Law Journal,* (2), 239–256.

Stephen, K. (2018). Societal impacts of a rapidly changing Arctic. *Current Climate Change Reports, 4*(3), 223–237. doi: https://doi.org/10.1007/s40641-018-0106-1.

Stroeve, J.C., Mioduszewski, J.R., Rennermalm, A., Boisvert, L.N., Tedesco, M., Robinson, D. (2017). Investigating the local-scale influence of sea ice on Greenland surface melt. *The Cryosphere, 11*(5), 2363–2381. doi: https://doi.org/10.5194/tc-11-2363-2017.

Sum, N.-L., & Jessop, B. (2013). Competitiveness, the knowledge-based economy and higher education. *Journal of the Knowledge Economy, 4*(1), 24–44. doi: https://doi.org/10.1007/s13132-012-0121-8.

Tanana Chiefs Conference (TCC). (2019). *Tanana Chiefs Conference.* TCC. https://www.tananachiefs.org/

Tanana Chiefs Conference (TCC). (2020). *Village Planning & Grant Writing.* TCC. https://www.tananachiefs.org/services/village-planning-grant-writing/

Tester, F., & Kulchyski, P. (2011). *Tammarniit (mistakes): Inuit relocation in the Eastern Arctic, 1939-63.* UBC press.

Tol, R. S. J. (2009). The economic effects of climate change. *Journal of Economic Perspectives, 23*(2), 29–51.

Tse, S. M., Weiler, H., & Kovesi, T. (2016). Food insecurity, vitamin d insufficiency and respiratory infections among inuit children. *International Journal of Circumpolar Health, 75*(1), 29954.

United Nations Educational, Scientific and Cultural Organization (UNESCO). (2013). *Creative Economy Report 2013 Special Edition: Widening Local Development Pathways.* http://www.unesco.org/culture/pdf/creative-economy-report-2013.pdf

Veijola, S., & Strauss-Mazzullo, H. (2019). Tourism at the crossroads of contesting paradigms of Arctic development. In M. Finger & L. Heininen (Eds.), *The global Arctic handbook* (pp. 63–81). https://doi.org/10.1007/978-3-319-91995-9_5

Wang, N., Degeorges, D., & Forsvarsakademiet. (2014). *Greenland and the new Arctic: Political And security implications of a statebuilding project.* RDDC Publishing House.

Wang, X., Jiang, D., & Lang, X. (2017). Future extreme climate changes linked to global warming intensity. *Science Bulletin, 62*(24), 1673–1680. doi: https://doi.org/10.1016/j.scib.2017.11.004.

Warwick, M. (2019, September 5). It's not so grim up North as fast Internet reaches the Arctic. *TelecomTV.* https://www.telecomtv.com/content/satellite/its-not-so-grim-up-north-as-fast-internet-reaches-the-arctic-36235/

Weber-Pillwax, C. (2004). Indigenous researchers and indigenous research methods: Cultural influences or cultural determinants of research methods. *Pimatisiwin: A Journal of Aboriginal & Indigenous Community Health, 2*(1).

Wexler, L., Chandler, M., Gone, J. P., Cwik, M., Kirmayer, L. J., LaFromboise, T., & Allen, J. (2015). Advancing suicide prevention research with rural American Indian and Alaska native populations. *American Journal of Public Health, 105*(5), 891–899.

World Economic Forum (2015). Arctic Investment Protocol: Guidelines for Responsible Investment in the Arctic. Global Agenda Council on the Arctic. http://www3.weforum.org/docs/WEF_Arctic_Investment_Protocol.pdf

Wuttunee, W. (2004). *Living rhythms: Lessons in aboriginal economic resilience and vision.* McGill-Queen's Press — MQUP.

Young, J. C. (2019). The new knowledge politics of digital colonialism. *Environment and Planning A: Economy and Space, 51*(7), 1424–1441. doi: https://doi.org/10.1177/0308518X19858998.

Young, J. C. (2017). *Encounters Across Difference: The Digital Geographies of Inuit, the Arctic, and Environmental Management* [Doctoral dissertation, University of Washington]. Semantic Scholar.

3 The cool economy

Technological innovation and the prospects for a sustainable Arctic economy

Ken Coates and Carin Holroyd

Introduction

Over the past 20 years, governments have become preoccupied with the commercialization of scientific and technological innovations as the cornerstone of the "new economy" (Schwab and Davis, 2018). National innovation policies have directed billions of dollars to basic and applied research, funding efforts to bring new products and services to market and supporting systematic efforts to train citizens to capitalise on opportunities. Around the world, the combination of advanced electronics, mass digitisation, high-speed wireless communications, robotics, artificial intelligence, biotechnology, and a wide variety of other technologies are transforming the global economy. Millions of jobs have been lost or reconfigured as a result of technological innovation in everything from manufacturing to banking, health care monitoring, and communications. Estimates of the medium-term impact of technology-based job loss call for between 25% and 50% of existing jobs to be displaced through technological innovation (Brynjolfsson & McAfee, 2014; David, 2015; Ford, 2015; Frey & Osborne, 2017). Obviously, other jobs will emerge as a result of the scientific and technological developments, but more of these new positions will be in the research laboratories, high-end factories, and design studios than in the field. Most of the job losses will be in traditional manufacturing, service firms, and retail operations, including in rural and remote regions (Lundvall, 1992; Nelson & Rosenberg, 1993).[1]

The emergence of the innovation-based economy changed government approaches to regional and national economic development. Commercial innovation has been focused on the triple-helix model of academic/research, government, and business collaboration, and on a simplified innovation system that involved an expansion of highly trained personnel, government investment in high risk, high return research, and specialised commercialisation efforts. These investments were believed to produce new businesses, employment, and societal prosperity. This approach underpinned Japan's late 1990s innovation economy, the rapid growth of South Korea, Taiwan, Singapore, and such North American regional success stories as Silicon Valley, the Seattle-Microsoft sector, and the Toronto and Waterloo Region

DOI: 10.4324/9781003172406-3

in southern Ontario (Bramwell & Wolfe, 2008; Bramwell, Nelles, & Wolfe, 2008; Collins, 2001; Holroyd & Coates, 2012, 2015; Kenney, 2000; Niosi & Bas, 2003; Pempel, 1987). Only a few northern regions, Oulu (Finland's center for Nokia's operations) and the Luleå region being the best example, fit into the new economy model (Inkinen & Suorsa, 2010; Rantakokko, 2012).

Various national governments spent billions of dollars on the promotion of the commercialisation of science and technology, building on and fueling a dramatic expansion in emerging technologies that have transformed the global economy.[2] Little of this investment capital made its way into the Arctic. The intensely competitive innovation economy focused on major universities and polytechnics, highly coordinated urban eco-systems, and existing high-technology companies (Etzkowitz, 2002, 2003; Etzkowitz & Leydesdorff, 2000; Etzkowitz & Zhou, 2017; Gibbons, 1994). The transition from the traditional approaches to the imperatives of the new science-based era left little place for rural and remote regions around the world, including agricultural zones and economies based on mining and forestry. Indeed, some of the technological developments, including robotics, long-distance control of machinery, digital surveillance, improved communications, and e-commerce and e-finance systems, changed and disrupted the traditional Arctic and sub-Arctic economies (Hall, 2020; Hall, Leader & Coates, 2017).

The potential for building an innovation economy in the North

Economically, the Arctic is a surprisingly complicated place. Overall, people in the region have high incomes, based largely on the wages paid in the natural resource economy where the workforce is transient and often from the south. A larger group of public servants generally earn less money but still have sizeable incomes and stable employment. A transient population of military personnel does not earn high incomes but receive housing and other benefits. Indigenous peoples generally have the lowest annual incomes in the Far North, although the gap is significantly smaller in Scandinavia than in Canada, Alaska, and Russia. The private sector commercial economy is more diverse, with those businesses connected to the resource sector following the boom and bust cycles that are the norm in the industry. Tourism generally pays less well and is often seasonal in nature.

Across the Circumpolar world, local and national governments, companies, and individuals have worked throughout the post-World War II era to create a more sustainable foundation for their economy. The effort has emphasised the urgent need to address Indigenous poverty, to create economic opportunity in small and remote communities, to improve overall prosperity, and to provide more stability across society. For decades, efforts across the Circumpolar world emphasised natural resource development, major infrastructure projects, and the provision of government services. In most jurisdictions, although less so in Alaska and Russia, transfer payments from governments to northerners make up a significant part of the regional economy. The standard formulas, based on government and private sector infrastructure projects, emphasised small business development but did not change the overall commercial trajectory in the region.

Several Arctic areas are actually extremely technologically advanced. The military establishments, particularly in Siberia (Russia) and Alaska (United States) are as sophisticated as any on earth (Farish, 2013; Osherenko & Young, 2005; Sergunin, 2015). Their military activities are economically extremely important; indeed, Alaska has, since the start of the Cold War in the 1950s, been sustained by a high and continued level of military spending, especially through the Elmendorf Airbase near Alaska and Fort Wainwright Army Base close to Fairbanks and numerous surveillance sites across the state focused on containing the Russian threat (Spohr, Höring, Cerioli, Lersch, & Soares, 2013). (In the 1950s and 1960s, the Americans also paid for the construction of an elaborate network of radar stations, particularly the Distant Early Warning Line (D.E.W), across the North American Arctic, providing an important economic boost to numerous isolated Indigenous communities (Fritz, 2010).) Similarly, there is a large and growing military presence across northern Russia that provides an important foundation for the regional economy and society (Bondar, 2015; Flake, 2014). In neither location, however, has the substantial military expenditure become the base for a technological eco-system, even though military investments have, for decades, been integral to the commercialisation of science and technology in the superpower states.

Other northern locations also have the advanced technological infrastructure. Svalbard, the Norwegian/international archipelago in the High Arctic, is typically seen as a northern tourism platform, but it has enormously powerful data collection systems focused on the Northern environment. The island also holds the world's seedbank, which stores seeds from every commercially viable crop globally (Grydehøj, 2020). Inuvik, Northwest Territories, a post-oil and gas boom small town near the Arctic Ocean, has superb satellite and space monitoring technologies, as does the University of Alaska-Fairbanks' geophysical laboratories[3] and Sweden's space center in Kiruna, Norrbotten (Gareis & Mercer, 2015). The University of Alaska-Fairbanks, the North American Arctic's most research-intense institution, has supported commercialisation in sectors as diverse as musk-ox rearing, northern mining, fishing and oil and gas exploration, but is only now attempting to make inroads in high-technology commerce (Jingfors & Klein, 1982). Only in Kiruna has the scientific enterprise begun to be directly connected to a plan for commercialisation and regional economic development.

However, the majority of Arctic and sub-Arctic isolated communities, especially those without road access, have almost no scientific infrastructure. Even basic cell phone and Internet coverage is inadequate and exceptionally expensive. Iqaluit, the capital of Nunavut and a town of over 7,000 people in Canada's Eastern Arctic, only recently secured average

Internet connectivity (Infrastructure Canada, 2019). Ironically, its neighboring capital, Nuuk in Greenland, has been connected to the Canadian fiber optic network for several years (McGwin, 2018). The Government of Canada declined the opportunity to build a spur line from St. John's, Newfoundland, to the Nuuk fiber optic cable. At present, there are several serious proposals for a fiber optic line across the Canadian Arctic, enhanced satellite-based Internet, and a variety of other Circumpolar connectivity efforts. If successful, then only Russia, the provincial North in Canada and remote parts of Alaska would be left as the most Internet and cell phone deficient regions in the Circumpolar world. The problems go beyond the Internet, however. Most remote communities rely heavily on diesel-generated electricity, which is both expensive and unreliable. In addition, few of these remote settlements have individuals with advanced technological abilities, including the skills needed to keep Internet-enabled devices fully operational.

Access to high-speed, reliable, and reasonably priced Internet is clearly critical to the future of the new economy in northern regions. Across the Arctic and sub-Arctic Scandinavia, almost all of the communities have competitively priced, high quality Internet and cell phone services. Almost the entire region has significantly better Internet services than the territorial capitals in Canada and remote regions across North America and Russia. The larger centers in Greenland, likewise, have high levels of connectivity. This means that these areas have access to the global Internet retail world and to such comprehensive entertainment sites as Netflix, Hulu and the like. The availability of high-speed, dependable Internet also means that Arctic businesses, artists, and content providers reach international markets and audiences (Delaunay, 2014). However, where high-speed Internet has been provided, including northern Scandinavia, urbanised parts of the provincial Norths in Canada and the major cities in Alaska, there has not so far been a surge in entrepreneurship and commercial innovation that can be connected to the development of digital infrastructure.

What also matters in terms of e-commerce and regional economic development is the balance between the loss of regional business to outside vendors—in the North American Arctic and sub-Arctic the retail "leakage" is severe—and the increased sales for northern sellers because they can reach international consumers quickly and reliably. Through e-commerce, businesses in small, isolated communities have access to global markets, reachable through website-based business systems. But at the same time, northern consumers have thousands of stores, including mega-retailers like Amazon. com, Alibaba, and Rakuten, that offer more products at lower prices than local retailers (Warf, 2011). Ironically, national governments have been subsidizing postal rates for northerners, which means that e-commerce firms can get supplies to Arctic communities at a lower cost than regional stores can. In Iqaluit, Nunavut, as an example, consumers have been ordering cartons of toilet paper, delivered cheaply by mail, undercutting local businesses (Rohner, 2019).

At present, the Arctic is not well-positioned to capitalise on the technology-rich economy. With a few partial exceptions (including the universities in Tromsø, Bodo, Umeå, Luleå, Oulu, Rovaniemi, Fairbanks, and Anchorage), the Arctic lacks the research facilities and the highly skilled personnel necessary to develop marketable innovations and create sustainable innovation eco-systems. Northern entrepreneurs often struggle to find risk capital and the local markets needed to allow them to scale-up operations. Most Arctic education systems are not currently focused on producing students well-versed in science, and technology, engineering. Even more fundamentally, smaller towns and even some of the larger communities do not have the Internet connectivity and technological infrastructure needed to build robust new economy operations. A small number of knowledge producers—filmmakers, creative personnel, musicians, and others—have capitalised on digital delivery systems to expand their commercial operations but these initiatives have yet to accumulate into a significant foundation for a sustainable northern economy.

While there will be substantial technology-based disruptions, are likely, successful adaptations by educational institutions, companies, regions, and governments could transform the Arctic. The flexibility, creativity, and responsiveness of technology-based companies could bring new solutions to bear on everything from energy supplies, environmental monitoring, local food production, on-site medical support, advanced education, and professional services. At present, southern companies and agencies have been extending their reach into the Far North, upsetting regional economies. Over time, regional firms can both produce Arctic-centered solutions to northern problems and opportunities and use modern communication technologies to serve customers and markets in the south and around the world. If the Arctic is passive in the face of the new economy, the region will endure substantial economic disruptions and, potentially, a demographic and commercial hollowing out. With a coordinated, Arctic-wide innovation agenda focused on responding to regional needs and opportunities and building out to service northern, rural and remote markets, the Far North could use the new economy to launch an era of job creation, business development, and prosperity. In the end, as will be argued herein, the more engaged outcome is unlikely, with significant implications for the Far North.

Over the last century, the North has attracted fanciful notions about how to jump-start the regional economy. In the past few decades, only a few have worked: oil and gas in the Russian North, Prudhoe Bay oil in Alaska, diamonds in the Northwest Territories, iron ore in Nunavut, North Sea oil off Norway, Santa Claus in Rovaniemi, Finland, and the massive hydro-electric developments in northern Quebec (Gouvernement du Québec, 2019; Simard, 2017; Brun et al., 2017). In the high-tech sector, notable developments are few in number, including Luleå's impressive transformation into Europe's server farm center and Skellefteå's planned battery mega-factory. This is not for a lack of effort in some communities, particularly in Norway. The massively subsidised Svalbard Island has an impressive science and technology-based economy but is not a model that can be replicated in many other locations. The combination of extensive satellite connectivity and a large international Arctic science capacity in Longyearbyen and a series of former mine sites has provided Svalbard with the foundations for a high-income academic and research-focused economy. Troms County in Norway, led by the University of Tromsø: Norway's Arctic University, has made a concerted effort to build an innovation-centered economy, with some significant achievements in the blue (ocean) economy, including off-shore oil and gas extraction (Hörnström, 2007; The Arctic Institute, 2019). Nord University in Bodø is deeply committed to the promotion of the new economy in the Arctic, focusing primarily on the promotion of entrepreneurship and commerce. Considerably further south, in Trondheim, Norway, the oil and gas economy has spawned a substantial innovation eco-system that has built a more sustainable system beyond the natural resource sector.

Northern Sweden has made a concerted effort to make the transition from the traditional natural resource economy to an expanded base that capitalises on new technologies. Skellefteå, a medium-sized city built on the mining industry, spent two decades trying to re-invent itself for the twenty-first century. In 2017, Northvolt concluded a lengthy inter-city competition and selected Skellefteå as the city for one of the world's largest battery factories. The plant, slated to open in 2023, will produce lithium-ion batteries for the growing electric car market. If the 3.8 billion Euro factory develops as forecast, Skellefteå will require several thousand skilled workers for the main plant and subsidiary operations. This one commercial operation will double the size of the community and provide a level of sustained new economic activity that is quite unique in the sub-Arctic.

Other parts of northern Sweden have undertaken similar transitions. Umeå, on the southern fringe of northern Sweden, has built creative and commercial environmental and renewable energy sectors around the city-owned utilities and the research capabilities of the University of Umeå (OECD, 2020). The long-time mining city of Kiruna, site of the wealthiest iron ore properties in the world, has focused its twenty-first century efforts on space industries, incorporating advanced education and detailed investigations into commercial aspects of space research. Arjeplog, a small natural resource town, reinvented itself as the world's most important center for winter automobile testing, inverting its summer economy into a winter-focused one. Luleå has become the posterchild for sub-Arctic new economic development, building off the strengths of the Luleå University of Technology, and producing hundreds of highly qualified personnel in the region. A satellite campus in Skellefteå focuses on gaming; a connection with the city's economic development office supported the emergence of several start-up gaming companies. The University's presence, plus the important combination of a winter climate and abundant an inexpensive hydroelectricity, made Luleå a highly competitive site for massive server farms. It became, in short order, one of Europe's most important server hosts, managing large portions of the continent's data. More than any other place in the Circumpolar world, northern Sweden demonstrates that sub-Arctic regions can use their natural resource wealth to build a diverse and robust economy that is internationally competitive and connected to the imperatives, demands, and profits of the high-technology economy (Holroyd & Coates, n.d.).

The Yukon's economy relies heavily on government spending, based through sustained and reliable transfers from the Government of Canada,[4] provides a foundation for a high income, high-skilled government economy. The mining sector has been historically important in Yukon since the Klondike Gold Rush in the late nineteenth century, but the territory has struggled to find stability after a strong boom through the 1960s. The industry rebounded in the 2010s, largely behind resurgence in the Dawson City and Mayo region. The Yukon Government's innovation efforts are locally significant but fall far short of creating either a regional eco-system or connecting to a Circumpolar innovation movement. The transition of Yukon College to Yukon University is more symbolic than transformational, for there is not a large investment of new money in the change. The funding of a local incubator, likewise, is in part a repurposing of funds previously spent through Yukon College. But together with a major commitment to high-speed Internet connectivity throughout the territory, the Government of Yukon has continued to support efforts to build the knowledge economy locally. At present, however, the territorial effort has improved the Whitehorse entrepreneurial culture but has had significantly less impact across Yukon (Coates & McPhee-Knowles, n.d.). The Yukon's effort, however small, is by far the most consistent and sustained in northern Canada, but it is overshadowed by the degree to which federal and territorial spending remains the solid foundation of the territorial economy.

The Government of Yukon understands that the boom and bust mining sector is not a strong foundation for the region; the Northwest Territories and Nunavut, in contrast, remain firmly committed to the natural resource economy. There is a small and active Nunavut Research Centre based in Iqaluit, but its most successful venture involves facilitating the work of southern academic institutions, researchers, and students. The Northwest Territories has the most robust effort to localise research activity in the Circumpolar world, mostly involving the hiring and training of local staff to support the work of outside scientists. Efforts to convince the Government of Canada to invest substantially in Arctic science peaked with the establishment of the Canadian High Arctic Research Station (CHARS) based in Cambridge Bay. CHARS, instituted by Prime Minister Stephen Harper in 2012, attracted initial investments of some $190 million and a $7 million annual budget, which was the cornerstone of the Conservative Party's campaign to enhance the northern economy. CHARS, located in the Central Arctic and far from the economic centers in the North, was focused on natural and Arctic science research and was not intended to be an innovation facility. As such, the CHARS investment continued the Government of Canada's pattern of seeing the Far North more in scientific than economic terms, and not recognizing that such initiatives did more to support southern researchers and institutions than these did to promote and enhance northern business.[5]

The Circumpolar World has seen some significant "new economy" achievements. The greatest advances occurred in the oil and gas and mining sectors, including off-shore exploration and development off Norway, the use of drones in mineral exploration in Yukon, and in the government-commercial interface in Arctic search and rescue. Greater changes lie in the offing, through the development of automated trucks (which will accelerate job losses in the Arctic) and the introduction of remotely controlled mining processes (already being used in the Kiruna iron ore mine and potentially disruptive of the workforce requirements for future mining projects). The eventual application of 3D printing could change northern commerce dramatically, as could the extension of digital business implementations.

Inverting innovation

Many of the future innovations, almost certainly developed in the South, especially in education and health care, have comparatively little impact on northern business. At the same time, these same innovations could reduce employment in the Far North—with virtual systems replacing on-site professional staff—and therefore further harm the northern economy. More dramatically, communities are adapting renewable energy systems, including solar power, wind and geothermal energy, to northern settings. These are important innovations, with many upsides, but one important effect could be the further reduction of jobs in the North. Russia, unfettered by environmentalist interventions, has moved rapidly to implement a range of nuclear power options, including a floating nuclear power plant. Implementations of food factories, as at Mayo, Yukon, are providing small-scale proof of concept with important benefits in terms of food security and local production. Scientific and technological innovation, delayed in the Far North, has the potential to both disrupt and improve economic realities in the North. It is yet unclear the relative benefits of rapid technological change in the Circumpolar World. It is obvious that, while significant technological contributions to the Arctic quality of life are already in progress, the net effect could well cause job losses and a weakening of the northern economy overall.

To sustain a dramatic improvement and stabilisation of the northern economy through scientific and technological innovation, the Circumpolar World needs to actually invert innovation. The current approach to the innovation economy serves southern universities, major urban centers, and southern-based companies more than northern commerce and the Arctic economy. Developments in the South, aimed for southern consumers and economies, the argument went, would filter into the North, even though they were not designed for the region, in the region, or by business and researchers in the region. Indeed, almost all the innovations were implemented without a conversation with northerners or even with an evaluation of the potential impact of disruptive technologies on northern peoples.

Inverting innovation requires repositioning the scientific and technological enterprise to focus on the needs of remote and northern regions. Many of the arguments about contemporary infrastructure, for example, focus on the need to bring the Arctic areas up to national and international standards by importing southern models into the North rather than on building region-specific technology models. Northern institutions and northern businesses can be empowered to develop new technologies that could be used, first and foremost, to address northern liabilities, needs, and interests. Inverting innovation would, likewise, require a change in emphasis from the standard investments in traditional science and university research and a new focus on addressing regional social problems, a strategy that has recently found favor in Japan (Fujisawa et al., 2015). The effort would require major changes in the Far North. It would require an upgrading of education systems, particularly in the smaller communities and the post-secondary institutions, to build the foundations to long-term engagement in scientific and technologies discoveries. A new approach would require, further, the upgrading of Arctic Internet and technological systems, major changes in the policy of sub-Arctic and Arctic governments. Using the often-under-utilised power of government procurement, regional authorities will have to work with local, regional and Circumpolar businesses to identify northern-centric products and services that could, if properly developed, find Circumpolar markets. The innovation empowered North would include widespread introductions of alternative energy, major changes in housing design and construction, improvements in educational delivery, quality and suitability for the Arctic, more accessible and engaged government services, and dramatic changes in wildlife and resource management.

This kind of a transition will not be cheap or easy and is certainly not assured. Countries and regions around the work are making comparable efforts to respond to the opportunities presented by the innovation economies. There is no shortage of demand for venture capital and highly skilled personnel. Business opportunities have to be identified and this is limited by the small, marginal, and widely scattered northern consumer and industrial markets. It is important to note that the largest commercial ventures connected to the "new economy," including the Northvolt battery plant in Skellefteå, Sweden, and the large server farms and related enterprises around Luleå, were both designed to capitalise on standard northern advantages (including winter and inexpensive hydroelectricity). National governments could do a great deal more to shift innovation expenditures to focus on local needs and opportunities and to investigate, as the United States has done so successfully over the years, commercial possibilities arising out of military activities. The possibilities are limited here, given the current small scale of military activity in the region, although rapid militarisation of the Russian Arctic could result in a 1950s-type surge in military-based economic development.

The Circumpolar World does not have a lot of advantages in the current highly competitive innovation environment. Indeed, a review of the standard factors of regional economic development—available highly skilled personnel, strong local post-secondary institutions with substantial research capacity, access to investment capital, a smart and wealthy consumer base, ready access to international markets, update-to-date infrastructure, and a strong entrepreneurial culture—reveals significant deficiencies in the Far North. Indeed, the smaller communities across the Arctic, already suffering from the diseconomies of smallness, isolation, and limited educational and commercial prospects, are much more likely to be disrupted by the "new economy" than stimulated by it. In the major cities, governments (including the armed forces) dominate the local economy and can usually crowd out the demand for workers, capital, and resources by the innovation sector. Conversely, these government-dominated centers could, as Whitehorse, Yukon, is attempting to do, become nodes of innovation activity.

As the Circumpolar World looks to position itself within the global innovation economy, the region and its residents face crucial questions about how they want to proceed. They could continue along the current, half-hearted path of engagement in the commercialisation and science and technology. There will be occasional achievements, primarily those of individual companies and cities, like Oulu, Kiruna, Luleå, and Skellefteå, that make themselves attractive to local business and international investors. They can, like Tromsø and Bodo in Norway, emphasise connections to the local resource economy, in this instance fishing and oil and gas exploration, and mobilizing local researchers and businesses to advance regional innovation. Laurentian University (in the city of Sudbury in northern Ontario) has collaborated with the government and business to create a strong regional innovation system connected to mining, but even this technically impressive effort has produced few jobs and new businesses.

Both of these approaches have proven somewhat economically successful and have stabilised local economies in important and long-term ways. In each of these cases, leading investments by governments in post-secondary education and research capabilities have been mobilised to support regional business operations. Investments in post-secondary institutions are no guarantee of commercial success, though, as seen by the uneven economic developments connected to the major campuses of the University of Alaska (Fairbanks, Anchorage, and Juneau), the University of Northern British Columbia, (Prince George, B.C.), Lakehead University (Thunder Bay, Ontario), University of Lapland (Rovaniemi, Finland) and the smaller colleges and university colleges across the Far North. Post-secondary institutions help societies in many ways, including by giving a boost to the local economy through the presence of students, faculty, and staff and by stabilizing the regional society. But, unfortunately, the opening or expansion of a new institution does not ensure a burst of scientific and technological innovation and the expansion of the economy (Kudtuashove & Sorokin, 2019). The effort to create a pan-Arctic institution, the University of the Arctic, produced small successes but not the region-integrating transformation desired (Poelzer, 2017).

There are other options available to northern regions seeking to connect to the global innovation effort. Northern communities could, recognizing the realities of the international innovation economy, realise that they will not be truly competitive and retain their focus on the current combination of government spending and natural resource development, specializing on monitoring and shaping the adoption of new technologies in the North. In this scenario, local businesses and government would assess and attract innovative technologies, adjusting training and employment systems to ensure that local workers benefit as much as possible from the new circumstances. Regions taking this approach would highlight the ability of emerging technologies to lower the expense of Arctic production, improve mineral exploration, and reduce societal costs through active engagement with the introduction of new products, including through the permitting, licensing, and regulation of science-based innovations.

Arctic regions could be more assertive in promoting an innovation-based economy. They would have to adapt the current model, which starts with an emphasis on basic research and early stage development, to suit the comparative shortcomings of the Arctic innovation environment. In this instance, Arctic authorities would leave much of the basic research to southern and urban institutions, continuing to focus their efforts on northern eco-system research. Innovation efforts would emphasise attracting external researchers and companies to bring compelling technologies to the North to ascertain their commercial suitability in the Arctic. The commercial effort would focus on testing and adapting emerging technologies to northern conditions, working with local business to see if northern consumers will accept the new product or service and marketing the items to other Circumpolar and remote regions. This latter approach, which is used by individual entrepreneurs and businesses, particularly in the mining sector, has not been adopted by regional governments as the foundation for the economy.

The search for a strong, stable, and sustainable Arctic economy remains a cornerstone of national and regional government efforts in the Far North. The Arctic's long-standing reliance on natural resource development has left the region subject to the vagaries of boom and bust cycles. Although commentators rarely see government as a core element in the economy, the reality is that Arctic governments (especially if military spending is included) provide stability to northern economies. These regions are all heavily subsidised—Russia removed many of its financial supports after the collapse of the Union of Soviet Socialist Republics, leading to a sharp decline in the

Siberian economy—and enjoy considerable stability for reasons of strategic importance, a commitment to social justice on reasons of regional equity and Indigenous equality, and the assertion of national sovereignty (Mote, 2018).

National and regional governments the world over have approached innovation with a combination of apprehension and enthusiasm. There is growing political and administrative awareness of the potential of the "Second Machine Age" or the "Fourth Industrial Revolution" to transform economies and cause substantial disruptions in the worlds of work and business (Brynjolfsson & McAfee, 2014; Schwab, 2018; Spencer, 2017). Technology-based services such as electronic banking, automated mines, online entertainment, e-commerce, and digital health systems are displacing northern workers and undermining the viability of Arctic companies. But at the same time, advanced manufacturing and 3D printing technologies have the potential to address many Arctic supply and construction issues.[6] The technological future is filled with potential and uncertainty. The Arctic's discussion of the next economy has been only tangentially engaged in these conversations, even though the technological disruptions have already been felt in many sectors. The reality is that times of technological disruption can distort and challenge existing economies, often resulting in substantial changes for good and for ill. The early decades of the twenty-first century are proving to be such a time for much of the world, especially the North.

Conclusion

Given the exceptionally high cost of being competitive in high technology fields and serious deficits in infrastructure and highly qualified personnel, the chances are slim that the North will be able to capitalise on the commercial advantages of the age of science and technology. There will be individual companies that do well and a small number of Arctic-based "lone eagles,"[7] high-value professionals who chose the North for lifestyle reasons and who work globally. Technological intrusions will continue, particularly in the natural resource sector; this could result in substantial northern job loss and workplace redefinition. Technological advances will make the Northern economy different, just as demand for more ecologically sustainable development could reduce the importance of traditional natural resource development, at least outside Arctic Russia. It is not yet clear which economic sectors, beyond government operations and seasonal international tourism, have sufficient scale to contribute substantially to the building of northern prosperity. Technology-driven economic growth, while still attracting a great deal of government attention, has rarely worked well outside of major metropolitan areas. There is little reason to believe that the Arctic will stand out as an exception to this rule.

At present, there are few reasons to be overly optimistic about the innovation future, and therefore the pursuit of a sustainable economy, in much of the Far North. The region generally lacks the fundamentals of a truly innovative economy, save for a few centers. Northern Sweden and Troms County in Norway are perhaps the prototypes of an adaptable, competitive sub-Arctic or Arctic economy. Northern Scandinavia is doing comparatively well in this regard, buoyed by continued resource development and some new economy successes. At present, however, the Circumpolar world is not well-prepared for the rapid, technology-based, and science-informed transformation. The world appears to be shifting toward a city-state economy, with economic opportunity focused in the larger centers with the most research-intensive university, college and institute environments. With the natural resource economy in a state of dramatic flux, the disruptive potential of the innovation economy could lead to a further weakening of the Arctic and sub-Arctic.

The age of science and technology has been surrounded by hyperbole and unchecked enthusiasm as people are now discovering, as jobs are lost and companies are displaced. In the Far North, disruptions are emerging faster than tangible economic benefits. While small improvements and individual corporate successes are both happening and likely, the macro-economic prospects are not overly promising. Preparations for the new economy require a reinvention of public education, a reformatting of government-business relations, and a whole economy approach to adapting emerging technologies to improve the quality of life in the Far North.

Notes

1. On the global innovation-based economy, see Kao, John. *Innovation nation: How America is losing its innovation edge, why it matters, and what we can do to get it back*. Simon and Schuster, 2007; Zhou, Lazonick, and Sun (Eds.) *China as an innovation nation*. Oxford University Press, 2016; Holroyd and Coates. *Innovation nation: science and technology in 21st century Japan*. Springer, 2007; Senor, and Singer. *Start-up nation: The story of Israel's economic miracle*. Random House Digital, Inc., 2011.
2. The gambles do no always work. For an excellent reminder of this reality, see Wong, Joseph. *Betting on biotech: Innovation and the limits of Asia's developmental state*. Cornell University Press, 2011.
3. The website of the Geophysical Institute is https://www.gi.alaska.edu/
4. The annual federal transfer to the Yukon Territorial Government amounts to approximately $1.2 billion per year.
5. Information on CHARS can be found at https://www.canada.ca/en/polar-knowledge/CHARScampus.html.
6. On 3D printing and house construction, see https://www.conferenceboard.ca/e-library/abstract.aspx?did=9948.
7. For an interesting Alaskan commentary see Odasz, Frank (1999).

References

The Arctic Institute. (2019, April 1). *The blue economy potential of Alaska and North Norway (AlaskaNor)*. The Arctic Institute. https://www.thearcticinstitute.org/blue-economy-potential-alaska-north-norway/

Bondar, C. S. (2015). Russian Militarization of the Arctic. In *Proceedings of the International Scientific Conference "Strategies XXI": The Complex and Dynamic Nature of the Security Environment* (Vol. 2) (pp. 12–18). Centre for Defence and Security Strategic Studies (CDSSS).

Bramwell, A., Nelles, J., & Wolfe, D. A. (2008). Knowledge, innovation and institutions: Global and local dimensions of the ICT cluster in Waterloo, Canada. *Regional Studies, 42*(1), 101–116.

Bramwell, A., & Wolfe, D. A. (2008). Universities and regional economic development: The entrepreneurial University of Waterloo. *Research Policy, 37*(8), 1175–1187.

Brun, A. et al. (2017). Le Plan Nord: enjeux géopolitiques actuels au regard des "plans nord" passés. *Recherches sociographiques, 58*(2), 297–335.

Brynjolfsson, E., & McAfee, A. (2014). *The second machine age: Work, progress, and prosperity in a time of brilliant technologies.* WW Norton & Company.

Coates, K., & McPhee-Knowles, S. (n.d.). The territories: Inverting innovation. In P.W.B. Philips, & D. Castle (Eds.), *Scientific and technological innovation in Canada.* University of Toronto Press.

Collins, S. W. (2001). Academic research and regional innovation: Insights from Seattle, Washington. *Industry and Higher Education, 15*(3), 169–178.

David, H. (2015). Why are there still so many jobs? The history and future of workplace automation. *Journal of Economic Perspectives, 29*(3), 3–30.

Delaunay, M. (2014). The Arctic: A new internet highway. *Arctic Yearbook*, 503–510.

Etzkowitz, H. (2002). *MIT and the rise of entrepreneurial science.* Routledge.

Etzkowitz, H. (2003). Innovation in innovation: The triple helix of university-industry-government relations. *Social Science Information, 42*(3), 293–337.

Etzkowitz, H., & Leydesdorff, L. (2000). The dynamics of innovation: From national systems and "Mode 2" to a triple helix of university–industry–government relations. *Research Policy, 29*(2), 109–123.

Etzkowitz, H., & Zhou, C. (2017). *The triple helix: University–industry–government innovation and entrepreneurship.* Routledge.

Farish, M. (2013). The lab and the land: Overcoming the Arctic in Cold War Alaska. *Isis, 104*(1), 1–29.

Flake, L. E. (2014). Russia's security intentions in a melting Arctic. *Military and Strategic Affairs, 6*(1), 99–116.

Ford, M. (2015). *Rise of the robots: Technology and the threat of a jobless future.* Basic Books.

Frank, O. "Alaskan professional development: Lone eagles learn to" teach from any beach!"." *THE Journal (Technological Horizons in Education)* 27.4 (1999): 90

Frey, C. B., & Osborne, M. A. (2017). The future of employment: How susceptible are jobs to computerisation? *Technological Forecasting and Social Change, 114*, 254–280.

Fritz, S. A. (2010). *DEW line passage: Tracing the legacies of Arctic militarization.* University of Alaska Fairbanks.

Fujisawa, Y., et al. (2015). A study of social innovation concepts: A Japanese perspective. *Japan Social Innovation Journal, 5*(1), 1–13.

Gareis, J., & Mercer, A. (2015). Celebrating the 50th anniversary of the Inuvik Research Laboratory. *Arctic, 68*(1), 132–139.

Gibbons, M. (Eds.). (1994). *The new production of knowledge: The dynamics of science and research in contemporary societies.* Sage.

Gouvernement du Québec. (2019). *Plan Nord.* Société du Plan Nord. https://plannord.gouv.qc.ca/en/

Grydehøj, A. (2020). Svalbard: international relations in an exceptionally international territory. In K. Coates & C. Holroyd (Eds.), *The Palgrave handbook of Arctic policy and politics* (pp. 267–282). Palgrave Macmillan.

Hall, H. M. (2020). Innovation, new technologies, and the future of the Circumpolar North. In K. Coates & C. Holroyd (Eds.), *The Palgrave handbook of Arctic policy and politics* (pp. 117–132). Palgrave Macmillan.

Hall, H., Leader, J., & Coates, K. (2017). Introduction: Building a Circumpolar innovation agenda. *Northern Review, 45,* 3–10.

Holroyd, C., & Coates, K. (2007). *Innovation nation: Science and technology in 21st century Japan.* Springer.

Holroyd, C., & Coates, K. (2012). *Digital media in East Asia: National innovation and the transformation of a region.* Cambria Press.

Holroyd, C., & Coates, K. (2015). *The global digital economy: A comparative policy analysis-student edition.* Cambria Press.

Holroyd, C., & Coates, K. (n.d.). *Pathways to a new northern economy? Community economic development strategies in Norrbotten, Sweden.* Unpublished research paper.

Hörnström, L. (2007). *Regional policy and regionalisation in the Northern peripheries-the case of Troms county.* Nordic Local Government Research Conference, Gothenburg.

Infrastructure Canada. (2019, August 19). *Nunavut residents to benefit from high speed internet.* Newswire. https://www.newswire.ca/news-releases/nunavut-residents-to-benefit-from-high-speed-internet-806159319.html

Inkinen, T., & Suorsa, K. (2010). Intermediaries in regional innovation systems: High-technology enterprise survey from Northern Finland. *European Planning Studies, 18*(2), 169–187.

Jingfors, K. T., & Klein, D. R. (1982). Productivity in recently established muskox populations in Alaska. *The Journal of Wildlife Management, 46*(4), 1092–1096.

Kao, J. (2007). *Innovation nation: How America is losing its innovation edge, why it matters, and what we can do to get it back.* Simon and Schuster.

Kenney, M. (2000). *Understanding Silicon Valley: The anatomy of an entrepreneurial region.* Stanford University Press.

Kudtuashove, E. V., & Sorokin, S. E. (2019). The "third mission" in the Arctic universities' development strategies. *Social and Economic Development, 34,* 17–34.

Lundvall, B. (1992). *National systems of innovation: Towards a theory of innovation and interactive learning.* Pinter Publishers.

McGwin, K., (2018, November 13). No matter where you live in Greenland, you can now get a 4G signal. *Arctic Today.* https://www.arctictoday.com/no-matter-live-greenland-can-now-get-4g-signal/

Mote, V. L. (2018). *Siberia: Worlds apart.* Routledge.

Nelson, R. R., & Rosenberg, N. (1993). Technical innovation and national systems. In R.R. Nelson (Eds.), *National innovation systems: A comparative analysis* (pp. 3–21). Oxford University Press.

Niosi, J., & Bas, T. G. (2003). Biotechnology megacentres: Montreal and Toronto regional systems of innovation. *European Planning Studies, 11*(7), 789–804.

OECD. (2020). The circular economy in Umeå, Sweden. *OECD Urban Studies.* https://doi.org/10.1787/4ec5dbcd-en.

Osherenko, G., & Young, O. R. (2005). *The age of the Arctic: Hot conflicts and cold realities.* Cambridge University Press.

Pempel, T. J. (1987). The unbundling of "Japan, Inc.": The changing dynamics of Japanese policy formation. *Journal of Japanese Studies, 13*(2), 271–306.

Poelzer, G. (2007). The University of the Arctic: From vision to reality. *Northern Review, 27,* 28–37.

Rantakokko, M. (2012). Smart city as an innovation engine: Case Oulu. *Elektrotehniski Vestnik, 79*(5), 248.

Rohner, T. (2019, December 13). *Why people in Canada's remote Arctic capital are obsessed with Amazon Prime. The Guardian.* https://www.theguardian.com/world/2019/dec/13/canada-iqaluit-amazon-prime

Schwab, K. (2018). *The fourth industrial revolution.* Currency.

Schwab, K., & Davis, N. (2018). *Shaping the future of the fourth industrial revolution.* Currency.

Senor, D., & Singer, S. (2011). *Start-up nation: The story of Israel's economic miracle.* Random House Digital, Inc.

Sergunin, A. (2015). *Russia in the Arctic: Hard or soft power?* (Vol. 149). Columbia University Press.

Simard, M. (2017). Le nord québécois: un plan, trois régions, neuf défis. *Recherches sociographiques, 58*(2), 263–295.

Spencer, D. (2017). Work in and beyond the second machine age: The politics of production and digital technologies. *Work, Employment and Society, 31*(1), 142–152.

Spohr, A. P., Höring, J. S., Cerioli, L. G., Lersch, B., & Soares, J. G. S. (2013). The militarization of the Arctic: Political, economic and climate challenges. *UFRGS Model United Nations Journal, 1,* 11–70.

Warf, B. (2011). Contours, contrasts, and contradictions of the Arctic internet. *Polar Geography, 34*(3), 193–208.

Zhou, Y., Lazonick, W., & Sun, Y. (Eds.) (2016). *China as an innovation nation.* Oxford University Press.

4 The potential of art and design for renewable economies in the Arctic

*Timo Jokela, Glen Coutts, Ruth Beer,
Svetlana Usenyuk-Kravchuc, Herminia Din,
& Maria Huhmarniemi*

Introduction

In this chapter, we discuss some of the ways in which art and design practices might help to support and develop renewable economies in the Arctic.
We use the term Arctic Art and Design (AAD) to refer to contemporary art,
design, and media productions aiming to contribute to renewable economies and sustainable development in the particular context of the North and
the Arctic. The sustainable development of the Arctic is defined in a variety of ways for different purposes and occasions (Fondahl & Wilson, 2017;
Gad, Jacobsen, & Strandsbjerg, 2019; Stephen, 2018; Tennberg, Lempinen,
& Pirnes, 2019). The dimensions of sustainability in this context include cultural and social sustainability, which means that the contemporary renewable productions must respect cultural diversity and heritage and must be
produced in collaboration with local inhabitants, so that the economic benefits are shared with the region. Various aspects and examples of AAD have
been studied in the research projects conducted in the Arctic Sustainable
Arts and Design (ASAD) network at the University of the Arctic (ASAD,
2019; Jokela & Coutts, 2018). In this chapter, we use this concept to describe
art, crafts, design, and cultural productions that transmit the heritage of
Arctic nature and culture. It is not limited to Indigenous art, instead, it also
covers non-Indigenous arts and their liminal productions such as indus-
trially produced craft-based products (Jokela, Huhmarniemi, & Hautala-
Hirvioja, 2019). We use the concept of AAD to highlight the view of art,
design, and crafts as interwoven with one another and as an integrated part
of the eco-social culture in the North (Härkönen, Huhmarniemi & Jokela,
2018; Jokela, 2017). The idea follows the concept of *duodji*, which sees Sámi
art, craft, and design as the union of expression, production, and way of
living (Guttorm, 2015). In addition, the concept of AAD carries the idea
of applying arts to societal and economic needs (Huhmarniemi & Jokela,
2019; Jokela, 2013) and combining the methods of socially engaged art
and service design (Härkönen & Vuontisjärvi, 2018; Jokela & Tahkokallio,
2015). AAD as the creative renewable economy can be seen as one of the
Arctic models of smart specialisation on green economy and as a model to

DOI: 10.4324/9781003172406-4

aim to economic and social resilience in rapidly changing Arctic regions (Giacometti & Teräs, 2019; Woien, Kristensen, & Teräs, 2019).

As industries traditionally associated with the Arctic region, such as large-scale resource extraction and global exploitation of finite natural resources, are increasingly seen as unsustainable, we applaud the move toward more sustainable business practices. We argue that the creative industries, and art and design in particular, can play a central role in developing new, more sustainable business opportunities that benefit the economy while preserving and promoting more local, place-based, and renewable business practices. The underpinning philosophy of AAD is closely related to that of the cultural and creative industries (Hesmondhalgh, 2007) or, as described by Howkins (2001), the creative economies. According to Howkins (2001, pp. 88–117), the creative economy comprises advertising, architecture, art, crafts, design, fashion, film, music, performing arts, publishing, research and development, software, toys, and games, TV and radio, and video games. Writing almost a quarter of a century ago, Landry and Bianchini (1995, p. 4) contended that "the industries of the twenty-first century will depend increasingly on the generation of knowledge through creativity and innovation." The economic impact of the creative industries has been measured on worldwide by United Nations Conference on Trade and Development (UNCTAD, 2018), and in Nordic Arctic countries (Olsen et al., 2016) and has been found to exceed that of the driver industry. The creative economy report by United Nations Conference on Trade and Development (UNCTAD, 2018) concluded: "The creative economy is recognized as a significant sector and a meaningful contributor to national gross domestic product. It has spurred innovation and knowledge transfer across all sectors of the economy and is a critical sector to foster inclusive development." In this report, both creative goods or products and creative services are subsumed under the term "creative economy." The discourse of potential of the creative economy started, and still is, mainly connected to urban cities, centers, and innovation hubs.

Few articles and studies have characterised the processes of creative economy development in the Arctic and hardly any of them specialised to the field of art and design (Nordic Councils of Ministers, 2018; Olsen et al., 2016; Petrov, 2014, Petrov, 2016, Petrov, 2017). The discussion is often focused on the challenges creative economies face under limiting factors such as the exploitation of nature at the core of the Arctic economy, population decline, high economic costs due to long distances, and the globalisation of the Arctic region. For example, crafters in Lapland are said to be lifestyle entrepreneurs and microentrepreneurs who give priority to artistic work and hesitate to step into business-oriented work, for example, with the tourism field (Kugapi, Huhmarniemi & Laivamaa, Forthcoming, 2020). Opportunities for renewable economies are hardly discussed, and the material and cultural heritage of the Arctic, commonly connected with Indigenous crafts and the skillful use of natural materials, has not been recognised as having the potential to make significant contributions to the economy, although we argue that this is now changing.

Our AAD model focuses on the economic potential of renewable natural and cultural resources that are plentiful in the Arctic. The notion of *ecosystem services* (ES) is central to our thinking (Milcu, Hanspach, Abson, & Fischer, 2013). ES focus on the use of nature, its conservation, and its social, cultural and economic relation to the Arctic. In addition to ES, such as water, wood, fibers, and food provisions and their use as new renewable bioeconomy (Teräs et al., 2014), the concept also includes *cultural ecosystem services* (CES), that is, the "non-material benefits obtained through spiritual enrichment, cognitive development, reflection, recreation, education, and aesthetic experiences" (Millennium Ecosystem Assessment, 2005, p. 4). Even when these cultural values are included in ES typologies, cultural, experiential, and other non-material values have generally received less attention compared to monetary and ecological values. Only a few studies, among other Chan, Satterfield, and Goldstein (2012) and Daniel et al. (2012), have paid attention to the cultural aspects of CES.

In this study, we will fill the gap by investigating the potential of art and design as renewable economies using the concepts of CES, and place-making as our theoretical and practical framework. Drawing on some of our own research and development projects, we present four cases that illustrate the ways in which AAD can be a contributor to a creative renewable economy. Our examples are from Alaska (United States), Canada, Finland, and Russia. We will share our experiences and findings and suggest future lines of enquiry.

Cultural ecosystem services and place-based development as conceptual framework

The authors of this chapter have been collaborating on the research and development of wide-ranging themes in AAD for many years. The ASAD research and development network, which was founded in 2011, has been instrumental the development of this innovative work (ASAD, 2019). Since its inception, ASAD has sought to "identify and share contemporary and innovative practices in teaching, learning, research and knowledge exchange in the fields of arts, design and visual culture education" (ASAD, 2019; Jokela & Coutts, 2018b). The organisation is one of the thematic networks of the University of the Arctic that aim to "foster issues-based cooperation within networks that are focused but flexible enough to respond quickly to topical Arctic issues" (University of the Arctic, 2019).

Among ASAD members, there has been emerging an interest in CES, and we believe that in the rapidly changing Arctic, discussions on CES can be used to weigh the balance between the use of nature, its conservation, and the social, cultural, and economic relations in the Arctic. According to the study by Milcu et al. (2013), mobilising CES as binding elements between social and ecological conceptual constructs is the core idea of the sustainability ideal. Hearnshaw and Cullen (2010) point out that thoroughly accounting for CES will be helpful in balancing primarily economic considerations and facilitate a more inclusive socio-ecological approach by exploring the interactions between social, ecological, and economic processes.

We know that in a global world, all cultural values in art and design may have little direct dependence on ecosystems, but we argue an especially significant relationship between ecosystems and the fulfillment of human needs can be demonstrated in Arctic CES.

We see that CES offer a theoretical and practical framework to think differently about the ways in which creative entrepreneurs may collaborate with communities using local natural, cultural, and social resources. Such collaboration, of course, must be undertaken in a way that is sensitive to and respectful of the unique nature, culture, and heritage of the Arctic and takes into account megatrends such as climate change, globalisation, and urbanisation (Nordic Council of Ministers, 2011; Stephen, 2018.)

In our thinking about AAD, CES are closely connected to place-based strategy, which is also known as place-making and can also be understood as an economic development strategy. It is the practice of using places and a community's capacities to make economic progress (Milone & Ventura, 2010; Vodden, Gibson, & Baldacchino, 2015). Building on existing strengths, this approach focuses on CES and the unique features of particular places to boost existing businesses and create new ones and even attract new investment. According to Daniels, Baldacchino, and Vodden (2015), place-based strategy is a reaction to conventional top-down, single-sector, national-stage development projects. Thus, place-making can also be understood as an identity policy of remote, rural and peripheral places that are centers for their inhabitants.

We are also familiar with the criticism of utilising CES in art and design and in the discussion of the creative economy. Spiritual and aesthetic cultural values are not best captured by instrumental or consequentialist thinking and they are grounded in conceptions of nature that differ from the ES conceptual framework (Cooper, Bardy, Steen, & Bryce, 2016). The different attitudes toward the use of CES can sometimes be quite passionate, since they are tightly bound to human values and behavior as well as social and cultural institutions and economic and political organisations. In the Arctic, they are also bound to Indigenous and non-Indigenous issues and relations.

An important factor is the effect of the diversity of ES to the diversity of Indigenous and non-Indigenous cultures in the North and the Arctic. Ecosystems influence the types of social relations that are established in particular cultures. The social relations in fishing societies, for example, differ in many respects from those in nomadic reindeer herding or non-Indigenous agricultural societies. According to Stephen (2018), the climate crisis has caused changes in harvesting, hunting, and fishing cultures, which has had a wider impact on cultural identities and traditional knowledge. In other words, the climate crisis has brought about changes in the ecosystems and has had effects on socio-economic and political realities, which has affected the cultures and self-understanding of Arctic Indigenous populations. We argue this is the case in many non-Indigenous communities in the North and the Arctic as well.

Besides material and social relations, Indigenous cultures of the Arctic add spiritual and religious dimensions and values to ecosystems. This calls for a certain cultural sensitivity in approaching Arctic art and design activities. Most often, it is commercial design productions and items that represents identities (for example, clothing) that cause heated discussions on cultural appropriation and exploitation. Visual symbols such as patterns and orna-ments have significance in the continuation of cultures and even the shar-ing of world views (Joy, 2019; Kramvig & Flemmen, 2019; Minnakhmetova, Usenyuk-Kravchuk, & Konkova, 2019; Schilar & Keskitalo, 2018). Thus, seeing Indigenous cultural traditions as an economic resource can cause tensions (Olsen et al., 2019 forthcoming; Smith, 1999). However, if members of Indigenous peoples themselves are participating in the transformation of tradition into contemporary and economic products, then there is less or no criticism. For example, the Sámi people were invited to collaborate in the production of the Disney film *Frozen 2*, which depicts Sámi culture.

In our review of the studies and articles dealing with CES, we have seen that the majority of these articles have been published in ecological jour-nals. From an art and design point of view, this may be a partial explanation for the rather vague discussion of the creative industries or art and design in most of the examined articles. On the other hand, however, art history research has shown that even when the concept of CES is not in use, nature still provides a rich source of inspiration for arts, design, craft, media, and architecture, especially in the Arctic (Mäkikalli, Holt, & Hautala-Hirvioja, 2019). The way nature inspires artists, can be seen to CES.

Besides art, many people find aesthetic, expressive, and emotional values in various aspects of ecosystems. For example, people value landscapes that are known for their beauty or the "sense of place" that includes locally and culturally significant stories and heritage (Hølleland, Skrede, & Holmgaard, 2017; Lindhjem, Reinvang, & Zandersen, 2015). These can all be connected to renewable creative industries because they affect where people choose to go to spend their leisure time or improve their well-being. CES plays an important role in nature, ecological, and cultural tourism and recrea-tion, which is often supported, represented, advertised, and made know by means of the creative industries (de la Barre et al., 2016; Müller & Viken, 2017; Rantala et al., 2019). When considering creative services as art and design products and goods, we see many opportunities for collaborations between art and design and tourism as renewable economies in the Arctic.

The following four case studies will demonstrate how art and design can play a role in cultural sensitive place-based renewable economies based on CES around the North and the Arctic.

Isuma: A lens to Canada's North

Canada is known for its abundance of natural resources and empty wilderness, seemingly ready for the taking. However, basing the economy on resource extraction such as mining and fossil fuel-related industries so prevalent in the Arctic is not sustainable. In order to protect the balanced ecosystems of both humans and non-humans, there is a growing need for the development of alternative and diverse economies more aligned with sustainability, in relation to both the environment and the well-being of communities and culture in the North (Schott, 2016). In many examples, CES literature refers more to recreational or touristic values, rather than a deep engagement with what the concept of culture means (Ihammar & Pedersen, 2017). Meanwhile, in Igloolik, Nunavut, the artist collective Isuma demonstrates the potential of cultural initiatives and renewable economies, using the concepts of ES, particularly CES, and place-making, while honoring their material and cultural heritage (Big River Analytics, 2017).

As Canada's first Inuit video-based production company, Isuma has a surprisingly long history (Kunuk, 2019). Co-founder Zacharias Kunuk began exploring the possibilities of the film almost three decades ago, when he used the profits from selling his traditional soapstone carvings to purchase his first video camera. This technology introduced him to new possibilities of storytelling, highlighting the life and landscape of the North, and in 1990 Kunuk and his collaborators, Paul Apak Angilirq and Norman Cohn, created Isuma Productions.

Since its inception, Isuma has produced a number of feature films, documentaries, TV series, and short films, including *Atanarjuat: The Fast Runner* (2001), *The Journals of Knud Rasmussen* (2006), *Before Tomorrow (Le jour avant le lendemain)* (2008), and *One Day in the Life of Noah Piugattuk* (2019), based on the landscape and stories of the Canadian Arctic. In 2008, they launched IsumaTV, a collaborative multimedia knowledge-sharing platform for Indigenous filmmakers and media organisations, and in 2012 they introduced Digital Indigenous Democracy, an innovative platform linking communities and presenting politics and legal issues within a culturally accessible framework. Based on knowledge and skill sharing, these initiatives feature stories of heritage and contemporary life. They promote interest in the land and people that contribute to the development of CES and related renewable economies through sharing with audiences and preserving for themselves the experiences of society and cultures in the North.

Isuma's films and other media projects are in oral Inuktitut, and like many AAD projects, encourage cultural preservation and resilience. This provides the community Elders with the ability to understand and appreciate the films, while strengthening the development and revitalisation of language and culture. The content on IsumaTV is accessible in over eighty languages, including Indigenous languages such as Cree, Ojibwe, and the Northern Athabaskan languages of the Na-Dene peoples. This contributes to the restoration of vulnerable languages which are intrinsic to cultural identity and knowledge.

To further this outreach and communication, Isuma has instituted an innovative digital service that brings media access, otherwise unavailable, to small, isolated communities in the Canadian North (Leask, 2016). Isuma productions and other creative Indigenous media activities have fostered interest, especially among northern youth, in educational and training opportunities that support these growing and economically impactful cultural initiatives, and in developing professional practices in video and film production, broadcasting, and related industries. Media education is provided by not-for-profit educational initiatives such as Wapikoni (2019) and Our World (2018) that work with local schools and hold workshops in remote Indigenous communities, encouraging youth to follow their passions while giving them practical skillsets to pursue their ambitions.

As Isuma becomes increasingly recognised in the Canadian North, it is also making an impact internationally. Isuma's *Atanarjuat: The Fast Runner* won Best First Feature Film at the 2001 Cannes Film Festival, and the production company has been recognised by six Genie Awards and numerous other international film awards. In 2019, the collective was selected to represent Canada in the prestigious Venice Art Biennale (Canadian Art, 2017; Sandals, 2019). In addition to the regional economic benefits of the film industry, the stories of the people who have been underrepresented and isolated from national and international conversations are now on the world stage, sharing culture and values, building resilience, and creating and strengthening international and intercultural alliances.

The changing landscape of the Arctic, a theme in many of Isuma's films, is a global topic of growing concern. For the Indigenous residents of the Canadian Arctic, this issue literally hits close to home: living off the land is not only an essential source of sustenance, but also necessary for the preservation of culture, language, and knowledge passed down from one generation to another. Isuma initiatives and the positive responses to their work demonstrate the importance of representing different knowledge systems, worldviews, and attention to cultural sensitivity within the concept of CES (Chan et al., 2012; Lepofsky et al., 2017).

Lapland snow and ice design and art in Finland

The snowy landscape has an important role to play in the local customs and traditions, identities, and cultures in the Arctic. The use of snow and ice in tourism has a long history in Finland, Lapland, and can be found in many forms, starting with snow-related sports. Snow and ice can be seen as ES and winter traditions as CES.

The concept of winter art was introduced in 2003 to describe the artistic features and phenomena related to winter aesthetics in Lapland, considering the cultural changes and opportunities related to the winter: "One manifestation of this change is the brisk increase in winter festivals, winter theatres, snow and ice sculpting events and snow architecture. At their best these phenomena can be called winter art" (Jokela, 2003, p. 7). Since then, art and design at the University of Lapland and the cold climate engineering department at the Lapland University of Applied Sciences have collaborated with local businesses on a project entitled Lapland Snow Design. The collaboration aimed to create new knowledge, innovations, and practices for the tourism and business sectors that utilised thinking and competence in snow and ice construction technologies (Jokela, 2014, 2019).

Long-term development brought artists, designers, and companies together in order to develop new high-quality products in cooperation with universities and businesses. At the same time, creative services based on the winter ecosystem were developed. The objective was to develop an internationally competitive product to leverage Lapland's versatile expertise of snow-related technologies and applications in the different services of the tourism industry. Different types of snow and ice environments, collaborative design methods, and marketing concepts were implemented during the project. Besides the development of new design and implementation methods for winter art, another goal was to build regional teamwork capacity and boots place-based thinking. The aim was to develop new and more efficient organisations to expand the applications of winter art in the business sector. One of the key outcomes of the project was the creation of a cluster-type regional network of experts that were able to apply the products in domestic and foreign contexts. The practices were designed to be flexible, so that they could be applied in a variety of environments and tailored to meet the needs of customers in different kinds of services.

Combining tourism and the development of culture-oriented creative industries, winter art, and snow design have contributed to the creative renewable economies in Lapland. Between eight and ten large-scale snow hotels with sleeping rooms, restaurants, bars, chapels, showrooms, and so on and several smaller-scale tourism facilities are built in Lapland every year (Jokela, 2014, 2019). Compared to Sweden, where the Jukkasjärvi Ice Hotel was first built in 1990 and still remains the only ice hotel in the country, in Finland, knowledge and skill in the renewable use of snow and ice was disseminated throughout the region to improve the competence of the local people through educational, participatory, and place-based activities (Gelter & Gelter, 2013).

Crow day, sharing traditions in Siberia, Russia

Today, the vast areas of the Russian North are turning into a complex site of conflict between Indigenous people, local non-Indigenous inhabitants, state-owned and private monopolies (extractive industries), and even tourists (Pashkevich, 2013). In order to ground the very idea of ethical and culturally sensible tourism in the environmentally and culturally fragile setting of the Russian Far North, the researchers and designers from the Arctic Design School (USUAA), propose a novel understanding of a tourist destination. It is a "laboratory" where innovative solutions to short-term existence in the extreme environment are generated and shaped together by tourists and local inhabitants. As global warming advances, these solutions can be further applied to facilitate adaptation to severe conditions and the development of new lifestyles for longer term visitors and non-Indigenous settlers across the Arctic regions.

In evidence for this vision, this case study presents designers' engagement with the Indigenous cultural heritage—not ignoring but learning from it, and not blindly preserving traditions but keeping them alive and available for the present and future (Nugraha, 2012). It is centered on an experiment within the AAD educational model: as a part of a master's course entitled "Regional Design" at USUAA, a student named Alexandra Nikolaeva developed a project, "A Hybrid Tradition: A designer's renewal of the traditional Crow Day," a context-sensitive adaptation of the identically named festival of the Ob Ugric people of Western Siberia. The main task was to redefine tourism as a form of mutually beneficial engagement with the land and the people, with an emphasis on inclusive participation. Seen through the CES lens, this case reveals the distinctive ability of the AAD approach to "wrap up" the protection and appreciation of the cultural and spiritual heritage into the "gift pack" of a memorable tourist experience.

Crow Day represents the end of the so-called Winter Year and the arrival of the long-awaited spring, (Golovnev, 1995). The design exploration into the context and structural elements of the celebration began with an investigation of historical materials such as publicly available collections, in-house publications, and catalogues of the archives and museums.

Diverse visual data provided insights into the complexity of human-environment interactions within the traditional culture of Northern and Arctic inhabitants. At the first stage of data analysis, the rites and ceremonies associated with the traditional Crow Day celebration were divided according to their sacredness, which determined their potential for public accessibility and tourist involvement. The three resulting groups were: (1) entirely sacred rituals that are performed by the community, for example, young girls or old women, and cannot be joined or even observed by others; (2) partly sacred rituals where spectators are allowed, but participation is restricted; and (3) open or public rituals where everyone can join the celebration. Accordingly, there were three interactive situations identified: a tourist as a participant, a deliberately invited spectator, and an occasional witness.

At the ideation stage, a cultural basis for design interpretation was proposed: to link the Northern Crow Day with the widely recognised and interculturally relevant celebration of New Year. The design outcomes included essential attributes for new and old rituals, such as a stylised New Year Tree, thematic food, a carnival with masks and costumes and a culminating ceremony of making a wish.

In terms of practical implications, the outcomes, namely the designed objects and celebration scenarios, can contribute to shaping the region's multicultural identity by disseminating environmentally and culturally appropriate "best practices" or "know-how" that originated in the heart of the Arctic. In the long run, contextually relevant design explorations can inform the process of developing and inhabiting remote Arctic or Northern territories.

A tourist memento—creating place-based sustainable souvenirs in Alaska

Traditionally, education for tourism has been provided by tour operators, while universities teach or conduct research on tourism from a more theoretical perspective. Given the sensitivity of the Arctic environment and the speed with which it is changing, it is incumbent on Northern universities to become more involved in the process of transmitting knowledge, raising public awareness, and encouraging stewardship of the Arctic. Tourism is an ideal mechanism for this effort.

In order to encourage renewable economies, promote stewardship and raise awareness of the Arctic, a collaborative approach to provide tourism "packages" is proposed.[1] These interdisciplinary solutions are needed to increase knowledge and engagement about sustainable tourism in the Arctic. A model for a designed sustainable tourism should include (1) citizen-engaged environmental observation, (2) place-based sustainable art, and (3) outdoor recreation and leadership.

Citizen-engaged Environmental Observation is a type of knowledge co-production that has received increasing interest as more people become affected in some way by Arctic environmental change (Alessa et al., 2015). Citizen science has proved to be an indispensable means of combining scientific, environmental research with education and public engagement. It has significant potential for engaging the tourism industry that provides a unique platform from which to conduct research in remote Arctic locations. Participants engaged in a sustainable tourism program are well-positioned to collect observational data on environmental change (de la Barre et al., 2016).

Place-based Sustainable Art creates artwork using local materials in the context of place and environment. As Hicks and King (2007) have observed, "Art education is well situated to address environmental problems that emerge at the point of contact between nature and social life." By creating place-based sustainable souvenirs (art), participants can become more fully engaged. Compared to purchasing imported tourist souvenir products, this creative experience promotes positive memories and a sense of deeper connection and meaning. It can be an effective methodology for systematically and purposefully developing art projects about the environment and sustainability.

Outdoor Recreation Leadership enhances the health and wellbeing of people and communities (Brymer, Cuddihy, & Sharma-Brymer, 2012;

Godbey, 2009; Gobster and Buchner, 2010). Outdoor recreational activities are designed to create a learning environment inspiring a passion for guardianship for our ecosystem. It focuses on life-long learning opportunities, stewardship of resources, and collaborative teamwork. A designed sustainable tourism program is fundamental to establishing this resiliency.

However, this project is an experimental "makerspace" introducing a unique approach to sustainable tourism in the Arctic—focusing on being environmentally and culturally responsible while appreciating nature and promoting conservation. We believe that participants will build skills and knowledge in basic biological sciences, understand environmental issues and develop wilderness travel proficiencies. Making a "place-based souvenir" as a part of the sustainable tourism experience will promote a sense of appreciation for the Arctic's natural environment and play a critical role in terms of renewable economic activity in the region.

Artification of the Arctic tourism and cultural revitalisation

Through analysing the case studies and juxtaposing them with a review of the current literature, we can demonstrate some key points of potential and challenges for art and design in the development of renewable economies in the North and the Arctic.

When using CES, the creative economy has both commercial and cultural value. The AAD model, as the intersection of art and design practices with planning and production, can be used effectively when designing and producing renewable goods and creative services often connected with responsible and sustainable tourism in the North and the Arctic. We recognise that tourism in the Arctic is characterised by a process that Naukkarinen (2012) has described as *artification*. This refers to situations and processes in which something not originally regarded as art is transformed into something that resembles art or is influenced by artistic ways of thinking and acting. According to Naukkarinen (2012), this phenomenon can be found in business, wellness and healthcare services, and academic education and research. We argue that Arctic tourism is a scene for artification when creative and learning tourism are developed is in relation to CES and through AAD. For example, this could include nature and northern lights photography tours; wintery, snow and ice experience environments; learning tourism makerspace or slow food design production as well the revitalisation of Indigenous cultures through film productions or festivals (de la Barre, & Broucher, 2013; Gelter & Gelter, 2013; Jokela, 2014; Jokela et al., 2020; Leask, 2016;Urry & Larsen, 2011).

In AAD, CES are understood not as simple products of nature that are utilised for particular economic benefits, but rather, as relational processes that people actively create and express through interactions with cultural ecosystems (Fish, Church, & Winter, 2016). In advancing this viewpoint, AAD approximates to Chan et al.'s (2012) understanding of CES as experiences and capabilities that arise from human–ecosystem relationships. We argue that through creative industries like art and design, these relationships can be transformed into renewable economies.

Reports from around the world have demonstrated that creative industries generate income through trade and intellectual property rights and create new opportunities, particularly for small- and medium-sized enterprises. Even though the importance of and interdependence between creative economies and cultural services have been consistently recognised, they are often characterised as subjective and difficult to quantify in monetary terms. It is evident that their potential for future development is underestimated by national and regional decision-makers and officials responsible for regional development.

When transformed into renewable AAD products and services, CES must always contribute to the satisfaction of human needs and wants, which necessarily involves subjective considerations. Besides economic value, while being subjective, CES also benefit human capacities by facilitating knowledge, social, and cultural development and, in the Arctic case, revitalisation of local Indigenous and non-Indigenous traditions. We argue that responsible artification can take place through the development of novel renewable products and services through AAD.

From artification toward Arcticfication: A risk or an opportunity

Without involving Northern and Arctic people as collaborators, we are faced with the obvious danger of reducing Arctic ecosystems to an exotic resource that benefits external parties rather than Arctic inhabitants. While the North and the Arctic are culturally rich and diverse, *Arcticfication* is the tendency to present the Arctic as a cold and snowy destination devoid of human activity. Arcticfication has been reinforced by tourism marketing, as presentations of such magnificent landscapes can trigger touristic demand (Rantala et al., 2019). As Chartier (2018) has described, the phenomenon has deep roots in Western art and scholarship, where the Arctic was historically marginalised as the "Imaginary North" —as an empty and horizontal landscape rather than a multi-ethnic, multi-cultural and multi-lingual space with a rich cultural history and diverse living traditions. At the same time, Arcticfication is also the social process that has created, on the one hand, new geographical images of Northern Europe as part of the Arctic and, on the other hand, new social, economic, and political relations (Müller & Viken, 2017). In the development and implementation of novel AAD-based renewable economic practices, Arcticfication presents an opportunity to introduce these innovations to decision-makers and into the larger social and political discussions on the future of the Arctic and the world.

Today, as insiders, many artists and designers in the Arctic have the agency to reflect and depict the transformations, nature, and culture of the region (Huhmarniemi & Jokela, forthcoming, 2020). In addition, more research is being conducted on creative industries and the use of art and design in areas such as tourism, which is a growing economic field in the Arctic (Huhmarniemi & Jokela, 2019; Huhmarniemi et al., 2021; Kugapi, Huhmarniemi & Laivamaa, 2020; Miettinen, Sarantou & Kuure, 2019). As our study has shown, AAD and place-based development utilising CES as an economic development strategy are particularly relevant in the North and the Arctic today. When communities, artists, and designers in remote and rural places commit to place-making as a method of economic development, the dual benefit of commercial and cultural development will stimulate the region's prosperity and well-being.

Responsible utilising of CES in AAD calls creative capacity building

Our study has shown that there are various opportunities for innovative applications of AAD in remote and peripheral areas. We agree with Petrov (2014) and Vodden et al. (2015) in arguing that innovation in the creative economy is not restricted to cities and innovation hubs only, but there are certain challenges in the Arctic. According to studies, the Arctic needs to generate more human capital by investing in its people to keep them in the region (Karlsdóttir & Junsberg, 2015; Karlsdóttir et al., 2017; Petrov 2016, 2017). The advent of what is often referred to as the "knowledge economy" necessitates the enhancement of human skills and creativity, which will be a key to the next stage of the development process toward AAD as the creative renewable economy. This calls for novel models for educating artists and designers for the Arctic. Artists with traditional artists training may lack the will and skills to work as entrepreneurs and producers of services, or they don't have enough specific knowledge about the Arctic to apply their skill to particular northern circumstances (Huhmarniemi & Jokela, 2019; Kugapi, Huhmarniemi & Laivamaa, forthcoming, 2020).

We refer that as drivers of the Arctic creative economy, art and design higher education institutions and universities can lay the ground for the formation of multidisciplinary and interprofessional creative clusters, like the Arctic Design Cluster in Rovaniemi, Lapland. The cluster is built around the research and education of the Faculty of Art and Design to boost the regional economy by implementing processes where art, design, creative services, and CES are combined with place-making to exploit existing strengths of the region.

Conclusion

Through the case studies and literature review, this chapter has filled the notable gap in the research connected to art, design, and CES. We argue that identifying, analysing, and using CES as the potential for creative renewable industries; particularly AAD can play an important role in the future of the Arctic in terms of sustainable economy. Merging CES and AAD with a place-making strategy is a way to exploit the existing strengths of communities to create renewable economies in the rural and remote areas of the Arctic.

As our study has demonstrated, there is no single way to implement the CES approach in the creative economy and art and design. Implementation necessarily depends on local, regional, social and cultural conditions. As the concept of CES is subjective and always linked to society and culture, it is necessary to understand the specific conditions they are operating in. Both identified trends: Artification of Arctic tourism and cultural revitalisation practices and Arcticfication as a risk or an opportunity for AAD should be recognised and utilised in responsible way when implementing CES in creative renewable economy in the Arctic. Therefore, art and design as renewable economies must be implemented through culturally sensitive and place-based strategies to respond to the challenges and ensure sustainability in the North and the Arctic. Higher art and design education have an important role to secure creative human capacity and promotion of sustainable future in Arctic.

Note

1. The concept is co-developed by Audrey Taylor, Ph.D., Assistant Professor of Environmental Studies, Department of Geography and Environmental Studies; Herminia Din, Ph.D., Professor of Art Education, Department of Art; and Timothy Miller, Director, Department of Health, Physical Education and Recreation at the University of Alaska Anchorage.

References

Alessa, L., Kliskey, A., Pulsifer, P., Griffith, D., Williams, P., Druckenmiller, M. … Jackson, L. (2015). Best practices for community-based observing: a national workshop report. Arctic Observing Summit. http://www.arcticobservingsummit.org/sites/arcticobservingsummit.org/files/Alessa%20et%20al%20-%20CBONReport_DRAFT%20FINAL-updated-2016-03-07.pdf

ASAD. (2019). *Arctic sustainable arts and design thematic network*. ASAD Network. http://www.asadnetwork.org/

de la Barre, S., & Broucher, P. (2013). Consuming stories: Placing food in the Arctic tourism experience. *Journal of Heritage Tourism, 8*(2-3), 37–41.

de la Barre, S., Maher, P., Dawson, J., Hillmer-Pegram, K., Huijbens, E., Lamers, M. … Stewart, E. (2016). Tourism and Arctic observation systems: Exploring the relationships. *Polar Research, 35*(1), 24980.

Big River Analytics. (2017). *Impact of the Inuit Arts economy*. Indigenous and Northern Affairs Canada, Government of Canada. https://www.rcaanc-cirnac.gc.ca/eng/1499360279403/1534786167549

Brymer, E., Cuddihy, T. F., & Sharma-Brymer, V. (2012). The role of nature-based experiences in the development and maintenance of wellness. *Asia-Pacific Journal of Health, Sport and Physical Education, 1*(2), 21–27. doi: https://doi.org/10.1080/18377122.2010.9730328.

Canadian Art. (2017, December 13). Inuit Art Collective Isuma to Represent Canada at 2019 Venice Biennale. *Canadian Art*. https://canadianart.ca/news/isuma-venice-biennale-canada-pavilion-2019

Chan, K., Satterfield, T., & Goldstein, J. (2012). Rethinking ecosystem services to better address and navigate cultural values. *Ecological Economics, 74*, 8–18.

Chartier, D. (2018). *What is the "Imagined North"?*. Presses de l'Université du Québec.

Cooper, N., Bardy, E., Steen, H., & Bryce, R. (2016). Aesthetic and spiritual values of ecosystems: Recognising the ontological and axiological plurality of cultural ecosystem "services". *Ecosystem Services, 21*(B), 218–229. doi: http://dx.doi.org/10.1016/j.ecoser.2016.07.014.

Daniel, T. C., Muhar, A., Arnberger, A., Aznar, O., Boyd, J. W., Chan, K. M. A. … von der Dunk, A. (2012). Contributions of cultural services to the ecosystem services agenda. *Proceedings of the National Academy of Sciences, 109*(23), 8812–8819. doi: http://dx.doi.org/10.1073/pnas.1114773109.

Daniels, J., Baldacchino, G., & Vodden, R. (2015). Matters of place: The making of place and identy. In K. Vodden, R, Gibson, & G. Baldacchino (Eds.), *Place peripheral: Place-based development in rural, island, and remote regions* (pp. 23–40). ISER Books.

Fish, R., Church, A., & Winter, M. (2016). Conceptualising cultural ecosystem services: A novel framework for research and critical engagement. *Ecosystem Services, 21*(B), 208–217. doi: https://doi.org/10.1016/j.ecoser.2016.09.002.

Fondahl, G., & Wilson, G. N. (Eds.). (2017). *Northern Sustainabilities: Understanding and addressing change in the circumpolar world*. Springer Nature.

Fullerton, C. (2015). Arts, culture, and rural community economic development: A Southern Saskatchewan case study. In K. Vodden, R, Gibson, & G. Baldacchino (Eds.), *Place peripheral: Place-based development in rural, island, and remote regions* (pp. 180–210). ISER Books.

Gad, U., Jacobsen, M., & Strandsbjerg, J. (2019). Introduction: Sustainability as a political concept in the Arctic. In U. P. Gad & J. Strandsbjerg (Eds.), *The politics of sustainability in the Arctic: Reconfiguring identity, space and time* (pp. 1–18). Routledge.

Gelter, H., & Gelter, J. (2013). An innovation lost. The Ice Dome Concert Hall Project in Piteå. In L. Lindeborg, & L. Lindkvist (Eds), *The value of arts and culture for regional development: A Scandinavian perspective* (pp. 252–266). Routledge.

Giacometti, A., & Teräs, J. (2019). *Regional economic and social resilience: An exploratory in-depth study in the Nordic countries*. Nordregio. doi: http://dx.doi.org/10.6027/R2019:2.1403-2503

Gobster, P. H., & Buchner, D. M. (2010). Healthy outdoor recreation: An integrated approach to linking physical activity with wellness goals. In L. Payne, B. Ainsworth, & G. Godbey (Eds.), *Leisure, health and wellness: Making the connections* (pp. 437–446). Venture Publishing.

Godbey, G. (2009). Outdoor recreation, health, and wellness: Understanding and enhancing the relationship. *RFF Discussion Paper No. 09–21*. doi: http://dx.doi.org/10.2139/ssrn.1408694

Golovnev, A. V. (1995). *Govoryashhie kul'tury* [Talking cultures]. URO RAN.

Griffith, D. (2014). *Imagining natural Scotland*. Creative Scotland Publications

Guttorm, G. (2015) Contemporary Duodji—A personal experience in understanding traditions. In Jokela T. and Coutts G. (Eds.), *Relate North: Art, heritage and identity*, (pp. 60–77). Lapland University Press.

Hesmondhalgh, D. (2007). *The cultural industries* (2nd ed.). SAGE Publications.

Hearnshaw, E., & Cullen, R. (2010, August 26-27). *The sustainability and cost-effectiveness of water storage projects on Canterbury rivers: the Opihi River case paper* [Paper presentation]. The NZARES Conference, Nelson, New Zealand.

Hicks, L. E., & King, R. J. H. (2007). Confronting environmental collapse: Visual culture, art education, and environmental responsibility. *Studies in Art Education*, *48*(4), 332–335.

Howkins, J. (2001). *The creative economy: How people make money from ideas*. Penguin.

Huhmarniemi, M., & Jokela, T. (2019). Environmental art for tourism in the Arctic: From handicraft to integrated art and reform on artists' skills. *Synnyt/ Origins*, 1(2019), 63–80. https://wiki.aalto.fi/pages/viewpage.action?pageId= 151504259

Huhmarniemi, M. & Jokela T. (2020). Arctic arts with pride: discourses on Arctic arts, culture and sustainability. *Sustainability* 12(2), 604. https://doi.org/10.3390/ su12020604

Huhmarniemi, M.; Kugapi, O.; Miettinen, S. & Laivamaa, L. (2021). Sustainable Future for Creative Tourism in Lapland. In N. Duxbury; S. Albino & C. Pato Carvalho (eds.), *Creative Tourism: Activating Cultural Resources and Engaging Creative Travellers* (pp. 239–253). Cabi International.

Härkönen, E., Huhmarniemi, M., & Jokela, T. (2018). Crafting sustainability. Handcraft in contemporary art and cultural sustainability in Lapland. *Sustainability*, *10*(6). doi: https://doi.org/10.3390/su10061907

Härkönen, E., & Vuontisjärvi, H. (2018). Arctic art & design education and cultural sustainability in Finnish Lapland. In T. Jokela & G. Coutts (Eds.), *Relate North: Practising place, heritage, art & design for creative communities* (pp. 86–105). Lapland University Press.

Høllcland, II., Skrede, J., & Holmgaard, S. (2017). Cultural heritage and ecosystem services: A literature review. *Conservation and Management of Archaeological Sites*, *19*(3), 210–237.

Jokela, T. (2003) Introduction. In H. Huhmarniemi, T. Jokela & S. Vuorjoki, S. (Eds.), *winter art. Statement on winter art and snow construction* (6–11). University of Lapland.

Jokela, T. (2013). Engaged art in the North: Aims, methods, contexts. In T. Jokela, G. Coutts, M. Huhmarniemi, & E. Härkönen (Eds.), *Cool: Applied visual arts in the North* (pp. 10–21). University of Lapland.

Jokela, T. (2014). Snow and ice design innovation in Lapland. In E. Härkönen, T. Jokela, & A. J. Yliharju (Eds.), *Snow design in Lapland: Initiating cooperation* (pp. 180–181). University of Lapland.

Jokela, T. (2017) Art, design, and craft interwoven with the North and the Arctic. In M. Huhmarniemi, A. Jónsdóttir, G. Guttorm, & H. Hauen, (Eds.), *Interwoven* (pp. 4–11). University of Lapland.

Jokela, T. (2019). Arts–based action research in the north. In *Oxford Research encyclopedia of education*. Oxford University Press.

Jokela, T., & Coutts, G. (Eds.). (2018a). *Relate North: Art and design education for sustainability*. Lapland University Press.

Jokela, T., & Coutts, G. (2018b) The North and the Arctic: A laboratory of art and design education for sustainability. In T. Jokela & G. Coutts (Eds.), *Relate North: Art and design education for sustainability* (pp. 98–117). Lapland University Press.

Jokela, T., Coutts, G., Huhmarniemi, M., & Härkönen, E. (Eds.). (2013). *Cool. Applied Visual Arts in the North.* Publications of the Faculty of Art and Design of the University of Lapland C 41. http://urn.fi/URN:ISBN:978-952-484-638-7

Jokela, T., Huhmarniemi, M., & Hautala-Hirvioja, T. (2019). Preface. *Synnyt 1/2019 special issue on Arctic Arts Summit,* 6–12. https://wiki.aalto.fi/pages/viewpage. action?pageId=151504259

Jokela, T., Huhmarniemi, M., & Paasovaara, J. (Eds.). (2020). *Soveltava taide ja luontokuvaus. [Applied visual art and nature photography].* Lapland University.

Jokela, T., & Tahkokallio, P. (2015). Arctic design week: A forum and a catalyst. In T. Jokela & G. Coutts (Eds.), *Relate North: Art, heritage & identity* (pp. 118–139). Lapland University Press.

Joy, F. (2019). Sámi cultural heritage and tourism in Finland. In M. Tennberg, H. Lempinen, & S. Pirnes (Eds.), *Resources, social and cultural sustainabilities in the Arctic* (pp. 144–162). Routledge.

Karlsdóttir, A., & Junsberg, L. (Eds.). (2015). *Nordic Arctic youth future perspectives.* Nordregio.

Karlsdóttir, A., Olsen, L., Harbo, L., Jungsberg, L., & Rasmussen, O. (2017). *Future regional development policy for the Nordic Arctic: Foresight analysis 2013–2016.* Nordregio.

Kugapi, O., Huhmarniemi, M. & Laivamaa, L. (2020). A Potential treasure for tourism: Crafts as employment and a cultural experience service in the Nordic North. In A Walmsley, K. Åberg, P. Blinnikka, G.T. Jóhannesson, G.T. (eds.), Tourism Employment in Nordic countries: Trends, practices, and opportunities (pp. 77–99). Palgrave Macmillan.

Kramvig, B., & Flemmen, A. B. (2019). Turbulent indigenous objects: Controversies around cultural appropriation and recognition of difference. *Journal of Material Culture, 24*(1), 64–82.

Kunuk, Z. (2019.). Zacharias Kunuk. *Isuma.* https://www.isuma.tv/members/ zacharias-kunuk

Landry, C., & Bianchini, F. (1995). *The Creative City.* Demos.

Leask, J. (2016). IsumaTV Builds Innovative Digital Systems to Share High-Def Streaming Video in Low Speed Remote Communities. *First Mile.* https://firstmile.ca/ isumatv-builds-innovative-digital-infrastructure-to-share-high-def-streaming-video-in-remote-communities

Lepofsky, D., Armstrong, C. G., Greening, S., Jackley, J., Carpenter, J., Guernsey, B. ... Turner, N. J. (2017). Historical ecology of cultural keystone places of the northwest coast. *American Anthropologist, 119*(3), 448–463. doi: https://doi.org/10.1111/aman.12893.

Lindhjem, H., Reinvang, R., & Zandersen, M. (2015). Landscape experience as a cultural ecosystem service in a Nordic context: Concepts, values and deci-cion-making. *TemaNord, 2015,* 547. doi: http://dx.doi.org/10.6027/TN2015-549

Miettinen, S., Sarantou, M., & Kuure, E. (2019). Design for care in the peripheries: Arts-based research as an empowering process with communities. *NORDES Nordic Design Research, 8.* https://archive.nordes.org/index.php/n13/article/view/467

Milcu, A. I., Hanspach, J., Abson, D., & Fischer, J. (2013). Cultural ecosystem services: A literature review and prospects for future research. *Ecology and Society, 18*(3), 44.

Millennium Ecosystem Assessment. (2005). *Ecosystems and human well-being: Synthesis.* Island Press. http://www.millenniumassessment.org/documents/document. 356.aspx.pdf

Milone, P., & Ventura, F. (Eds.). (2010). *Networking the rural: The future of green regions of Europe.* Van Gorcum.

Minnakhmetova, R., Usenyuk-Kravchuk, S., & Konkova, Y. (2019). A context-sensitive approach to the use of traditional ornament in contemporary design practice. *Synnyt/origins, Special issue on Arctic Arts Summit* 1, 49–62. https://wiki.aalto.fi/pages/viewpage.action?pageId=151504259

Müller, D. K., & Viken, A. (2017). Toward a de-essentializing of indigenous tourism? In A. Viken & D. K. Müller (Eds.), *Tourism and indigeneity in the Arctic* (pp. 281–289). Channel View.

Mäkikalli, M., Holt, Y., & Hautala-Hirvioja, T. (Eds.). (2019). *North As a meaning in design and art.* Lapland University Press.

Naukkarinen, O., (2012) Variations on artification. *Contemporary Aesthetics, Special Volume 4 (2012) ARTIFICATION.* https://digitalcommons.risd.edu/cgi/viewcontent.cgi?article=1189&context=liberalarts_contempaesthetics

Nordic Council of Ministers. (2011). Megatrends. *TemaNord, 2011, 527.* http://norden.diva-portal.org/smash/get/diva2:702166/FULLTEXT01.pdf

Nordic Councils of Ministers. (2018). *Arctic business analysis: Creative and cultural industries.* Nordisk Ministerråd. doi: http://dx.doi.org/10.6027/ANP2018-708

Nugraha, A. (2012). *Transforming tradition.* Unigrafia Aalto University.

Olsen, L., Berlina, A., Jungsberg, L., Mikkola, N., Roto, J., Rasmussen, R., & Karlsdottìr, A. (2016). *Sustainable business development in the Nordic Arctic* [Nordregio working paper 2016: 1]. Nordregio.

Olsen, K. O., Abildgaard, M. S., Brattland, C., Chimirri, D., de Bernardi, C., Edmonds, J. ... Viken, A. (2019). *Looking at Arctic tourism through the lens of cultural sensitivity. ARCTISEN – A transnational baseline report.* University of Lapland.

Our World. (2018). *Our World: Youth, Film, Culture.* https://www.ourworldlanguage.ca/

Pashkevich, A. (2013). Tourism development planning and product development in the context of Russian Arctic territories. In R. H. Lemelin, P. T. Maher, & D. Ligget (Eds.), *From talk to action: How tourism is changing the polar regions* (pp. 41–60). Lakehead University.

Petrov, A. N. (2014). Creative Arctic: Towards measuring Arctic's creative capital. In L. Heininen, H. Exner-Pirot, & J. Plouffe (Eds.), *Arctic yearbook 2014: Human capital in the North* (pp. 149–166). Northern Research Forum.

Petrov, A. N. (2016). Exploring the Arctic's "other economies": Knowledge, creativity and the new frontier. *The Polar Journal, 6*(1), 51–68.

Petrov, A. N. (2017). Human capital and sustainbale development in the Arctic: Towards intellectual and empirical framing. In G. Fondalh & G. N. Wilson (Eds.), *Northern Sustainabilities: Understanding and addressing change in the circumpolar world.* Springer.

Rantala, O., de la Barre, S., Granås, B., Þór Jóhannesson, G., Müller, D. K., Saarinen, J. ... Niskala, M. (2019). *Arctic Tourism in Times of Change: Seasonality.* Tema Nord. http://norden.diva-portal.org/smash/get/diva2:1312957/FULLTEXT01.pdf

Sandals, L. (2019). Zacharias Kunuk Speaks on Isuma's Venice Biennale Project. *Canadian Art.* https://canadianart.ca/news/zacharias-kunuk-speaks-on-isumas-venice-biennale-project

Schilar, H., & Keskitalo, E. C. (2018). Ethnic boundaries and boundary-making in handicrafts: Examples from northern Norway, Sweden and Finland. *Acta Borealia, 35*, 29–48.

Schott, S. (2016). The changing face of economic development in the Canadian North. *Open Canada*. https://www.opencanada.org/features/changing-face-economic-development-canadian-north

Smith, L. T. (1999). *Decolonizing methodologies: Research and indigenous peoples.* Zed Books.

Soini, K., & Birkeland, I. (2014). Exploring the scientific discourse on cultural sustainability. *Geoforum, 51*, 213–223.

Stålhammar, S. S., & Pedersen, E. (2017). Recreational cultural ecosystem services: How do people describe the value? *Ecosystem Services, 26*(Part A), 1–9. doi: https://doi.org/10.1016/j.ecoser.2017.05.010.

Tennberg, M., Lempinen, H., & Pirnes, S. (Eds.). (2019). *Resources, social and cultural sustainabilities in the Arctic.* Routledge.

Teräs, J., Lindberg, G., Johnsen, I. H. G., Perjo, L., & Giacometti, A. (2014). *Bioeconomy in the Nordic region: Regional case studies.* Nordregio.

Stephen, K. (2018). Societal impacts of a rapidly changing Arctic. *Current Climate Change Reports, 4*, 223–237.

Urry, J., & Larsen, J. (2011). *The tourist gaze 3.0.* Sage.

United Nations Conference on Trade and Development (UNCTAD). (2018). *Creative economy outlook: Trends in international trade in creative industries.* UNCTAD. https://unctad.org/en/pages/PublicationWebflyer.aspx?publicationid=2328

University of the Arctic. (2019). Thematic Networks and Institutes. *University of the Arctic.* http://www.uarctic.org/organization/thematic-networks/

Wapikoni. (2019). Mission, Values, and Objectives. *Wapikoni.* http://www.wapikoni.ca/about/who-are-we/mission-values-and-objectives

Vodden, K., Gibson, R., & Baldacchino, G. (Eds.). (2015). *Place peripheral: Place-based development in rural, island, and remote regions.* ISER Books.

Woien, M., Kristensen, I., & Teräs, J. (2019). The status, characteristics and potential of smart specialisation in Nordic regions. *Nordregio Report, 2019*, 3.

5 Touring in the Arctic

Shades of gray toward a sustainable future

Patrick T. Maher, Gunnar Thór Jóhannesson, Trine Kvidal-Røvik, Dieter K. Müller, & Outi Rantala

Introduction

Throughout this chapter, we will use the lens of the "destination" to look at the way tourism has developed, and continues to develop, in the Arctic. In Canada this includes Yukon, Northwest Territories and Nunavut, Iceland as a whole nation, Northern Norway (the counties of Troms and Finnmark, plus the county of Nordland), Sweden (Norrbotten Västerbotten county and in some cases the region of Swedish Lapland) and Finnish Lapland. Over time, these destinations have seen some tremendous changes in tourism, as shown by the numbers in Tables 5.1 and 5.2; alongside other Arctic jurisdictions:

In tourism studies, we often talk about destinations, but as Morgan, Pritchard, and Pride (2011, p. 4) acknowledge, destination is a concept which is "variously used by marketers and professionals (as a geopolitical system with its own Destination Management Organizations) and by sociologists and geographers (as a socio-cultural construction)." In line with this description, a place only becomes a destination through the narratives and images communicated by its tourism promotion material (Morgan et al., 2011).

Tourism and the production of place images have become an important aspect of modern societies and the image of the Arctic matters to the overall sustainability of tourism in Northern areas. That is, sustainable societies, based on regional resources, require that tourism "is balanced with the development of inclusive and democratic places for people living in the Arctic" (Rantala et al., 2019, p. 19).

Understanding the image of the Arctic, rests on an understanding of how place meanings are created. Place can be understood as a commodity to be consumed, and representations tied to place are important to this consumption (Andersson, 2010; Urry, 1995). Thus, tourism is not necessarily that different from other industrial sectors—the consumption is simply taking away experiences, perhaps with tangibles such as photographs and other souvenirs, versus physically removing trees or minerals.

What a place is, and how a place comes to be seen, will depend on a variety of conditions, "ranging from local institutional contexts and interactions, to

DOI: 10.4324/9781003172406-5

Table 5.1 Early 2000s, estimated Arctic tourist numbers (data from 2001-2010; modified from Maher, 2013)

Country/Region/Province	Tourist Numbers (Estimates)	Sources/Notes
USA (Alaska)	• 1,631,500	• Summer 2006 data for all out-of-state visitors
Canada		
• Yukon	• 8049	• 2004 data – covers only the Northern Yukon tourism region
• Northwest Territories	• 62,045	• 2006-2007 data for all non-resident travellers to the entire territory
• Nunavut	• 9,323	• 2006, summer only
Greenland	• 33,000 (air arrivals) • 22,051 (cruise arrivals)	• Data reported in 2011
Iceland	• 277,900	• Data reported in 2002
Svalbard (Norway)	• 29,813	• AECO personal communication, August 2010; 2009 cruise visitors arriving from overseas
Norway (Finnmark county)	• 2,420,959	• Data from 2002
Sweden (Norrbotten county)	• 1,700,000	• Data from 2001 tourist overnight stays
Finland (Finnish Lapland)	• 2,117, 000	• 2006 data for the number of registered tourist overnights
Russia	• Estimated at a few tens of thousands and growing steadily	• Actual data difficult to obtain

specific situations concerning economic and social life as well as narratives and symbols available" (Granås, 2009, p. 119). Places compete in attracting residents, businesses, and visitors, and as Morgan, Pritchard, and Pride (2011, p. 3) state, "A place with a positive reputation finds it easier to vie for attention, resources, people, jobs, and money; a positive place reputation builds place competitiveness and cements a place as somewhere worth visiting."

Of particular relevance to tourism, place images create expectations and demands among potential customers and collaborators, which the local businesses and communities must relate to. At the same time, representations (while potentially produced with intentions such as creating attractive destination brands) are also key to inhabitants—their sense of place and identification as a local to a certain area. Amundsen (2012, p. 140) points out that, "tensions between definitions of what should be offered to tourists and

Table 5.2 Most recent, estimated Arctic tourist numbers (updated from Maher, 2017)

Country/Region/Province	Tourist Numbers (Estimates)	Sources/Notes
USA (Alaska)	• 2,242,900	• https://www.commerce.alaska.gov (Accessed March 2019); November 2018 report for October 2016-September 2017 data for all out-of-state visitors
Canada		
• Yukon	• 334,000	• http://www.tc.gov.yk.ca (Accessed March 2019); 2017 estimated total overnight visits to the entire territory
• Northwest Territories	• 112,530	• http://www.iti.gov.nt.ca (Accessed March 2019); 2017-2018 total visitors (leisure and business) to the entire territory
• Nunavut	• 16,750	• http://nunavuttourism.com; 2015 exit strategy – non-resident visitors
Greenland	• 90,025	• http://www.tourismstat.gl (Accessed March 2019); 2017 Greenland Tourism Statistics Report
Iceland	• 2,224,074	• http://www.ferdamalastofa.is (Accessed March 2019) 2017 international visitors to Iceland (via flights and ferries)
Svalbard (Norway)	• 158,248	• https://en.visitsvalbard.com (Accessed March 2019); 2018 Visit Svalbard statistics for overnight stays in Longyearbyen
Norway (Nord Norge)	• 2,794,973	• Innovation Norway, 2017 report (available from https://assets.simpleviewcms.com/simpleview/image/upload/v1/clients/norway/Key_Figures_2017_pages_9b3f82d5-43f4-4fe9-968c-7a85a36704b2.pdf); 2017 data for all foreign overnight stays and overnight stays from domestic tourists from other regions
Sweden (Swedish Lapland)	• 2,947,969	• 2018 total guest nights (Accessed August 2019 from https://www.swedishlaplandvisitorsboard.com/en/the-destination/statistics/)
Finland (Finnish Lapland)	• 3,100,000	• https://yle.fi/uutiset/osasto/news/record_number_of_tourists_visit_finnish_lapland_in_2019/11186729 (Accessed January 2020); 2019 data for the number of registered tourist overnights
Russia	• 500,000	• Tzekina (2014)

how local places should develop thus involves a range of perspectives and actors and it is likely that this continues to be a source of dispute."

This speaks to a dynamic that is important to address when discussing overall sustainability of Arctic societies; and across societies in different regions. Tourism scholars situated in the northern areas around the world are asking how tourism can be developed so that it "strengthens communities and makes them better places to live in," something which further "begs the question of how to find the balance between economic, social and environmental sustainability" (Rantala et al., 2019, p. 40), the essence of a green economy, which when tied to the marine environment, as much of the Arctic is, is also a blue economic approach.

An important success criterion for the tourism industry will be to provide the right experience to the right visitor. For this to happen, the image of the Arctic alongside realities of small communities must be addressed. Many regional actors assume that the publicity of a place will lead to increased number of tourists, investors or inhabitants (Falkheimer, 2006), and this may not always be the right trajectory. So how are Arctic places positioned at the moment? As a green option for development, versus yet another exploitative/non-renewable one. The next section of this chapter will outline the background contexts against which tourism is overlaid in each of the five nations (and sub-regions).

Background

Canada

Canada's Arctic, specifically the jurisdictions of Yukon, Northwest Territories and Nunavut, together cover over 3.5 million km². This is roughly 40% of Canada's landmass; yet the total population is less than 1% of the total Canadian population (115,000). As a result, the diversity of tourism products is extensive due to shear geography, yet the capacity to develop a broad sector is limited. Much of the tourism in Yukon has historically been linked to travelers transiting the Alaska Highway, with United States travelers being the primary supply as they make their way from the southern States to their Northern frontier. Yukon has a strong history of Indigenous presence in its tourism industry (see Hull, de la Barre, & Maher, 2017), which also extends to the creative success of non-Indigenous cultural tourism. In the Northwest Territories, hunting and fishing has long been the draw; alongside paddling journeys and more recently the sparkle of viewing the Aurora Borealis and diamond mines. Nunavut, which until 1999 was a part of the Northwest Territories, has no external road access, and thus has seen increased access by expedition cruise ships and as such a far larger dependence on marine tourism (see Johnston, Dawson, & Maher, 2017). Nunavut was created through a land claim process and as a result is heavily invested in tourism that shares a variety of Inuit traditional activities, such as carving, kayaking, dogsledding, drum dancing, etc.

Iceland

Iceland has a long history as a destination for travelers and explorers, but only during the last decades of the twentieth century has tourism started to develop at an exponential pace (compare Tables 5.1 and 5.2). The population of this cold-water island state is less than 360,000; of which approximately 60% lives in the capital region of Reykjavík, on the southwest corner of the island (Statistics Iceland, 2020). Iceland barely touches the Arctic Circle, yet it is very much part of the same popular imagery surrounding other circumpolar destinations. The travelogues of many explorers and adventurers who visited Iceland in the seventeenth and eighteenth centuries described the island as a place of natural wonders, a narrative that was further boosted by its "island-ness,"—the sense of distance, isolation, separateness, tradition, and "otherness" (Jóhannesson, Huijbens, & Sharpley, 2010). This image is still partly sustained today in Iceland's marketing material for tourism.

Norway

Northern Norway, which consists of the counties of Troms Finnmark, and Nordland makes up a large area of Norway's mainland (35%), but the population is only approximately 486,000 people (less than 10% of Norway's overall population). Statistics shows the importance of tourism as an industry in Northern Norway. In 2018, the industry supported NOK 19.3 billion in economic activity, and employed 17,242 people. At those levels, the industry accounted for 7.1% of 2018 employment in Northern Norway. In comparison, the employment shares or other primary industries (including fish processing) was 6.2% and other industries such as petroleum development and mining was 5.7% (NHO, 2019). Importantly, at the same time as visitor numbers are increasing, expectations of professionalism among industry players are increasing.

Recently, public attention has also begun to focus on the growth in winter tourism in the region, which a recent national tourism strategy describes as the most significant change in tourism over the last few years (Innovation Norway, 2021). This has clearly changed "the conditions for the tourism industry and influenced social life in the villages and towns most strongly affected by this increase" (Rantala et al., 2019, p. 21). Even though the Summer season is the biggest season for tourism in Northern Norway, there is an important increase in Winter tourism, a growth linked with long-term initiatives to develop the region into a. year-round destination for international tourism (Innovation Norway, 2021). Winter has increased its "market share" of total international visits throughout the year from 9 to 30% over the last decade, and in 2016/2017 more British and Asian visitors came to Northern Norway during the Winter than during the Summer season (NHO, 2017).

Sweden

The Swedish North is historically the home region of the Indigenous Sami; however, in the nineteenth century, the state and industry identified the area as a rich source of natural resources, such as timber and minerals (Sörlin, 1988). At the turn of that century, tourism was identified as a part of the industrial mix of the region (Müller, Byström, Stjernström, & Svensson, 2019). Yet it was not developed into a core industry until recently.

For 100+ years, tourism functioned as an alternative and complementary livelihood during bust periods in the traditional natural resource industry cycle (Müller, 2013a). Today, tourism in the Swedish North is experiencing a boom period. In this context, new products and new seasons have emerged meeting the increasingly global demand for northern tourism. In the footprints of the Icehotel, established in the early 1990s, more winter tourism products are being developed; including dog-sled tours and aurora borealis chasing Thus, commercial overnight stays during the winter season are growing almost twice as much as during summer (Tillväxtverket, 2018).

Despite the steadily growing figures, tourism in the Swedish North is still small-scale compared to many other parts of the world. The region has only 520,000 inhabitants or 5% of the Swedish population. Within the region, approximately a third of the guest nights are related to international tourists (Tillväxtverket, 2018). A majority of these guest nights occur in Umeå and Luleå, the two counties' coastal capital cities, where a majority of the small population is concentrated. In the inland areas, particularly places accessible by airplane, there is some positive development. Kiruna, the home municipality of the Icehotel, and the primary spots for aurora observations, takes a dominant position (Müller, 2011).

Finland

Finnish Lapland is often regarded as a peripheral area of Finland, since it covers 30% of the area of Finland, but only 3% of the Finnish population lives in Lapland. The area is seen as rich in material resources for forestry and mining, but also rich in exotic imaginaries for the tourism industry. The current reliance on these traditional industrial sectors brings turnover and employment to the county, but the tourism sector has steadily increased its importance during the last three decades. When considering the local community perspectives, highlighting peripherality, exoticness and resource-richness seems too straightforward. We are currently witnessing an overwhelming human influence upon the Earth (for example, Crutzen, 2002) —and in line with that the need to bring up alternative perspectives in the era of environmental crises. It should be highlighted that Lapland has biodiversity rich areas, know-how on multiple uses of the forests—e.g. superfood companies, and lively creative industries. Thus, the vision of the county of Lapland is to be the world's cleanest county in 2040, based on "Arcticness," openness and smartness. The vision is to be achieved by applying sustainable practices and smart technologies into the use of the resources—accompanied by a high level of digitalisation (Hyry et al., 2017).

Policy and development

As can be seen in the background context, there are many similarities between these Arctic tourism destinations. They rely on a few unique factors: large landscapes (at different scales) that attract visitors because they are so different to the tourists' regular city landscapes; very small populations; peripheral constraints (perceived or real); and historical narratives of exoticism and marginalisation. This next section will examine some policies and development trajectories across borders. Common themes include the dependence on transport (specifically aviation), seasonality, the role of the periphery against the core, sustainability of the system, access and environmental protections, and the realities of the workforce.

Aviation

In Iceland, aviation is the precondition for large scale tourism on the island. Soon after World War II, two Icelandic airlines started to operate international flights and established route networks connecting the island to Europe and North America. These companies later merged under the name Icelandair. Icelandair, continued to develop a hub and spoke system, making effective use of the location of Iceland in the middle of the North Atlantic, connecting various destinations in Europe and North America. Icelandair has been the major driver of tourism development in the country and still holds a key position in that regard. Other Icelandic-based airlines have operated for some periods, but have struggled to survive. Most recently, WOW air, established in 2011 went bankrupt in spring 2019. It had operated with a similar hub and spoke system as Icelandair. Many international airlines have also operated routes to and from Iceland in recent years, especially during the high season (22 additional airlines during summer 2019).

On a far smaller scale, airlines are a critical piece of the tourism infrastructure for many areas of Arctic Canada. Nunavut relies on airlines to bring visitors to the territory, with no road access linking it to the rest of the country; and the linkages of small regional (largely domestic) airlines across the three territories is a necessity. These airlines have been owned and/or managed by the territories themselves and specific Indigenous groups, more so than the major airline players (Air Canada and WestJet), thus they have unique community connections and expectations.

Although aviation is critical to some destinations, in other areas there is the recognition that we need to move beyond the development of flight connections and airports. In Finnish Lapland, the focus is now on regional accessibility by train. This links to the impacts of tourism being recognised, and actors such as the Responsible tourism network of Lapland, having more visibility than in the past.

Seasonality

The most recent tourism strategy for Finnish Lapland highlights year-long sustainability (Sievers, 2020), and in doing so prioritises tourism development that increases the amount of tourism during the snowless seasons. One reason for focusing on snowless seasons is the aim to balance the impacts of tourism on the local environments and societies away from the busy winter season.

While tourism originally developed in the summer in Finnish Lapland, since 1980s the winter has been the high season. There were plans already afoot in the 1950s to build a "Christmas land" in Rovaniemi, but it was not until mid-1980s that the plans were actualised and the Santa's Village Christmas tourism destination was built at the Arctic Circle (Ilola, Hakkarainen, & García-Rosell, 2014). The winter season continues to be the most important season in most of Lapland, with the month of December being the most popular, both in terms of overnight stays and passenger traffic. The lowest number of overnight stays in 2017 was in May, with 63,000, whereas in December there were 465,000 overnight stays registered. The months from January to April form the second peak season after Christmas tourism, and the summer and autumn months from June to September the third season. According to a survey conducted among entrepreneurs from northern Finland, northern Norway and northern Sweden, tourism entrepreneurs see May as the most problematic period for developing tourism and autumn season as the most potential one for the development of year-around tourism (Rantala et al., 2019, p. 25).

In Iceland, tourism is easily characterised by seasonality. More than 90% of all tourists enter the country via Keflavík international airport, close to the capital Reykjavík; and during the summer months of June, July and August. There is some growth in tourist arrivals during the winter, but this is an interesting trend in light of the often uncertain and harsh weather conditions at that time. Although it has led to considerably less seasonality, particularly in the capital region and along the south coast of the island, i.e. the areas most easily accessible during wintertime. A new challenge is now the significant regional differences in tourism within the country, which is not likely to change in foreseeable future.

Tourism in Northern Norway has also been characterised by seasonality, and while this might work well for some businesses in Northern Norway, for instance "Indigenous entrepreneurs, who may rely on seasonal engagement in tourism to make the entrepreneurship fit into the annual life cycle of their Indigenous community" (Rantala et al., 2019, p. 33), particularly reindeer herding. It also brings about some challenges as well, with year-round tourism said to better enable "larger companies to deal with environmental issues, and to recruit competent staff, who demand full-time position" (Rantala et al., 2019, p. 32).

Returning to Finnish Lapland, the strong seasonality of tourism has negatively impacted local communities and environments due to the pressure of a single high peak (Rantala et al., 2019). There are new practices being developed to mitigate the impacts of a peak season; i.e., in Rovaniemi a social worker has been hired—since 2013—to work with foreign tourists in the regional hospital from November to April. The social worker enables hospital employees to concentrate on serving local people and also enables hospital to get payment back from their services (previously local tax money has been used to take care of tourists in the hospital). Seasonality has also strengthened the image of tourism work as low-skilled and precarious (Rantala et al., 2019), which has led to labor shortages during the high season. At the same time, the seasonal nature of tourism has enabled communities "to take a break" from tourists and formed a basis for lifestyle entrepreneurship (Rantala et al., 2019). This is present in Canada's Yukon too, with lifestyle entrepreneurs working on their own time. A particular example is how dog sledding kennels can focus on their racing and training at some times of the year, and tourism endeavors at others.

Peripherality

Paulgaard (2008, p. 56) puts it well when she says, "the branding of the place and the people within the field of tourism represents the local culture in accordance with the hierarchical understanding of the distinction between centre and periphery." Müller (2015, p. 149) adds that "while the periphery position *can* represent a practical challenge of distance, it is not necessarily the physical distance that can be seen as the challenge, but rather the symbolic distance embedded in such a center-periphery construct." Müller (2015, p. 149) expands that to point out,

> "it should be noted that the Arctic is not a remote destination. It is in fact surrounded by major demand markets in North America and Europe, and is in fact much closer to these markets than other popular destinations like Southeast Asia and Australia. Hence, it is not the physical distance that makes the area remote, but rather the cognitive perception of a different climate and ecosystem. Still, traveling in the Arctic can be expensive, but this is a consequence of limited market demand rather than physical distance."

Tourism in Northern Sweden has been promoted as an opportunity to create employment and stabilise communities in rural and northern peripheries (Müller & Brouder, 2014). Many tourism stakeholders in the North embrace this message (Lundmark & Müller, 2010). Governments promote the numerous national parks and nature reserves in the North as resources for tourism development and obviously, tourism is seen as an industry that can support a transition to a more environmentally friendly use of northern resources, too.

However, not everybody in the tourism industry embraces this idea. Instead, some stakeholders see nature protection and the regulations that follow along as a threat to business (Lundmark & Stjernström, 2009; Müller, 2013b). This applies not least when motorised transportation is included in the products. Furthermore, nature protection does not seem to have the promised positive impacts on employment (Byström & Müller, 2014; Lundmark, Fredman, & Sandell, 2010). A similar discussion takes place in Canada, where many parks and protected areas have been created in the Arctic—in the past due to a lack of population, and a center-based desire to protect a system of ecosystems. Nowadays, protection is a recognition of Indigenous land claims and overall stewardship (see Maher, 2012). Some Indigenous Sami entrepreneurs consider tourism as integrated part of their traditional activities and use the income from tourism to support their reindeer herding (Leu, 2019).

In Finnish Lapland, previous tourism strategies categorised different areas into strong tourism centers, and this categorisation was used for directing investments (Hakkarainen & Tuulentie, 2008; Regional Council of Lapland, 2007). This has led into tourism, which is driven by tourism centers (small cities) that are then complemented by peripherical attractions (Hakkarainen, 2017). In Arctic Canada, this is similar to the situation in each territory; hubs such as Whitehorse and Yellowknife act as the conduits to attractions elsewhere in the territories. Again, it is driven in part by aviation infrastructure, as noted earlier.

In Northern Norway, tourism growth is also unevenly distributed; Nordland County has considerably higher tourist numbers than the rest of Northern Norway. Finnmark County to the north east has tourism numbers that are less than half of those of Nordland. In Nordland, the summer season is a peak season, while in Troms County, located in the middle of the region, the summer and winter seasons have more equal numbers, and the development from 2012 to 2018 led to higher numbers during winter. In the town of Tromsø, the number of international commercial overnight stays during the Winter season increased from 18.000 in 2008 to 200.000 in 2018, and AirBnB comes on top of this (Jakobsen & Engebretsen, 2019). In the same period, the growth in winter tourism in the most northern part of the region was much more limited. With such different structures of the communities, we could perhaps say there is no "one size fits all" when it comes to tourism development in northern Norway.

In the past, there have also been one-sided media accounts that re-create the myths of Northern Norwegians, versus those in the south near Oslo and Bergen as "naïve and natural, living among the fjords and the fish" (Paulgaard, 2008, p. 51), and this has been claimed as the reason why young people cease to identify with the northern places in which they live. Speaking to this issue, Guneriussen (2008, p. 233) says,

"this region has been considered a backward, poor, weakly-developed and mostly pre-modern periphery in Norway, in need of state subsidies and regional development programmes in order to become 'modern'. Such a negative labelling of the region has been typical and not only by 'outsiders' (particularly representatives from national political, economic and cultural centres). It has also been an important part of the northerner's self-understanding or self-image. People in the north have habitually considered themselves subordinate in many respects. They felt that the modern centre in the south, with all its advanced technology, culture and economic power represented a higher level of development."

Similarly, Kraft (2008, p. 222), stated that "Northern Norway has traditionally been constructed according to a north-south axis, with 'south' as the centre of power and decision-making, and 'north' as a suppressed and exploited backyard. Related to this perspective of subordination and victimization, the people of the north have been imagined thorough a primitivist discourse, in contrast to a presumed modern, Western identity."

Despite these controversies, there seems to be a general agreement among stakeholders that tourism should continue to develop as an industry that contributes to sustain communities and labor markets in such peripheral regions. A recent government commission in Northern Sweden outlined ideas that clearly set tourism in the context of a green economy (SOU, 2017). In this context, a more sustainable transportation system including the availability of public transport in peripheral areas is among the proposed actions. This is remarkable considering the low population density; but could be a model to assist the ongoing core-periphery tension in many of the regions discussed here. Furthermore, since many of the tourism entrepreneurs in the region (Swedish Lapland) have been attracted to the region by outdoor activities and the related lifestyles themselves, they engage in adapting their activities to become environmentally friendly and sustain the resource base of their lifestyle and business (Carson, Carson, & Eimermann, 2018).

Sustainability

The focus on sustainability and on the need to balance the tourism sector's activities can be seen as a tremendous dilemma, resulting from the strong increase of the tourism industry during the last few decades, and the impacts of this increase. In Finnish Lapland there are some estimates that the actual number of the overnight stays may be 2.5 or 3 times larger than the 3.1 million reported because many visitors spend nights in private rental cabins and in AirBnB accommodations that are not being registered. Airbnb accommodations have increased rapidly during the last years. In Rovaniemi—the capital city of Lapland, there were a total of 136 AirBnB accommodations listed in March 2016, while in November 2017 they reached 500, and at the beginning of 2019 the number rose to almost 900. In comparison that is 14.4 AirBnB locations per 1,000 inhabitants in Rovaniemi, while the same number in the far larger capital city of Helsinki is 4.2 per 1,000 inhabitants (Retrieved January 13, 2020 from https://shareabletourism.com). This issue is considerable threat to sustainability in Iceland too, particularly Reykjavik.

The extensive increase of AirBnB accommodations and sharing economy has caused conflicts in the development of tourism—especially in Rovaniemi region. These conflicts include, e.g., the lack of clarity of rules and regulations regarding AirBnB accommodation in the city. However, the sharing economy—and especially the trend toward "living like a local" —has enabled the inclusion of new responsible areas into the agenda of tourism development in Finnish Lapland (Haanpää, Hakkarainen, & Harju-Myllyaho, 2018). The same situation, although under smaller circumstances, can be seen in Arctic Canada and Norway. Visitors may get stretched by the limited accommodation options available, and while they wish to have a "local" experience that causes tensions around commodification and country food security.

The workforce

In Iceland, tourism exports account for approximately 40% of foreign currency income and provides 8.7% of GDP in 2017 (Mælaborð ferðaþjónustunnar, 2019). Approximately 33.000 people or 16% of the labor force work in tourism, of which one third is migrant workers. Development has been driven by entrepreneurs operating within a weak organisational and regulation framework. The sector has been characterised by a few large firms and an abundance of micro and nano-sized companies.

The importance of tourism in the regional economy of Finnish Lapland is also significant, as the share of tourism in GDP was 5.7%, while Finland's national average was 2.5% (House of Lapland, 2020). In 2017, Lapland accounted for over 4,000 person-years of work in the tourism sector; up to 7,000 people when including also the seasonal workforce (House of Lapland, 2020). The turnover of the tourism industry was 630 million euros in year 2017, with 16% growth from the previous year.

With different county- and municipality-level strategies, development plans, and visions, the tourism sector is often expected to bring employment and income to peripherical communities. People in these communities are expected to develop different kinds of tourism related services and innovations—without offering them concrete tools (Hakkarainen, 2017). Hence, tourism does not inevitably bring the means "to save" the peripheral areas from outmigration, but little by little tourism has formed as one way to enhance employment in the villages in Finnish Lapland—for example by combining tourism work to reindeer husbandry, mining, car-testing, or agriculture (Hakkarainen, 2017). By combining different sources of livelihood, the seasonal nature of tourism has been mitigated for local conditions

Growth in tourism leads to increased use of the region's areas and services that are also used by locals. This places much stronger demands on strategic and comprehensive planning from both the authorities, the companies, and other actors in the tourism industry. Growth has led to an increased use of nature and public areas. In Northern Norway, tourism businesses and the communities are experiencing a paradoxical situation:

> "On the one hand the infrastructure presently available is too limited for further growth during the high season, while on the other hand because of the limited infrastructure, it is not viable to run tourism the whole year round"
>
> *(Rantala et al., 2019, pp. 38–39).*

Some international and regional organisations have raised concern for these potential impacts on the natural environment, wildlife, local residents, and Indigenous population, in the wake of the increased tourism in the region (Chen & Chen, 2016). One specific aspect of some of these discussions, deals with the Norwegian *allemannsretten/*"right to roam," which has relevance in terms of pressure on nature in northern areas. Even though public access rights have been at the core of discussions regarding use of land for tourism, "public interest in access has largely prevailed because at the heart of the Nordic conception of citizenship is a deeply embedded tradition of outdoors sporting and recreational activity, as embodied in a cultural sensibility towards *friluftsliv* (a simple life in nature) and *idrett* (purposeful outdoor sporting activity), underpinned by the notion of *allemannsretten*" (McNeish & Olivier, 2017, p. 290). If this issue remains unaddressed in tourism, it will potentially undermine the sustainable development of Northern communities in both Norway and Sweden.

Also, the core of Arctic tourism is made up by many small companies and lifestyle entrepreneurs that offer experiences and services for visitors. These small companies are often "based on lifestyle entrepreneurships that are strongly embedded in places, environments and communities" (Rantala et al., 2019, p. 30). In order for tourism industry to grow, and become increasingly professionalised, larger companies might be beneficial, however limited growth in tourism may be a better solution to the region, following the concept of carrying capacity that determines the optimal number of visitors to be hosted at given time and space (Chen & Chen, 2016).

Next steps

Across the destinations included in this chapter, it can be seen that tourism holds wide-ranging opportunities, as well as considerable impacts within society. In a relatively short period, 2000 onwards, there has been a change from consumption of resources to a service-based economy.

For example, in Iceland, the national economy, which has historically fluctuated in tandem with environmental conditions of the sea for fishing and market conditions for aluminum now sees tourism as one more pillar for the economy. Though tourism is also marked by fluctuations and volatility as recent downturn reflects. Iceland is a nature-tourist destination, with more than 90% of visitors saying that the natural environment of the country gave them the idea to travel to Iceland (Óladóttir, 2019). So it may be particularly prone to negative dialogue on overcrowding and environmental damage, except that overall, tourists state they are happy with their visit (Ferðamálastofa, 2019). In fact, the main source of complaints in Iceland is about pricing and expenses.

This is the dilemma for many destinations; cheap flights, packaged accommodation, and the negative impacts of tourism associated with overtourism (see Jóhannesson & Lund, 2019); versus true sustainability and responsibility. Travel choices are at an impasse. Today, sustainability is a global imperative, across all sectors and society as a whole. Blue and green economics dictate where we should be headed in Canada and the Nordic states. Long-term sustainable development of tourism in the Arctic relies on engaging in community development and caring for local environments (Rantala et al., 2019). There is a need to consider how to achieve a market mix that minimises travel emissions and pays respect to the planetary resource limitations in the Anthropocene (Gren & Huijbens, 2014).

Globalisation implies a spatial expansion of the tourism system that now increasingly includes long-haul travel to reach the destination, while the time spent at the destination seems to decline. The related emissions do not match the idea of a tourism industry that sells experiences of unspoiled nature and aspires to become part of a sustainable future.

For a number of years, scholars have challenged the image of Arctic places; in Northern Norway, the notion of being peripheral and wild has been challenged; implying that Northern Norway is about to become a new, vital and "dynamic area" in the nation, in Europe and even in a global context (Guneriussen, 2008, p. 233), but "myths are not easily deconstructed, even though they may not correspond to people's experiences in their daily life" (Paulgaard, 2008, p. 53). Also, the myths may not be negative for all, "the construction of a centre and periphery as asymmetrical counter-concepts has both positive and negative connotations" (Paulgaard, 2008, p. 52).

It is still a perceived reality that one needs to be "in touch with wilderness" in order to become "a healthy, natural human being" (Guneriussen, 2008, p. 242) and the "Arctic magic" is something that is seen to appeal to "modern actors who long for something extraordinary" (Guneriussen, 2008, p. 242). This is a tool for tourism in the Arctic to capitalise on. The North "has come to be conceived as something very attractive when viewed from within a modern and highly urbanized culture" (Guneriussen, 2008, p. 242) and "wilderness has become a prime attractor for various forms of tourism—a spectacle for modern spectators, something good and authentic with which to make contact" (Guneriussen, 2008, p. 242).

DMO's such as Visit Norway, discuss being attracted by the Midnight Sun, fishing opportunities, and picturesque landscapes; seduced by the Northern Lights and snow-related activities in the winter, such as husky rides and visiting ice hotels (Chen & Chen, 2016). This can be said of nearly all the regions covered in this chapter; then does it become superficial or even fake? Perhaps it can also be beneficial to the sustainability discourse; the next steps are "stay home" if you live in one of these areas, or is it "rather urgent that the Arctic government[s] and host communities contemplate the consequences of global warming on future tourism development and put forward appropriate policies and regulations to better anticipate and respond to ongoing climate transformations" (Chen & Chen, 2016, p. 5). Again, this is a balance of how one sees ongoing tourism development as part of blue or green economies or is it the antithesis of such.

In order to apply sustainable practices on development to tourism, wider discussion is needed regarding the future aims and directions; that move beyond simply increasing tourism. The strong increase of tourism has already had diverse implications—such as a need to hire extra personnel in a hospital, a need to invest more on the infrastructure of recreational areas that are used based on everyman's rights, and need to clarify regulations and rules related to sharing tourism between core centers and peripheral spaces, in all seasons.

Acknowledgments

This chapter is a product of the ongoing work of the University of the Arctic Thematic Network on Northern Tourism. The five authors (and their institutions) play an important core role in the larger 30+ partners of the network. Since 2008 the Network has been successful in receiving funding from a number of different sources—most recently the Nordic Council of Ministers. This chapter was written just prior to the COVID-19 pandemic, thus much of its outlook may have changed dramatically as a result of that global event.

References

Amundsen, H. (2012). Differing discourses of development in the Arctic: The case of nature-based tourism in Northern Norway. *The Northern Review, 35*(Spring 2012), 125–146.

Andersson, M. (2010). Provincial globalization: The local struggle of place-making. *Culture unbound, 2*, 193–215.

Byström, J., & Müller, D. K. (2014). Tourism labor market impacts of National Parks: The case of Swedish Lapland. *Zeitschrift für Wirtschaftsgeographie, 58*(2-3), 73–84.

Carson, D. A., Carson, D. B., & Eimermann, M. (2018). International winter tourism entrepreneurs in northern Sweden: Understanding migration, lifestyle, and business motivations. *Scandinavian Journal of Hospitality and Tourism, 18*(2), 183–198.

Chen, J. S., & Chen, Y.-L. (2016). Tourism stakeholders' perceptions of service gaps in Arctic destinations: Lessons from Norway's Finnmark region. *Journal of outdoor recreation and tourism, 16*, 1–6.

Crutzen, P. J. (2002). Geology of Mankind: The anthropocene. *Nature, 415*(6867), 23.

Falkheimer, J. (2006). When place images collide. Place branding and news journalism. *Geographies of communication. The spatial turn in media studies* (pp. 125–138). Nordicom.

Ferðamálastofa. (2019). Ferðaþjónusta í tölum - Ágúst 2019. https://www.ferdamalastofa.is/static/files/ferdamalastofa/talnaefni/ferdatjonusta-i-tolum/2019/agust-2019-isl.pdf

Granås, B. (2009). Constructing the unique: Communicating the extreme dynamics of place marketing. In T. Nyseth & A. Viken (Eds.), *Place reinvention: Northern perspectives* (pp. 111–126). Ashgate.

Gren, M., & Huijbens, E. H. (2014). Tourism and the anthropocene. *Scandinavian Journal of Hospitality and Tourism, 14*(1), 6–22.

Guneriussen, W. (2008). Modernity re-enchanted: Making a "Magic" region. In J. O. Bærenholdt, & B. Granås (Eds.), *Mobility and place: Enacting Northern European peripheries* (pp. 233–244). Hampshire, Burlington: Ashgate.

Haanpää, M., Hakkarainen, M., & Harju-Myllyaho, A. (2018). Katsaus yhteiskunnalliseen yrittäjyyteen matkailussa: Osallisuuden mahdollisuudet pohjoisen urbaaneissa paikallisyhteisöissä. *Finnish Journal of Tourism Research, 14*(2), 44–58.

Hakkarainen, M., & Tuulentie, S. (2008). Tourism's role in rural development of Finnish Lapland: Interpreting national and regional strategy documents. *Fennia, 186*(1), 3–13.

Hakkarainen, M. (2017). *Matkailutyön ehdot syrjäisessä kylässä.* Acta Universitatis Lapponiensis 357.

House of Lapland (2020). *Industry Brief: Tourism.* https://www.lapland.fi/business/tourism-industry-in-lapland/

Hull, J., de la Barre, S., & Maher, P. T. (2017). Peripheral geographies of creativity: The case for aboriginal tourism in Canada's Yukon territory. In A. Viken & D. Müller (Eds.), *Tourism and Indigeneity in the Arctic* (pp. 157–181). Channel View.

Hyry, M. et al. (2017). Lappisopimus. *Regional Council of Lapland Publications A47/2017.* http://www.lappi.fi/c/document_library/get_file?folderId=3589265&name=DLFE-32814.pdf

Ilola, H., Hakkarainen, M., & García-Rosell, J.-C. (2014). Johdanto Rovaniemen joulumatkailuun. In Ilola H., Hakkarainen, M. & García-Rosell, J.-C. (Eds.), *Joulu ainainen? Näkökulmia Rovaniemen joulumatkailuun* (pp. 10–22). Matkailualan tutkimus- ja koulutusinstituutin julkaisuja, The Lapland University Consortium.

Innovation Norway (2021). National Tourism Strategy 2030. https://assets.simpleviewcms.com/simpleview/image/upload/v1/clients/norway/Nasjonal_Reiselivsstrategi_engelsk_red_c59e62a4-6fd0-4a2e-aea1-5e1c6586ebc6.pdf

Jakobsen, E. W., Engebretsen, B. E. (2019). Ringvirkningsanalyse av reiselivet i Tromsø. Menon Economics Report 26.04.19. https://www.menon.no/wp-content/uploads/2019-127-Troms%C3%B8-2.pdf

Johnston, M. E., Dawson, J., & Maher, P. T. (2017). Strategic development challenges in marine tourism in Nunavut. *Resources, 6*(3). http://www.mdpi.com/journal/resources/special_issues/polar_tourism

Jóhannesson, G. T., Huijbens, E., & Sharpley, R. (2010). Icelandic Tourism: Past directions – future challenges. *Tourism Geographies, 12*(2), 278–301.

Jóhannesson, G. T., & Lund, K. A. (2019). Beyond overtourism: Studying the entanglements of society and tourism in Iceland. In C. Milano, J. M. Cheer, & M. Novelli (Eds.), *Overtourism: Excesses, discontents and measures in travel and tourism* (pp. 91–106). CABI.

Kraft, S. E. (2008). Place-making through mega-events. In J. O. Bærenholdt & B. Granås (Eds.), *Mobility and place: Enacting Northern European peripheries* (pp. 219–232). Ashgate.

Leu, T. C. (2019). Tourism as a livelihood diversification strategy among Sámi indigenous people in Northern Sweden. *Acta Borealia, 36*(1), 75–92.

Lundmark, L. J., Fredman, P., & Sandell, K. (2010). National parks and protected areas and the role for employment in tourism and forest sectors: A Swedish case. *Ecology and Society, 15*(1), 19.

Lundmark, L., & Müller, D. K. (2010). The supply of nature-based tourism activities in Sweden. *Tourism, 58*(4), 379–393.

Lundmark, L., & Stjernström, O. (2009). Environmental protection: An instrument for regional development? National ambitions versus local realities in the case of tourism. *Scandinavian Journal of Hospitality and Tourism, 9*(4), 387–405.

Maher, P. T. (2012). Expedition cruise visits to protected areas in the Canadian Arctic: Issues of sustainability and change for an emerging market. *Tourism: An International Interdisciplinary Journal, 60*(1), 55–70.

Maher, P. T. (2013). Looking back, venturing forward: Challenges for academia, community and industry in polar tourism research. In D.K. Müller, L. Lundmark, & R.H. Lemelin (Eds.), *New issues in polar tourism: Communities, environments, politics* (pp. 19–36). Springer.

Maher, P. T. (2017). *Tourism futures in the Arctic.* In K. Latola & H. Savela (Eds.), *The interconnected Arctic.* (pp. 213–220). Springer.

Mælaborð ferðaþjónustunnar. (2019). *Hagstærðir.* https://www.maelabordferdath-jonustunnar.is/is/hagstaerdir-1/hagstaerdir

McNeish, W., & Olivier, S. (2017). Contracting the right to roam. In K. Spracklen, B. Lashua, E. Sharpe, & S. Swain (Eds.), *The Palgrave handbook of leisure theory* (pp. 289–307). Springer.

Morgan, N., Pritchard, A., & Pride, R. (2011). *Destination brands: Managing place reputation.* Elsevier.

Müller, D. K. (2011). Tourism development in Europe's "last wilderness": An assessment of nature-based tourism in Swedish Lapland. In A.A. Grenier & D.K. Müller (Eds.), *Polar tourism: A tool for regional development* (pp. 129–153). Presses de l'Université du Québec.

Müller, D. K. (2013a). Hibernating economic decline? Tourism and labour market change in Europe's northern periphery. In G. Visser & S. Ferreira (Eds.), *Tourism and crisis* (pp. 123–138). Routledge.

Müller, D. K. (2013b). National parks for tourism development in sub-Arctic areas? Curse or blessing – the case of a proposed national park in Northern Sweden. In D.K. Müller, L. Lundmark & R.H. Lemelin (Eds.), *New issues in polar tourism: Communities, environments, politics* (pp. 189–203). Springer.

Müller, D. K. (2015). Issues in Arctic tourism. In B. Evengård, J. Nymand Larsen, & O. Paasche (Eds.), *The new Arctic* (pp. 147–158). Springer.

Müller, D. K., & Brouder, P. (2014). Dynamic development or destined to decline? The case of Arctic tourism businesses and local labor markets in Jokkmokk, Sweden. In A. Viken & B. Granås (Eds.), *Destination development in tourism: Turns and tactics* (pp. 227–244). Ashgate.

Müller, D. K., Byström, J., Stjernström, O., & Svensson, D. (2019). Making "wilderness" in a northern natural resource periphery. In E.C.H. Keskitalo (Ed.), *The politics of Arctic resources: Change and continuity in the" old North" of Northern Europe* (pp. 99–111). Routledge.

NHO. (2019). Nordnorsk Reiselivsstatistikk 2018. *Tromsø.* https://indd.adobe.com/view/78fbc36d-5c8b-4de1-a8f5-84b82f6ae92f

Óladóttir, O.Þ (2019). *Erlendir ferðamenn á Íslandi 2018 - Lýðfræði, ferðahegðun og viðhorf.* Ferðamálastofa.

Paulgaard, G. (2008). Re-centering periphery: Negotiating identities in time and space. In J. O. Bærenholdt & B. Granås (Eds.), *Mobility and place: Enacting Northern European peripheries* (pp. 49–59).

Rantala, O., de la Barre, S., Granås, B., Jóhannesson, G.Þ, Müller, D. K., Saarinen, J. ... Niskala, M. (2019). *Arctic tourism in times of change: Seasonality.* TemaNord, 2019,528.

Regional Council of Lapland, (2007). *Lapin matkailustrategia 2007-2010 (Lapland Tourism Strategy 2007-2010).*

Statistics Iceland. (2020). *Mannfjöldaþróun 2019 (Demographic development 2019).* Accessed at https://hagstofa.is/utgafur/utgafa/mannfjoldi/mannfjoldathroun-2019/

Sievers, K. (2020). *Lapin matkailustrategia 2020-2023 (Lapland Tourism Strategy 2020-2023).*

Sörlin, S. (1988). *Framtidslandet: Debatten om Norrland och naturresurserna under det industriella genombrottet.* Carlssons.

Tillväxtverket (2018). *Fakta om svensk turism 2018.*

Urry, J. (1995). *Consuming places.* Routledge.

6 The social economy and renewable resource development in Nunavut

Barriers and opportunities

Chris Southcott

Introduction

Nunavut and the rest of Canada's North has undergone tremendous social, cultural, and economic change over the past 60 years. Northern communities have experienced processes of development quite different from most other communities in Canada. These processes have resulted in many unique challenges. These challenges have been met by communities developing approaches that serve to assist these communities to ensure healthier and more sustainable futures. While increased attention is being devoted to the importance of non-renewable extractive developments in the region, these other approaches remain important and are being utilised to support renewable resource development in the region.

One of these new approaches is that of the social economy (Bouchard, 2011; Quarter, Mook, & Armstrong, 2017). It is a means of social action which seeks to empower communities by developing social capital and human capital capacity through assisting non-profit, voluntary, and co-operative organisations work more effectively in the interests of their communities. This chapter summarises research done to examine the potential of the social economy to assist Northern communities to deal with changes these communities are experiencing. It is an attempt to provide a "portrait" of the social economy in Nunavut and to outline the barriers and opportunities. It is based largely on the results of a survey of social economy organisations undertaken in Northern Canada in 2008.

Data from this survey shows that when compared to the other Northern territories, Nunavut has a much higher percentage of social economy organisations engaged in activities that elsewhere are provided by profit-oriented private sector organisations. Many of these are engaged in renewable resource development. It also shows that social economy organisations in Nunavut tend to be younger than those elsewhere in Canada and that Nunavut has a much smaller percentage of organisations that are legally registered as charitable organisations. Membership numbers for social economy organisation show healthy growth with a relatively high level of activity, although fewer of these organisations use volunteers than in the

DOI: 10.4324/9781003172406-6

other territories. In terms of challenges, obtaining and retaining paid staff is one of the most important along with decreasing government funding. In terms of general needs faced by social economy organisations, Nunavut respondents identified finding funding as the most important overall need followed by providing staff training and development.

Findings confirm that social economy organisations continue to be an important part of community developments in Nunavut and that they are a significant contributor to the renewable resource sector. Despite challenges, these organisations continue to be an important part of any attempt to resist an overdependence on extractive industry development in the region. While extractive industry development may continue to be important in Nunavut, a review of problems being experienced by social economy organisations suggest that communities could potentially use financial, training, and other benefits from non-renewable resource development to help social-economic organisations better contribute to renewable resource development.

The social economy and development in Nunavut

Until World War II, Nunavut was isolated from many of the changes occurring in the rest of Canada. The Inuit people of the region were able to maintain their traditional lifestyles: lifestyles both supported and challenged by the interests of the fur trade. This started to change with military development during the World War II. The establishment of military installations in the region introduced new influences and created infrastructure such as airbases that allowed for increased exposure to southern influences after the war.

Almost from the beginning of this change, there was a debate about how the "modernisation" of the North should be managed. The fur trading industry and the Canadian government first thought it best to keep the Indigenous population as isolated as possible from the forces of change (Damas, 2002). Continued dependence on their traditional activities was considered to be the best option for the Inuit peoples of the region. During the 1950s, the federal government reversed its policy on the issue and decided that as Canadian citizens, the Indigenous peoples of the North had a right to basic services such as education, health, and social services. These could best be provided by establishing permanent settlements for the Inuit people of the region.

Yet it was recognised that the North would not be simply an extension of the urban life of the southern parts of Canada. The communal and sharing culture that was the basis of traditional Inuit culture should be maintained by special approaches to development. Co-operatives and community economic development initiatives were highlighted (Lotz, 1982; MacPherson, 2000; Pell, 1990). With the assistance of the federal government, the Inuit people of Nunavut established co-operatives as the main vehicles for economic development in their communities. The people of the region tried to ensure that Inuit traditional ways and knowledge became part of the economic development of their communities by using community-based initiatives. With the negotiation and signing of new treaties, these alternative approaches became institutionalised by the Indigenous peoples themselves in their attempts to maintain traditional activities (Saku, 2002).

The economic institutionalisation of Inuit traditions through a preference for community-based or social economy, initiatives is seen in the Nunavut Economic Development Strategy (NEF, 2003), the first development strategy developed following the creation of Nunavut. Its guiding principles link IQ with "placing control of economic development in the hands of community members" and "integrating economic development activities with community efforts in the areas of community wellness, community learning and community governance." Social economy organisations are clearly identified as important agents of desirable economic development.

These ideals were reiterated in subsequent updates on the Strategy produced by the Nunavut Economic Forum (Impact Economics, 2005, 2008a). In 2007 and 2008, two separate studies were commission by the Nunavut Economic Forum to look at non-profit organisations and their role in the development of Nunavut. The 2007 study was the first to try and get a general idea of the situation of non-profits in Nunavut and the challenges that they face (Aarluk Consulting, 2007). As is noted later on in this report, the conclusions of the 2007 study are similar to those found in this report: non-profits are essential elements of Nunavut communities, but they are faced, by funding and human resource difficulties. Further analysis of the situation in 2008 confirmed and elaborated on the findings of the 2007 study but highlighted the important potential of the non-profit sector to improve well-being in Nunavut's communities (Impact Economics, 2008b).

The Government of Nunavut Report Card, evaluating the performance of Nunavut ten years after its formation, noted that improvements could be made to make Nunavut a better place to live (North Sky Consulting, 2009). In many ways, the Report Card was a reiteration of points made in the initial Nunavut Economic Development Strategy. It noted the need for greater involvement of people in their community and their government. Greater involvement would assist in ensuring a greater degree of self-reliance and ensure that more effective education and training, housing, poverty reduction, and cultural programs are developed and delivered. Although the Report Card did specifically state the usefulness of social economy organisations in making these improvements, the previously mentioned reports all noted the importance of community-based, non-profit and voluntary organisations in achieving greater community involvement.

The conditions of the social economy in the Canadian North

Indigenous traditions linked to the mixed economy, the role of the state, and dependence on natural resource exploitation can be expected to have an impact on the type, form, operation, and development of social economy organisations in the Canadian North. Each of these factors impact the social economy in different ways. It is not a simple matter of saying that this factor will have a positive impact or that factor will have a negative impact. The reality will be much more complex.

Indeed, we can discover initial clues to this complexity in the findings of the 2003 National Survey of Non-profit and Voluntary organisations. While publicly available data from this survey does not allow in-depth investigation of social economy organisations in the north, a 2005 report from this study did list some interesting statistics related to the situation of non-profit and voluntary organisations in the three northern territories (Statistics Canada, 2005). It should be pointed out that this data does not include all social economy organisations. In particular cooperatives, an important part of many communities in the North, were not included in the 2003 survey. The study did not allow for a comparison of Nunavut with the other regions of the Canadian North.

The study counted 851 organisations in the Territories. It is interesting to note that this was the highest percentage of social economy organisation per population in Canada. At 825 organisations per 100,000 population, the percentage was significantly more than the Canadian average of 508 per 100,000 population (Statistics Canada, 2005, p. 19). Only a minority of these organisations are Registered Charities. At 37%, this rate is the lowest in the country and significantly less than the national average of 56% of organisations that are Registered Charities (20). Not surprisingly, compared to the provinces, the Territories had the highest percentage of non-profit or voluntary organisations serving Aboriginal communities (20).

The study listed interesting financial characteristics of social economy organisations in the North. Organisations in the Territories had average revenues of $1.4 million. This was higher than the average of organisations in all other provinces in the country (Statistics Canada, 2005, p. 30). Compared to the provinces, social economy organisations in the Territories had the highest percentage of income from "Earned income" —fees for goods and services. This source comprised 57% of all income for these organisations in the north.

Data showed that social economy organisations in the North varied from other provinces by primary activity. The Territories had the highest percentage of organisations involved in Law, Advocacy, and Politics (Statistics Canada, 2005, p. 19). The region also had higher than average percentages of organisations involved in Arts and culture, Sports and recreation, Education and research, the Environment, and Business and professional associations and unions. The region had lower percentages of organisations involved in health, social services, development and housing, grant-making, fundraising, and voluntarism promotion, and religion.

The study also showed that social economy organisations in the Territories were most likely to report problems related to organisational capacity (Statistics Canada, 2005, p. 53). Interestingly the one capacity area where they did not have problems was obtaining board members. Northern organisations are also far more likely to report problems, such as difficulty providing training to board members (52% in the territories versus 34% in Canada); difficulty providing staff training and development (45% versus 27%), and, difficulty obtaining the type of paid staff the organisation needs (44% versus 28%).

Many of these conditions were investigated further in an inventory, or portrait, of the social economy in Northern Canada undertaken by the SERNNoCa research project (Abele & Southcott, 2016; Southcott, 2015). This attempt at a mapping of the social economy of the Yukon, Northwest Territories, Nunavut, Nunavik, and Labrador involved two stages. The first was a "census" of all social economy organisations in the Canadian North. A list of all possible social economy organisations with their main activity and location was constructed. The second was a questionnaire survey in order to uncover some of the basic characteristics of these organisations in comparison with other regions of Canada.

SERNNoCa researchers were very much aware that much social economy activity, especially in smaller northern communities that rely heavily on the mixed economy, is not undertaken by the formal organisations that this portraiture work was researching. Much of this activity is done in a much more informal manner that can only be studied using different techniques. Other research projects were undertaken by SERNNoCa to try and better understand this aspect of the social economy in the North (Abele, 2009; Natcher, 2009).

No single list exists for all social economy organisations in the North. As a result, an important first step in the portraiture process was the construction of this list. Before the list could be assembled, researchers had to decide on a definition of what constitutes a social economy organisation. The mapping exercise used in this project is based upon a broad definition of social economy that refers to activities that focus on serving the community rather than generating profits. The focus is on economic activities that are not primarily state-driven and not primarily profit-driven and include the traditional economies of Indigenous populations of the North. While a literature review of definitions was conducted, the project leaned most heavily on the definitions contained in Bouchard, Ferraton, and Michaud (2006).

Creating a list of social economy organisations in the Canadian North proved to be problematic. This was particularly the case with many Aboriginal organisations that undertake activities similar to social economy organisations but that are the products of treaties giving sovereign power to these communities. As it was pointed out by at least one questionnaire respondent, to include these organisations as a social economy organisation is to deny the legitimacy of these self-government initiatives. Provisional lists of social economy organisations were established in 2006 and 2007 to serve as the sample frame for the initial questionnaire survey. Table 6.1 shows the numbers for each region of the Canadian North for

Table 6.1 Provisional list of social economy organisations in Northern Canada

	First Nations & Inuit Organizations	Nonprofit and volunteer organizations
Yukon	40	482
NWT	153	475
Nunavut	35	309
Nunavik	19	46
Labrador	36	172

these initial provisional lists. These lists included all potential social economy organisations and as such it was recognised at the time that the actual census list of social economy organisations would be smaller.

The construction of the list of social economy organisations gave researchers quite a bit of information about Northern social economy organisations independent of that gathered from the questionnaire. An extensive amount of information about these organisations can be gathered unobtrusively, directly from the internet. As concerns the Canadian North as a whole, as of May, 2008 1,190 organisations were identified as being probable social economy organisations (Southcott and Walker, 2009).

The activities of social economy organisations in the Canadian North

Internet-based research done on the census allowed researchers to identify the main activity of all but 28 organisations. These results are shown in Table 6.2.

These figures show several important differences in the types of social economy organisations that exist in each of the territories. In looking at Nunavut, compared to the averages for the Territorial North, it has a much higher percentage of social economy organisations engaged in trade, finance, and/or insurance. This is due primarily to the importance of co-operatives in the retail trade sector in Nunavut compared to the Northwest Territories and especially the Yukon. Another important difference concerning the social economy in Nunavut is the relative absence of organisations engaged in law, advocacy, and politics. This can be partially explained by the fact that many of the national advocacy groups have not established branches in Nunavut. Finally, Nunavut has a larger than an average number of organisations that are business associations, professional associations, or unions. The main reason for this is the fact that each community in Nunavut has a hunters and trappers association, the existence of which is linked to the 1993 Nunavut Land Claim Agreement. As well, there are more arts and

Table 6.2 Social economy organisations in the territorial North by main activity

Activity	Nunavut		Northwest Territories		Yukon		Total Territories	
	Total No.	Pct of Total	Total No.	Pct of Total	Total No.	Pct of Total	Total No.	Pct of Total
Manufacturing, Processing and/or construction	1	0.3	0	0.0	2	0.4	3	0.3
Trade, Finance and/or Insurance	29	9.8	8	2.1	3	0.6	40	3.4
Development and Housing	30	10.2	19	5.0	25	4.8	74	6.2
Sports & Recreation, Tourism	39	13.2	32	8.4	128	24.8	199	16.7
Arts & Culture	44	14.9	33	8.7	82	15.9	159	13.4
Education and Research	7	2.4	12	3.2	13	2.5	32	2.7
Health	11	3.7	19	5.0	15	2.9	45	3.8
Social Services	45	15.3	49	12.9	80	15.5	174	14.6
Environment	7	2.4	21	5.5	27	5.2	55	4.6
Law, Advocacy and Politics	14	4.7	70	18.5	49	9.5	133	11.2
Grant-making, Fundraising and Voluntarism Promotion	3	1.0	7	1.8	8	1.6	18	1.5
Religion	15	5.1	49	12.9	43	8.3	107	9.0
Business Association, a Professional Association or a Union	48	16.3	34	9.0	41	7.9	123	10.3
Unknown	2	0.7	26	6.9	0	0	28	2.4
Total	295		379		516		1190	
Undetermined Aboriginal	0		21		31		52	

crafts business associations in Nunavut than in the other territories. This clearly shows the importance of social economy organisations for renewable resource development in Nunavut.

The 2008 questionnaire survey of Nunavut social economy organisations

The initial census served as the sampling frame for the 2008 questionnaire survey. In order to ensure that comparisons were eventually possible across Canada, the construction of the questionnaire was loosely based on a question-naire designed by social economy research networks in both Atlantic Canada

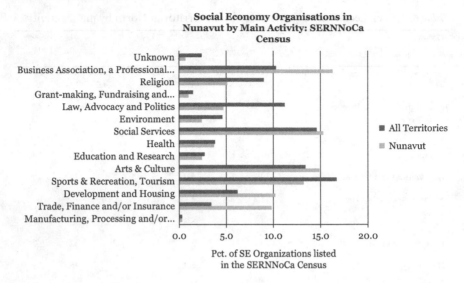

Figure 6.1 Social economy organisation in Nunavut by main activity: SERNNoCa
Census.

and Southern Ontario. Due to the remoteness of many communities, it was
decided to deliver the questionnaire using e-mail where possible and mail where
this was not possible. In Nunavut, the questionnaire was sent out to 285 organisations. Just under half, 127 were sent out by mail and the rest by e-mail.

Looking at the results for the three Territories, it can be seen that a total
of 153 questionnaires were returned from respondents identifiable as social
economy organisations. This represents a response rate of 13%. Looking at
each of the territories, the Yukon had a response rate of 14%, Nunavut had
a response rate of 13%, and the Northwest Territories of 11%. The relatively
low response rate from the questionnaire survey means that the results from
the survey may not be representative of all social economy organisations
in the Territories. At the same time, there was at least one indicator, which
suggests that the results could be fairly representative.[1] As noted above,
we do have main activity statistics for the entire territories. When these are
compared to the activity statistics for the questionnaire respondents, we
see that the statistics for the two groups are remarkably similar. As well,
the percentage of respondents from each territory is similar to the percentage of social economy organisations found in the census.

Types of organisations

Table 6.3 shows the types of organisations that responded to the questionnaire in both Nunavut and all three territories. We can see that the respondents in Nunavut had fewer non-profits, fewer voluntary organisations, and
more cooperatives.

Table 6.3 Structures of social economy organisations

Your organization is....	Nunavut	All Territories
a non-profit	81.6	91.5
a voluntary organization	34.2	52.3
a cooperative	13.2	7.8
a federation or association	28.9	30.1
organized as a worker cooperative	5.3	3.3
legally registered as a not-for-profit or charity	50	64.7
a subunit of a larger parent organization	31.6	26.1

Looking at the data more closely, we see that there were fewer non-profits because there were more cooperatives and because some organisations, such as Hunter and Trappers organisations, are unsure whether they were non-profit or not. The relative importance of organisations created in association with the Nunavut Land Claims Agreement (NLCA) is a unique aspect of the social economy in Nunavut.

There were significantly fewer voluntary organisations in Nunavut that responded to the questionnaire than in the rest of the Territorial North. This could be an indication that the formal voluntary sector in Nunavut is smaller than elsewhere in the North, but more research needs to be done before this can be stated as a certainty. Several organisations that stated they were not voluntary organisations also stated that volunteers are used in various activities.

Legal differences among non-profit and voluntary organisations may influence the types of activities they undertake and their ability to access resources. A key distinction is between those organisations that are registered charities and those that are not. To better understand the key characteristics of the social economy organisation in Nunavut and the rest of Northern Canada, it is important to understand whether most organisations are part of larger organisations, legally constituted as a non-profit organisation, or a registered charity. It is also interesting to find out the percentage of social economy organisations that are organised as worker cooperatives, but this is a question that is included primarily to allow comparisons with other regions of Canada.

Table 6.3 shows that approximately 30% of the respondents in Nunavut were organised as a federation or association, a similar percentage to respondents in all the territories. Just over 31% were a subunit of a larger parent organisation, a somewhat higher percentage than for the territories as a whole. One of the most interesting findings is that only 50% of respondents stated that they were legally registered as a not-for-profit or charity. This is significantly less than the percentage of respondents in all the territories. While this difference is partially explained by the larger number of cooperatives among the respondents in Nunavut, it is clear that fewer social economy organisations in Nunavut are accessing the advantages that may come from a legal not-for-profit or charitable status.

Age of organisations

The national survey of non-profit and voluntary organisations done in 2003 showed that in Canada as a whole most of these types of organisations have existed in communities for a long time. The average age of organisations was 29 years. Organisations in Canada's North are much younger than the national average, reflecting the particular historical development of the North. Data from the 2008 survey showed that the average age of respondent organisations in all three territories was 21 years. Comparing the three territories, we see that Nunavut has the newest organisations, with an average age of 16 years, followed by respondent organisations in the NWT (21 years) and the Yukon (24 years).

Looking at Nunavut's organisations more closely, we see that co-operatives are by far the oldest organisations. The average age of Nunavut's co-operative organisations was 34 years. Some Hunter and Trappers organisations, though reorganised following implementation of the NLCA, also reported that they had existed for 30 years or longer. Almost 40% of Nunavut respondent organisations have been in existence for less than ten years, while almost 75% have been in existence for less than 25 years.

The location of social economy organisations in Nunavut

Figure 6.2 lists the percentages of respondent organisations by their main community of operations. As stated above, because of the low number of respondents, these percentages are not necessarily an adequate representation of the location of all social economy organisations in Nunavut. Still, in analysing the findings of the questionnaire, survey it is important to determine whether the sample of respondents is reasonably representative of the general population.

Without a better census of social economy organisations for Nunavut, population statistics are a good guide to tell us whether the respondents are reasonably representative of Nunavut social economy organisations. According to the 2006 Census, Iqaluit represents only 21% of the population of Nunavut, yet they represent 47% of our respondents. Respondents from Iqaluit are therefore probably over-represented. Cambridge Bay represents 5% of the population of Nunavut compared to the 8% of respondents for

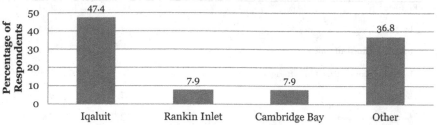

Figure 6.2 Main location of respondent organisations.

the questionnaire survey. Rankin Inlet has 8% of the population Nunavut as well as 8% of questionnaire respondents. Other communities in Nunavut represent 66% of the total population, yet only 37% of questionnaire respondents. organisations in these other communities are therefore probably underrepresented in our findings.

While Iqaluit is probably overrepresented, it is important to state that of the 18 organisations that listed Iqaluit as the main community operations, 7 also stated that they have operations in other communities. Only one organisation that listed their main community as outside Iqaluit, Rankin Inlet, or Cambridge Bay stated that they have operations in other communities.

Membership

One of the key strengths of social economy organisations is their members. Membership is sometimes restricted to certain individuals and at other times open to all. It is usually relatively easy to become a member of an organisation, and for this reason, the degree of involvement of members varies. An organisation that has an active membership has the potential to do much to both support the interests of the individual members and the community at large.

Not all organisations have members, but the majority does. The National Survey of Non-profit and Voluntary organisations showed that 76% of these organisations had members. Our survey showed that the percentage of respondent organisations in the Northern Territories with members was significantly higher at 86.3%. The percentage of respondent organisations in Nunavut with members was 89.5%.

For all three territories, 111 organisations were able to list the number of members, which totaled 56,901. This meant that respondent organisations had an average membership of 513. In Nunavut 26 respondent organisations were able to list the number of members, which totaled 29,683 for an average membership of 1,142. It is important to note that membership numbers

Table 6.4 Membership of respondent organisations by eligibility restriction

	All Territories	*Nunavut*
Restricted	38.6	47.4
by Age	5.9	10.5
by Ethnocultural affiliation	0	0
by First Nations/Metis/Inuit affiliation	7.2	15.9
by Religion	0.7	0
by Geographic area	7.2	2.6
by Workplace affiliation	2	2.6
by Gender	3.3	0

varied greatly based on several criteria, such as the geographic area covered by the organisation and whether the membership was voluntary or related to pre-determined criteria such as land claims. A probable reason that the numbers for Nunavut were so high was that several organisations included all Inuit in a given area of Nunavut.

Some organisations have restrictions on who can become members. Table 6.4 shows the percentage of respondent organisations with restricted membership eligibility and the main characteristics of these restrictions. In the territories as a whole, 39% of all organisations had membership restrictions. In Nunavut, this percentage was slightly higher at 47.4%.

When looking at the characteristics of these restrictions, we see that for all the territories, First Nation/Métis/Inuit affiliation and geographic area are the most important factors of restriction, while age and gender are also important. In Nunavut, by far the most important reason for membership restriction among questionnaire respondents was First Nation/Métis/Inuit affiliation. Almost 16% of questionnaire respondents stated their membership was restricted based on the criteria. Further examination shows that many of these organisations have restricted membership due to treaty requirements.

Level of activity

One indicator of the health of social economy organisations is whether their membership numbers are increasing or decreasing. Table 6.5 lists the growth or decline of memberships for respondent organisations from three years ago. Overall most organisations had membership numbers that were stable—71.2% in all the territories and 61.8% in Nunavut. Very few organisations showed decreases in the number of members. In all the territories, 8.3% showed decline, while in Nunavut, 2.9%, or one organisation, indicated their membership numbers were declining. Most declines were seen in the Yukon.

Table 6.5 Levels of activity

	All Territories	Nunavut
Change in number of members from 3 years earlier		
Decreased	7.2	9.7
Stayed the same	61.4	65.3
Increased	17.6	12.5
Percentage of Membership attending AGM		
0 to 25 %	23.5	21.1
26 to 50 %	13.1	7.9
51 to 75 %	15.7	26.3
76 to 100 %	21.6	18.4

Out of 11 organisations that indicated their membership was in decline, 7 were based in the Yukon. The highest percentage of organisations whose membership showed growth was found in Nunavut. For all the territories, 20.5% of organisations indicated that their membership had grown over the past three years, while in Nunavut, this percentage was significantly higher at 35.3%.

An indicator of whether social economy organisations are active or not is whether they have regular annual meetings. As well, an active, well-organised organisation that is responsive to its membership would generally, but not in all cases, have a quorum at its Annual General Meeting (AGM) and would have the organisation's financial report approved. Data from the 2008 survey shows that the large majority of social economy organisations that responded to the questionnaire have AGMs and that these meetings have quorum and approve the financial reports of the organisation.

Looking at the percentage of the membership that attends AGMs, we see in Table 6.5 that there is a great deal of variety. For the territories as a whole the largest single category of participation of members is 0 to 25%. For Nunavut, the largest category is that of 51% to 75%. The frequency of general membership meetings is another indication of whether organisations are active or not. For the territories as a whole, respondent organisations held an average of 4.3 meetings a year for the general membership. For Nunavut, the number of meetings was significantly less at 2.5 meetings a year.

Board activity and membership

Generally speaking, social economy organisations are run by a smaller group of individuals who are more involved in the guiding the activities of the organisations. This smaller group of individuals is generally known as a governing Board. Information about the activity and composition of Boards can help us better understand the nature of social economy organisations. As can be expected, organisations held more Board meetings than general membership meetings. Looking at the respondent organisations in all three territories, an average of 7.6 Board meetings were held a year. In Nunavut alone, this number was slightly less at 6.6 meetings a year. The average number of Board members for respondent organisations in all the territories combined was 6.8, while the number was slightly less in Nunavut at 6.4.

The survey respondents were asked if all their Board positions were filled or not. In both the territories as a whole and Nunavut 55% of the organisations responded that all the Board positions were filled, meaning that 45% had vacant positions. The vast majority of Board members are volunteers that receive no compensation for their participation. Of all the respondent organisations in all three territories, only 16.2% gave any sort of compensation to Board members. At the same time, there is a significant difference between Nunavut and the other territories in this regard. A much higher percentage of Nunavut organisation (35.2%) gave compensation to its Board members.

Figure 6.3 shows the characteristics of Board members for those organisations that responded to the questionnaire. It is interesting to note that 55% percent of Board members in Nunavut are either First Nation, Métis, or Inuit while the corresponding figure for all the territories in 31%. It is also interesting to note that while a majority of Board members in all the territories are female, only 35.2% of those in Nunavut are women.

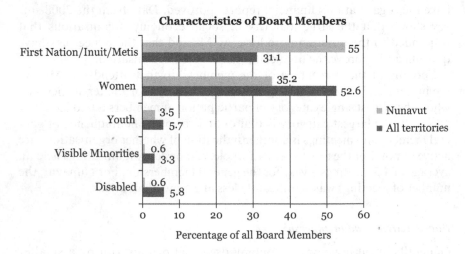

Figure 6.3 Characteristics of board members by percentage of all board members.

Volunteers and employees

One of the most significant differences between Nunavut and the other territories is that fewer social economy organisations in Nunavut use volunteers
for their activities. While 78.9% of respondents in all the territories use volunteers, only 55.3% of Nunavut respondents do.

In terms of the number of volunteers used by respondent organisations
each year, there is a lot of variation. Looking at the Territories as a whole,
most organisations used less than ten volunteers in a given year. The next
largest category is 11 to 20 volunteers. Over 14% of respondent organisations used over 50 volunteers a year. The percentages of Nunavut are also
included, but because of the small number of Nunavut respondents (22), the
percentages are not very reliable.

In both Nunavut and the other territories, approximately 46% of respondents reported that their organisations had no paid employees and therefore
issued no T4 slips. Of the remaining 54% of respondents, there was a significant difference between Nunavut and the rest of the territories in terms of
the average number of employees per organisation. The average number of
employees in respondent organisations in all the territories was 6.5, while in
Nunavut, the average was much higher at 12. Nunavut has a larger number
of smaller organisations employing less than 40 people, while the other territories have a large number of bigger employers.

Human Resource issues

Respondents were asked if they had human resource problems. The most
serious problems in this area relate to obtaining and retaining paid staff.
Figure 6.4 shows that just over 30% of respondents in all territories said

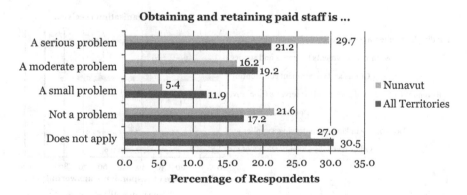

Figure 6.4 Percentage of respondents rating the statement "Obtaining and retaining paid staff is...."

it was an issue that did not apply to them. Generally, this is because these particular organisations do not have paid staff. Just over 40% stated that this was either a serious or moderate problem. In Nunavut, just under 30% stated it was a serious problem. Next in importance was training. Just under 38% of Nunavut respondents listed providing staff training and development as a serious or moderate problem against 22% who said it was not a problem.

Obtaining and retaining Board members does not seem to be that much of a problem in Nunavut. While 30% of respondents agreed that it was either a serious or moderate problem, 38% said it was not a problem. Providing training to board members is the least serious problem. In Nunavut, less than 30% of respondents claimed it was a serious or moderate problem compared to over 40% who said it was not a problem.

Finances

In addition to the many contributions that social economy organisations make to communities, they also represent an important economic presence. The NSNPVO Survey showed that many non-profit and voluntary organisations earn income by providing goods and services for a fee; some also depend substantially on governments—particularly provincial governments—for funding. Smaller organisations rely heavily on donations of money and in-kind donations of goods and services.

Figure 6.5 shows the sources of revenue for social economy organisations in the North according to the 2008 survey. For respondents in all the territories 87.8% received funding from another organisation such as a government, a foundation, or a corporation. In Nunavut, this total was slightly less at 78.4%. The largest single source for funding was from the government in the

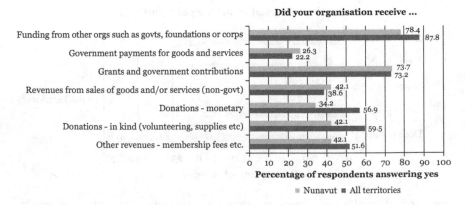

Figure 6.5 Graph of percentage of respondents answering yes to the statement "Did your organization receive...."

form of grants or contributions. In all the territories, 73.2% of organisation received funding from this source compared to 73.7 in Nunavut. It is notable that the NSNPVO Survey of 2003 showed a much lower reliance on government sources in the rest of Canada, where only 49% of non-profits and voluntary organisations received government funding. In comparing Nunavut organisations with those in the rest of the North we can see that a slightly higher percentage of Nunavut social economy organisations receive funding from government payments for goods and services, non-government sales of goods and services, and a lower percentage receive funding from donations, either monetary or in-kind, and other types of revenues.

Respondents were asked whether their organisation's revenues had either increased, decreased, or stayed the same over the past three years. Only 12% of all the respondents in the North stated that their revenues had decreased and in Nunavut only one organisation did. A large percentage, 48% in all the territories and 48.6 in Nunavut stated that their revenues had increased over the past three years. This is an indication that these organisations are continuing to be a dynamic part of the economic makeup of the Canada's North.

The respondents were also asked if a series of financial issues identified in previous research were a problem for their particular organisation. The results are displayed in Figure 6.6. Of all the issues listed, the most serious for Northern social economy organisations was reductions in government funding. In all the territories, 44.8% of respondents stated that it was

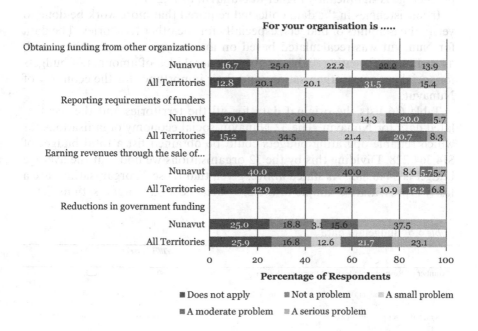

Figure 6.6 Graph of percentage of respondents rating the statement "For your organisation is..." for funding.

a serious or moderate problem. In Nunavut the percentage was even higher at 53.1%. In Nunavut, 37.5% of respondent organisations stated that reduction in government funding was a serious problem. Obtaining funding from organisations such as government, foundations or corporations was the second most important problem followed by the reporting requirements of funders. Earning revenues through the sale of goods and/or services was the least important problem for organisations in all the territories and in Nunavut.

Organisations were also asked if their organisation made a surplus profit last year from the sales of goods or services. There is a significant difference between the respondents in Nunavut and those in the other territories. Only 21.1% of respondents in all territories stated that they earned a surplus last year, whereas 38.3% of respondents from Nunavut did. While not many social economy organisations generate a surplus, respondents were asked what would happen if a surplus was generated. Clearly the most popular direction of distribution is back into the organisation—an option identified by slightly more than half of respondents, in both Nunavut and the other territories. The next most popular direction of distribution is to hold it in reserve for community benefit or in a community trust. In all the territories, 17% of the respondents favored this option while the percentage in Nunavut was slightly less at 13.2%. In all the territories, only 4.6% of the respondents indicated they would distribute the surplus to individual members, but this percentage is significantly higher in Nunavut at 13.2%.

Inconsistencies in the data collected required that more work be done to verify the amounts of budgets, especially for the other territories. The data for Nunavut was recalculated based on an analysis of revenue data and, as such, is sufficiently reliable to give a general idea of amounts of budgets for social economy organisations and their importance for the economy of Nunavut.

Table 6.6 lists the original data for all the territories and the recalculated data for Nunavut. The 32 Nunavut social economy organisations for which reliable operating budgets could be obtained list a total budget of $14,364,228. Dividing this by the 32 organisations we find that the average budget is $448,882. As stated above, given that these 32 organisations are a fairly representative sample of the 295 social economy organisations listed

Table 6.6 Budgetary data

	All Territories	Nunavut
Number of Respondents Listing Operating Budget	130	32
Total Operating Budget of Social Economy Organizations	$52,443,330	$14,364,228*
Average Budget per Organization	$403,410	$448,882*
Standard Deviation	$804,242	$824,237

in the 2008 version of the SERNNoCa census, we can state that the social economy organisations in Nunavut manage approximately $132 million in funds each year.

While government grants account for the largest amount of revenue at 46% of total revenue, sales of goods and services (33%) and memberships and subscriptions (145) are also important sources of revenue.

General needs

The respondents were asked a series of questions about issues related to their general needs. The first of these questions asked how much collaboration their organisation has with other social economy organisations such as non-profits, voluntary organisations, or co-operatives. Most organisations have some degree of collaboration. At the same time, there is less collaboration in Nunavut compared to the territories as a whole. In all territories, 31% of organisations collaborate a lot with similar organisations, while in Nunavut this percentage is only 17.6%

Respondents were asked the degree to which a series of issues was a problem for their organisation. Figure 6.7 lists the responses to these questions. Finding funding was clearly the most important overall need of the social economy organisations responding to the questionnaire. Of all the respondents, 55% listed it as either a moderate or serious problem. Only 15% said it was not a problem. The numbers for Nunavut were similar to the averages for all the territories.

Getting volunteers is the next most serious problems faced by the respondents in all the territories. Of all the respondents, 42% said it was either a serious or moderate problem. Just over 18% said it was not a problem. In Nunavut, the issue is less a problem than in the territories as a whole as just 23% stated it was a serious or moderate problem and 26% stated it was not a problem. In Nunavut 26% of respondents stated that getting volunteers was an issue that does not apply to them compared to only 11% in all the territories.

Providing staff training and development was the second next most important issue to respondents in Nunavut and the third most important for respondents in all the territories. Only 12% in both Nunavut and the territories as a whole said it was a problem that did not apply to them. In Nunavut a slightly higher percentage of respondents stated it was not a problem, but 47% of Nunavut respondents stated it was a serious or moderate problem.

Internal capacity in areas such as internal administrative systems, information technology, software, or databases was clearly not as important an issue as the previous three. This is especially the case in Nunavut were only 22% of respondent organisations listed it as a serious or moderate problem.

The least serious issue for most respondents was collaboration with other social economy groups. While very few organisations stated that it was not a problem that applied to them, only13% in all the territories said it was a

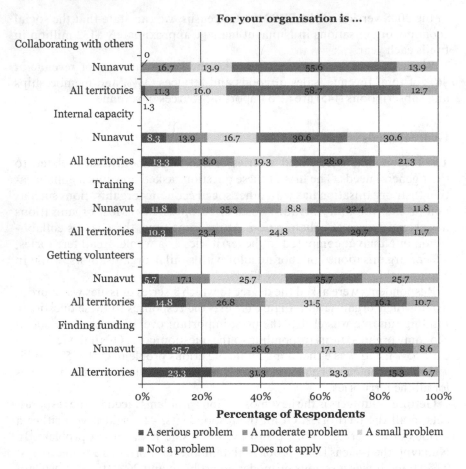

For your organisation is ...

Figure 6.7 Graph of percentage of respondents rating the statement "For your organization...."

serious or moderate problem. In Nunavut, no organisation listed it as a serious problem, and only 17% stated it was a moderate problem.

The survey also asked respondent a series of open-ended questions to allow respondents to mention issues that might not have been adequately dealt with in the questionnaire. Responses to these questions are listed below (Figure 6.8). The first asked respondents if they had any special research needs that could help their organisation better contribute to the well-being of their community.

Responses varied considerably, with some looking at general questions for Nunavut and other concerned with issues specific to their organisation.

Finally, respondent organisations were asked about concerns or problems that their organisation has had in the past. Funding surfaces as an important issue, as it was mentioned in 6 of 11 responses.

> **3. Does your organisation have any special needs in terms of research that could help your organisation better contribute to the well-being of your community and region?**
> - **if yes please specify**
>
> **Responses from Nunavut respondent organisations:**
>
> - number of working artists (full time and part time) and in what mediums; amount of income from selling art. How has the development of the regional mining activities affected arts production (i.e., Artists working for mining companies)
> - to expand our outreach into Nunavut, increase research opportunities related to landscape and the environment, and increase knowledge of the profession to Inuit, young people and anyone with an interest in the profession
> - population growth
> - coach training and recruitment
> - The members need to understand the importance of cooperating with one another in order to have a healthier organisation as a community
> - utilising compost to support local plants
> - market research to find new customers
> - We need to track our participants better, we need to understand the bottom line needs of the teachers and students in communities with respect to training in the world of skilled trades and technologies
> - Research is needed on number of visitors, conversion information, economic potential of the industry, and all types of marketing research
> - research for alternative education delivery
> - support to develop our own research capacity
> - language, training
> - research that supports non profits in their quest to do their work more effectively- this would include arguments for increased funding from government

Figure 6.8 Responses from Nunavut respondent organisations to the question "Does your organization have any special needs in terms of research that could help your organization better contribute to the well-being of your community and region?"

Summary and discussion

Few places in the world have gone through social change with the speed and intensity that the communities of Nunavut have. In the space of 50 years, the people of this region have had to adapt themselves from a traditional migratory hunting and fishing existence based on community co-operation and self-reliance that had served them for generations to an urban lifestyle characterised by dependence and new economic and social values such as individualism, competition, and industrialism. This transition has not been easy. The loss of much of their former self-reliance and the disparagement of their traditional culture has taken a toll on these communities. This is seen in the numerous social and health challenges that are often brought to public light.

Yet despite these challenges, it is not unfair to say that the people of Nunavut have been able to survive this transition and are now looking to change the power relations that characterised their adaptation processes in the past. Rather than adapting their cultural values and lifestyles to Western patterns, they are now increasingly adapting Western values and lifestyles to meet their needs. While this is most evident in the creation of Nunavut, it is also evident in the ability of the people of Nunavut to develop innovative community organisations, often based on traditional cultural values, that are designed to meet the needs of their community.

Elsewhere in Canada and the rest of the world, these innovative community organisations have been called social economy organisations in that they are separate from private and government and function as an economic organisation that is based on the values of service to a community to meet social needs rather than generate profits. When given the choice, the people of Nunavut have, since the beginning of the transition, given preference to these types of organisations. This is seen, starting in the late 1950s, in the use of co-operatives as the organisation of choice to deal with economic needs. More recently, communities in the region have started to develop a range of new social economy organisations to assist in the well-being and sustainable development of their communities.

The research summarised in this chapter is an attempt to provide a portrait of the social economy in Nunavut and to outline barriers and opportunities faced by these organisations. Previous research had indicated that these organisations were facing funding and human resource difficulties. The SERNNoCa survey undertaken during 2008 confirmed the findings of these earlier studies and has given us a better understanding of some of the characteristics of these organisations. Because of the importance of co-operatives in the retail and arts and crafts sectors when compared to the other Northern territories, Nunavut has a much higher percentage of social economy organisations engaged in activities that elsewhere are provided by profit-oriented private sector organisations. Social economy organisations in Nunavut tend to be younger than those elsewhere in Canada except for co-operatives and hunter and trapper organisations. The structures of organisations in Nunavut differ from elsewhere in the North in that Nunavut has a much smaller percentage of organisations that are legally registered as charitable organisations.

Membership numbers for social economy organisation show healthy growth with a relatively high level of activity. The composition of the governing Boards of social economy organisations in Nunavut differs slightly from those elsewhere in the North in that a higher percentage of Board members in Nunavut are Indigenous, but a lower percentage are female. Another key human resource characteristic of social economy organisations in Nunavut is that fewer of these organisations use volunteers than in the other territories. In terms of human resource challenges, while obtaining and retaining Board members is not a major problem, obtaining and retaining paid staff is.

Regarding finances, the survey results confirmed previous indications that social economy organisations represent an important economic force in Nunavut. Rough calculations based on questionnaire results indicate that social economy organisations in the territory manage approximately

$132 million in funds each year. Government funding, as is the case for social economy organisations throughout the North, represent a higher percentage of revenues then elsewhere in Canada. Compared to the other territories, Nunavut social economy organisations receive a higher percentage of revenues from the sales of goods and services and a lower percentage from donations. In general, revenues for these organisations are on the increase. Decreasing government funding is the most serious financial challenge being faced by organisations in Nunavut.

When asked about the relative importance of a series of general needs faced by social economy organisations, Nunavut respondents identified finding funding as the most important overall need. Providing staff training and development was the second next most important issue identified by respondents in Nunavut. The least serious issue for most respondents was collaboration with other social economy groups. When asked what research would benefit the social economy in the North, responses tended to note the importance of research on general social and economic issues of concern to Nunavut. Clearly, respondent organisations are concerned about the general well-being of their communities.

Generally, the research indicates that, despite challenges, the social economy is stronger and more important in Nunavut than elsewhere. This confirms the observations found in other research that the Indigenous traditions in Nunavut continue to influence the way communities view economic development (Abele & Southcott, 2016; Harder & Wenzel, 2012; Southcott, 2015). The desire for a larger role for co-operative and collaborative relationships in the economy likely stems from the role these relationships played in pre-contact activities and their continued importance for traditional subsistence activities (Natcher, 2009). While extractive activities, and profit-oriented organisations that are associated with these activities, are increasing in importance in Nunavut, it is unlikely that most communities would see these activities and their organisations as an end in themselves. They are seen by many as being necessary to provide communities with necessary jobs, training, and funding (Bernauer, 2011). At the same time, it is likely that communities would use these extractive industry benefits to support more sustainable and renewable forms of long-term economic development. Social economy organisations appear to be the preferred vehicle by which Nunavut communities would like to shape their economic future.

Note

1. As the survey was not based on random sampling, probability theory-based indicators of representivity could not be used.

References

Aarluk Consulting. (2007). Not-for-Profit Groups in Nunavut—A Review. *Nunavut Economic Forum.*

Abele, F. (2009). The state and the northern social economy: Research prospects. *Northern Review, 30*, 37–56.

Abele, F., & Southcott, C. (Eds.). (2016). *Care, cooperation and activism: Cases from the Northern social economy*, University of Alberta Press.

Bernauer, W. (2011). *Mining and the social economy in Baker Lake, Nunavut.* Northern Ontario, Manitoba, and Saskatchewan Regional Node of the Social Economy Suite. http://usaskstudies.coop/documents/social-economy-reports-and-newsltrs.

Bouchard, M. J. (Eds.) (2011). *L'économie sociale, vecteur d'innovation: l'expérience du Québec.* Presses de l'Université du Québec.

Bouchard, M. J., Ferraton, C., & Michaud, V. (2006). Database on social economy organizations: The qualification criteria. *Working Papers of the Canada Research Chair on the Social Economy*, Research Series no. R-2006-03.

Damas, D. (2002). *Arctic Migrants/Arctic villagers: The transformation of Inuit settlement in the Central Arctic.* McGill-Queen's University Press.

Harder, M. T., & Wenzel, G. W. (2012). Inuit subsistence, social economy and food security in Clyde River, Nunavut. *Arctic, 65*(3), 305–318.

Impact Economics. (2008a). *2008 Nunavut Economic Outlook.* Nunavut Economic Forum.

Impact Economics. (2008b). *Understanding Nunavut's Non-profit Sector.* Nunavut Economic Forum.

Impact Economics (2005). *2005 Nunavut Economic Outlook.* Nunavut Economic Forum.

Lotz, J. (1982). The moral and ethical basis of community development: Reflections on the Canadian experience. *Community Development Journal, 17*(1), 27–31.

MacPherson, I. (2000). Across cultures and geography: Managing co-operatives in Northern Canada. In I. Sigurdsson & J. Skaptason (Eds.), *Aspects of Arctic and sub-Arctic history*, pp.550–562. University of Iceland Press.

Natcher, D. (2009). Subsistence and the social economy of Canada's aboriginal North. *Northern Review, 30*, 83–98.

North Sky Consulting. (2009). *Qanukkanniq? The government of Nunavut report card: Analysis and recommendations.* Government of Nunavut.

Nunavut Economic Forum. (2003). *Nunavut Economic Development Strategy.* The Sivummut Economic Development Strategy Group.

Pell, D. (1990). Soft economics: Community organizations that can do something!. In J. Potvin (Eds.), *Community economic development in Canada's North*, pp. 145–167. Canadian Arctic Resources Committee.

Quarter, J., Mook, L., & Armstrong, A. (2017). *Understanding the social economy: A Canadian perspective.* University of Toronto Press.

Saku, J. (2002). Modern land claim agreements and Northern Canadian aboriginal communities. *World Development, 30*(1), 141–151.

Southcott, C. (Ed.). (2015). *Northern Communities working together: The social economy of Canada's North.* University of Toronto Press.

Southcott, C., & Walker, V. (2009). A portrait of the social economy in Northern Canada. *Northern Review, 30*, 13–36. Springer.

Statistics Canada. (2005). *Cornerstones of Community: Highlights of the National Survey of Nonprofit and Voluntary Organizations.* Ministry of Industry, Government of Canada.

7 An academic lead in developing sustainable Arctic communities

Co-creation, quintuple helix, and open social innovation

Martin Mohr Olsen

Introduction

In the pursuit of renewable economies and communities within the Arctic, it is essential that we keep in mind the regenerative properties and inherent potential for renewal contained within the region's institutions of higher learning. While colleges and universities in the periphery naturally tend to be smaller and less resourceful compared to their metropolitan counterparts—they are, more often than not, centers of specialised local and regional knowledge. Not only do institutions of higher learning located in remote regions often play a vital part in maintaining and developing local languages, practices, and cultures—they are also ideal incubators for the education and training of local agents of change capable of making a genuine impact on both local and regional issues. Realising that smaller institutions can often be protective and somewhat conservative, it is nevertheless of growing importance that they make an increased, systematic, and collaborative effort toward engagement in issues of sustainability on multiple operational levels in order to combat many issues facing the Arctic region; climate change and ecological challenges that affect the sustainability of communities, political, geopolitical, and securitisation issues, effects from resource extraction, impacts of increasing tourism, etc. (Arctic Council, 2016).

While it may be easy to tout the benefits and values of small, local, and specialised universities—it should also be acknowledged that universities in the periphery are fighting at least two distinct, yet interwoven, problems. On the one hand, Arctic universities often rely heavily or entirely on government subsidies that are often subject to prevailing political moods and often see allocations differ from year to year, making it hard to plan ahead and plan for long-term commitments or strategies. Located in smaller countries and communities, they are also often unable to attract major sources of external funding from benefactors or industry and establishment of large-scale research projects is similarly rare. As issues of sustainability facing the Arctic mentioned align with all three aspects of sustainability; economic, environmental, and social—the involvement of universities as key stakeholders is paramount for the development of much-needed solutions. At the

DOI: 10.4324/9781003172406-7

same time, unfortunately, such demands for solutions to often very complex issues can place an even greater burden upon the universities.

In this chapter, I will argue that smaller institutions of higher learning within the Arctic must play a greater role in tackling the issues facing the region in a more *practical* sense. They should work with geographically embedded knowledge in a real-work setting and focus on solutions relevant to the area and its stakeholders. However, I will also argue that for this to become a reality, changes to how many small Arctic universities currently operate must be made. What follows is an attempt to outline an operational framework that addresses the two problematics mentioned above; issues of resources and sustainability. The framework presented is an early attempt at a conceptual visualisation of all the different practical aspects that universities in the Arctic will need to consider systematically in order to minimise reliance on input resources in order to maximise their sustainability output. The framework is being developed for use by the new *Innovation Unit* at the University of the Faroe Islands and is under continual revision. In order to explain the basic framework, a point-by-point analysis of each step in the process will be given, outlining a theoretical basis and practical considerations.

Background and context

Before we get to the conceptual framework, there are two contextual aspects that should be clarified. First, some background on how this project is based on prior work and attempts to speak to some objective needs with Arctic academia, and secondly, a short description of the newly formed Innovation Unit at the University of the Faroe Islands, and their role in the project, will be given.

Initial insights

The chapter is an attempt to follow up on our recent review of emerging trends within the Arctic academic environment (for more, see, Blaxekjær et al., 2018). The first of these trends is a cross-sector demand for more innovative and entrepreneurial skills to be included in academic courses and curriculums. Where the main justification for entrepreneurship and innovation in an academic setting was initially economic growth—it can now be found in most academic fields (Chiu, 2012; Reffstrup & Kærn Christiansen, 2017). Second, there is a growing interest in going beyond the three main pursuits that are at the core of most university mission statements (teaching, research, and dissemination). This fourth emerging mission statement, at times, referred to as "co-creation of sustainability" (Trencher et al., 2014) is in many cases a way for universities to experiment with novel triple- and quadruple helix models that can facilitate cooperation, but has also allowed them to diversify from purely economic pursuits to a wider array of non-commercial purposes dealing with sustainability (Trencher et al., 2014; Rosenlund, Rosell, & Hogland, 2017). Third, the Sustainable Development Goals (SDGs) have also, however slowly, begun to take hold in the Arctic region. Places of higher learning, such as the University of the Faroe Islands, the University of Greenland, and local stakeholders in general, are noticing the benefits of taking a stance on including sustainability into their missions. In recent years, there has been a positive increase in the amount of focus given to the SDGs within Arctic universities. And, fourth, as is evidenced by the number of conferences and university collaborations with a focus on the Arctic and sustainability—there is a genuine window of opportunity for smaller regional universities to make their mark and take part in the growing interest in the region and the resources available. While we have seen an increasing interest in these trends over the last couple of years, it is important to note that for smaller Arctic universities, acting on and realising these interests can be a challenge. As has been mentioned above, a large number of Arctic universities are publicly funded, are lacking in resources and staff, are spread out over a vast geographic area where travel is expensive and cumbersome, and often lack networking possibilities and the patronage of large donors or collaborators.

The innovation unit[1]

Reacting to these emerging trends, the University of the Faroe Islands decided to initiate a response in the form of a small and lean project office that will be tasked with analysing and mapping best practices available to a relatively small and resource-weak institution. Novel to the University of the Faroe Islands, we were able to secure funding for a full-time member of staff that was not bound by any teaching responsibilities or office work. The Innovation Unit is headed by a coordinator not affiliated with any specific department or academic discipline, rather she answers to the rector and the board of directors. The unit is intended to be flexible and nimble in that it can incorporate members of staff for short-term projects as well as initiate long-term working groups for larger initiatives. As such, the name Innovation Unit was deliberately chosen in order to convey its *ad hoc* configuration and the non-physical placement within the conventional organisational hierarchy.

The main intent behind the Unit is to have it function as an auxiliary support on matters that would otherwise be beyond the scope of the duties of the university management staff and beyond the resources of the research staff. It is meant to facilitate and coordinate projects and initiatives that would previously likely have fallen into the boundary between two organisational camps, resulting in inaction. By adhering to the tenets of boundary spanning, the unit attempts to push a holistic agenda of maximising benefits on behalf of not only the university as a whole, but also the wider community it finds itself in (Tushman, 1977). It purposely looks to deal with complex projects that can be classified as "wicked issues," projects that cross academic disciplines and rely on the participation of multiple departments, civil society, municipalities, governmental agencies, and industry to solve (Williams, 2002).

The initial findings for the Innovation Unit reveal that there are four distinct, yet overlapping, foundational changes that should be made at an institutional level in order to meet these growing trends relating to sustainability. First, using the SDGs as primary guidelines in decision making and in the evaluation of the viability of new projects or existing modes of operation. Second, adding to the core mission statements of the organisation (teaching, researching, and knowledge transferal) to include a fourth mission statement of sustainable co-creation. Third, revising and formalising how the university deals with external stakeholders based on sound Helix models. Fourth, the implementation of modern innovation protocols such as Open Innovation and Social Innovation in order to foster local capacity building, dialogue, a shared sense of ownership, and mechanisms that can mitigate risk and help launch socially beneficial initiatives beyond the initial stages of prototyping and proof-of-concept.

A framework for the future

In an effort to outline the conceptual framework proposed for the Innovation Unit, what follows is an analysis of the constituent theoretical elements contained within it. First, I will argue for a systematic implementation of the Sustainable Development Goals. Secondly, I will give an historical summary of the university mission statement and then turn my attention to the arguments for a review of the beneficial addition of a fourth-mission statement based on sustainability co-creation as is outlined in Trencher et al. (2014). Third, I will give a quick overview of the Triple Helix approach and then focus in on Ranga and Etzkowitz (2013) analysis of Triple Helix Spaces, followed by a quick outline of the benefits and utility of quadruple- and quintuple-helix models as they are outlined in Carayannis and Campbell (2012) and Carayannis, Barth, and Campbell (2012). Fourth, I will provide some background relating to both Open and Social Innovation and then focus on how these two approaches could favorably be combined into the new concept of *Open Social Innovation* as has been outlined in Chesbrough and Di Minin (2014) and Martins and De Souza Bermejo (2014). I will then conclude that these four theoretical concepts (the SDGs, academic mission statements, Helix Models, and innovation frameworks), when combined into a framework, allow a small university such as the University of the Faroe Islands to put forward a set of guidelines that are both rigid in terms of adherence to sustainability and co-creation, while also being flexible in terms of scale and resources.

The conceptual operational framework, as seen in Figure 7.1, has been adapted and expanded upon based on the initial work done by Martins

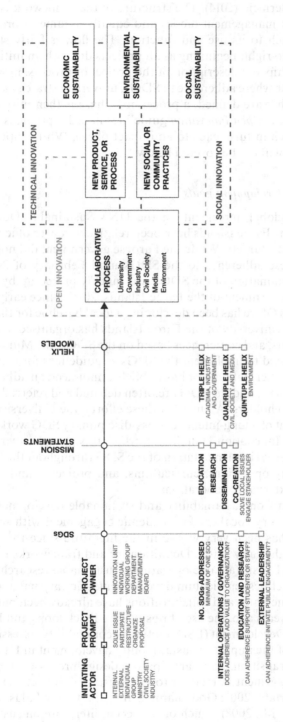

Figure 7.1 Conceptual operational framework.

and De Souza Bermejo (2014). The structure of the framework is based on common project management tactics and equally assumes a project management approach to its use and function. The flow is fairly straightforward, from left to right, beginning as all projects do, with an initiation that involves ownership and description of the project or process, moving on to an auditing phase where adherence to SDGs, university strategy and collaborative approaches are defined, a processing phase is then initiated where participants follow *Open Innovation* guidelines in order progress to an output phase—which in turn leads to an impact phase. While important, the initiation phase will not be covered below.

The sustainable development goals

Since the near-global agreement on the UN's Sustainable Development Goals in 2015, the Faroe Islands have been relatively slow to officially adopt the goals and their targets. While the Faroese government did not officially announce Faroese adherence to the SDGs until February of 2018, interest in- and dissemination of the SDGs was quickly picked up by a few key stakeholders (Government of the Faroe Islands, 2018). Since early 2018, the Prime Minister's Office has been developing a local baseline for the tracking of progress, the University of the Faroe Islands has organised a number of workshops, events, and conferences based on the SDGs, the Municipality of Tórshavn has voted to implement the SDGs as guidelines for future policy work, salmon farmer Bakkafrost has publicly announced an adherence to a number of key goals, and the SDG are often debated and referred to by politicians and stakeholders. Along with these efforts, the University green-lit the establishment of a long-intended cross-disciplinary SDG working group overseen by the Innovation Unit. The work of this group is two-fold: best practices for internal implementation of the SDG throughout the University in terms of daily operations, curriculums, and projects—and simultaneously through external collaborations.

Academic focus on sustainability and sustainable development are not new to HEIs. A very vocal cry for academic engagement with sustainability, in general, has been very visible in the literature since the 1990s and especially since the early 2000s. Definitions of and frameworks for sustainability within HEIs sustainable operations, sustainable research, environmental literacy, ethics, curriculum development, and internal and external multi-helix partnerships and collaborations have already been outlined and analysed many years ago, the use of positive feedback loops and Education for Sustainable Development (ESD) tactics, accountability, assessments and measurements of the impact of sustainability development in HEIs and the use of systems transitions and participatory design processes in relation to stakeholder engagement in relation to HEI sustainability, etc. (Lukman & Glavič, 2007; Wright, 2002; Godemann et al., 2014; Ferrer-Balas, Buckland, H., & de Mingo, M., 2008). Much of the recent literature on sustainability and sustainable development that has been published after the SDG's launch in 2015 is thematically preoccupied with addressing the Sustainable Development Goals head-on, but in many regards, the arguments and theoretical basis are much the same and build on previous works such as those mentioned.

For the Innovation Unit at the University of the Faroe Islands, the initial question with regards to the SDGs was very binary and simple: *"In doing this, are we working against the principles of any of the SDGs?"* If the operation or project could be argued to have a negative impact on any of the SDGs, they would be scrapped or reworked. However, in a broader sense, this simplistic way of filtering does nothing to further implementation or garner collaborative support for further development internally or externally, and the need for a proactive stance rather than a reactive one quickly became apparent. In the context of a larger operational framework, the first question we should be asking is: *"In doing this, are we furthering the principles of the SDGs – if so, how many and to what extent?"* Cross-referencing with a national baseline would, of course, also benefit an initial scoring as this is able to provide far better metrics on the benefits of a certain operation or project. If the answer is positive, we should (as is outlined in SDSN Australia/Pacific, 2017, pp. 10–30) ask ourselves how the operation can contribute to the SDGs through: internal operations and governance, education, research, and external leadership.

This type of early stage evaluation is typical to general project management along with risk assessment and delegation of ownership, but the really beneficial aspect to this type of scoring is that is lends itself to a speedier and more nimble learn-by-doing style of SDG implementation, not unlike Jeff Sutherland's SCRUM methodology (Sutherland, 2014). Rather than spending years developing local implementation guidelines that run the risk of being too unwieldy or dated—an early score-card with simple metrics outlining benefits and drawbacks that allow for easy reporting and quick iterations will allow for better results, the option for broader inclusiveness and collaborations, and an accumulative positive net effect in relation to expertise and know-how down the line. There are, of course, more elaborate and technical ways of assessing the implementation of the SDGs on an academic level. Laurent, A. et al.'s (2019) *SDG Assessment Methodology* is one such tool.

University strategy and mission statements

With a solid foundation ensuring adherence to the SDGs and considering their implications as they relate to internal operations, education, research, and external leadership, as seen above, an operational framework is slowly starting to take shape. With the initiating participant having covered ownership and the factoring in of sustainable development goals and targets, it is equally important that they experience there being a willing institutional support toward practical societal contributions within the organisation.

The three common missions that can be found in most university charters or strategies are education, research, and dissemination. While the University of the Faroe Islands and its staff generally strive to go beyond these three missions, there is no codified or systematic approach to any such operations. Arguing for a systemic approach to academic work that goes beyond these three common missions, I will be relying on Trencher, Yarime, McCormick, Doll, and Kraines' (2013) analysis of an emerging fourth mission they refer to as: *co-creation for sustainability*. In order to discuss this new type of academic mission, a very brief outline of the missions relating to teaching, research and dissemination will follow below before I return to an analysis. It is important to note that these activities in no way stand-alone within academia, each are pillars supporting the same roof and the same is true for any fourth pillar we might add. Trencher et al. (2013) illustrate this point very well in Figure 7.4 of their paper (Trencher et al., 2013, p. 168).

Teaching

Growing out of cathedral and monastics schools of late medieval Europe, the first institutions historically to be acknowledged as "universities" were the University of Bologna and the University of Paris, established in the eleventh and twelfth centuries, respectively. The driving force behind these and more than 100 universities that were established around Europe between the twelfth and the fifteenth century, was the expansion of the Catholic Church and the need for a systematised approach to instruct students in matters relating to the church; theology, law, and medicine (Arbo & Benneworth, 2007, p. 19). This first mission of teaching was to be the main objective of early universities well into and beyond the Enlightenment.

Research

Following the Enlightenment, a genuine demand for academic reform was raised in the established European academic institutions. As new types of schools and disciplines were developed in order to cater to social changes brought on by the Enlightenment, the French Revolution, the Industrial Revolution, and the growing field of natural sciences—the older institutions were increasingly seen as part of the *l'ancien régime* of days past. Following the carnage of the Napoleonic Wars, a renaissance was to hit the Prussian academic establishment that would cement the second mission of research as a cornerstone of academia. Brought on by the shuttering of a number of prominent Prussian universities during the wars and a renewed sense of nationalism following the wars, the establishment of the Berlin University in 1810 was guided by the "ideals of Bildung, academic freedom and the collective research process as its corner-stones" (Östling, 2018, pp. 23–9). This notion of *Bildung* has generally attributed to Wilhelm von Humboldt, and the transition from an academic environment focused purely on teaching, to one of teaching and research were to be codified in what became known as the Humboldtian Reform or Humboldtian Model (see Wittrock, 1993; Östling, 2018) which dictated a more holistic approach to academic education where students were given academic autonomy to pursue their interest and engage with the world through reason and self-determination. Within this ideal, the pursuit of knowledge was to be available to all and aim to change the world for the better independently of economic interests. In other words, knowledge for the sake of knowledge (see Östling, 2018; Anderson, 2004).

Dissemination and technical transferal

While the third mission encompasses academic dissemination to peers and the public in a variety of forms not covered by the first and second missions, the focus here is rather on the technical transferal of knowledge for use outside of the academic realm (E3M Project, 2012). The rise of this part of the third mission emerged from Vannevar Bush, 1945 conceptualisation of a "university-industrial complex" following the World War 2 (Bush, 1945). Bush's blueprint for academia was meant to "establish a link between university research and business innovation" in order to "advance economic well being" (Zomer & Benneworth, 2011, p. 83). As Clark (1998) and Etzkowitz (2002) point out, the connection between academia and industry within a third mission formulation dates back to the European agricultural universities and the American land-grant universities of the Industrial Revolution. However, Bush's university-industrial complex did not pick up speed until the 1980s, helped along with the passing of the Bayh-Dole Act of 1980 that allowed researchers to claim ownership of federally funded inventions and technical advancements through patents (Mowery, 2007 as cited in Trencher et al., 2013). This new law was instrumental in capitalising, commercialisation, and commodification of knowledge (Zomer & Benneworth, 2011, p. 84; Etzkowith, 1998, p. 826). This commodification, in turn, gave rise to the Entrepreneurial University (see for example: Clark, 1998; Etzkowitz, 1998; Etzkowitz, Webster, Gebhardt, & Terrad, 2000), where external funding often dictates the direction of research and applied sciences in order to maximise profits outside of the university. Led by MIT and Stanford—for the Entrepreneurial University, "identifying, creating and commercialising intellectual property have become institutional objectives [with the aim of] improving regional or national economic performance as well as the university's financial vantage and that of its faculty" (Etzkowitz et al., 2000, p. 313 as cited in Trencher et al., 2013, p. 151).

It should be noted here that technical transferal of knowledge in the form of patents or products might not be relevant to the majority of smaller Arctic universities. In cases where this commercial component is missing or lacking, the argument for an organisational focus on social issues such as sustainability would be a much simpler sell as that addition of the next mission we will deal with will show.

Co-creation for sustainability

As is argued by Trencher et al. (2013, pp. 156–157), the term "social contribution is a useful synonym for describing the core notion of the third mission." However, they go on to argue that "the idea of societal contribution is today widely perceived and promoted as being chiefly an economic contribution" championed by OECD efforts to "emphasize the economic benefits and gains in international competitiveness for governments when universities focus their third-stream activities on innovation transfer and spurring regional development" (Trencher et al., 2013, pp. 156–157). In their critique of the third-mission regime, Trencher et al. (2013) conclude that efforts to introduce the concept of sustainability and green innovation into the existing third mission has yet to produce much in the way of results—likely due to the fact that the majority of funding and knowledge transfer relates to medicine, biomedical- and computer research. Based on their analysis of the current state, their "position is that the potential of the third mission regime to function as a useful guiding concept or propelling force in the quest for low-carbon development and sustainable transformation of individual towns, cities, and regions is yet to be proven" (Trencher et al., 2013, pp. 156–157). They (Trencher et al., 2013, pp. 157–9) then argue for a fourth mission statement, one of co-creation for sustainability. They do not argue that this new mission should supplant the third mission, but rather supplement it in compounding the effects of all four mission statements by *transforming Entrepreneurial Universities into Transformative Universities* (Trencher et al., 2013, p. 169, emphasis is of the author) that weave together: teaching, research, dissemination, and co-creation for sustainability. This mission, they argue (Trencher et al., 2013, p. 158) is one that requires the transformative university to encompass a broad range of transdisciplinary sciences working together long-term on crucial issues that are place and stakeholder oriented. Further, they call for large-scale coalitions with multi-helix specialists and non-specialists utilising Open Innovation tactics in order to produce socially embedded knowledge and mutual learning. To clarify, let us take a look at Helix models and Innovation frameworks in turn.

Helix models

If the Berlin University of 1810 could be said to be Single Helix and the addition of industry to form the Entrepreneurial University of the 1980s could be said to be a Double Helix, then the addition of government as a third actor is what makes up the Triple Helix Model. Conceptualised by Etzkowitz (1993) and Etzkowitz and Leydesdorff (1995) in an effort to interpret the shift from a dualistic relationship between universities and industry in the industrial society, to the triadic relationship between academia, industry, and government in the modern knowledge society, the Triple Helix Model is commonly seen as the standard for modern academic cooperation.

Triple Helix

Often depicted as three intertwined strands in the style of the classic representation of DNA, the Triple Helix model is probably better illustrated in the style of a Venn diagram consisting of three circular spheres of influence; academia, industry, and government overlapping. Since Etzkowitz and Leydesdorff's early work of the 1990s, an abundance of theoretical development has pushed the conceptual framework of the model forwards immensely and we will not be able to gain a complete overview of the literature here. A majority of the early literature is concerned with the in-depth analysis of systemic and organisational interactions at fairly large scales where actors tend to be portrayed as entire universities, entire companies or entire governments—often ignoring the roles of individual actors such as members of staff or specialised working groups. While much of the early theoretical work on the topic provides a good foundational approach to how these three spheres of influence can come together, they do not, however, "provide an explicit analytical framework for conceptualizing Triple Helix interactions in an innovation system" (Ranga & Etzkowitz, 2013, p. 238) and the large-scale holistic focus on "'block' entities, without going deeper to the level of sphere-specific actors" (Ranga & Etzkowitz, 2013, p. 242) are not conducive to the Arctic conditions dealt with here. For the sake of the operational framework, I will be relying on Ranga and Etzkowitz's (2013) more fine-grained analysis of the Triple Helix systems approach below.

In very broad strokes, the conventional theoretical approach to the Triple Helix model is based on two complementary perspectives; one institutional and one evolutionary. Within the institutional approach, the way the three spheres of influence interoperate fall into three different configurations: *A statist configuration*, where the government acts as the main driver and planner for innovation and development. *A laissez-faire configuration* with limited governmental control is where the industry is the main driver and universities provide skilled workers. *And, a balanced configuration* where the institutions act jointly in formalised partnerships (Ranga & Etzkowitz, 2013, p. 239; Etzkowitz & Leydesdorff, 2000, p. 111). For our sake, my references to the Triple Helix model will refer to the balanced configuration where trilateral networks and hybrid organisations are possible (this is also similar to the notion of Triple Helix type 3 as noted by Etzkowitz & Leydesdorff, 1998, p. 197). The evolutionary perspective argues that universities, industry, and government exist and co-evolve within social systems where they are influenced by markets, technological advancement, environmental concerns, and so on. The interoperation between the stakeholders here relies heavily on two processes of communication—a functional and indirect one between science and markets, and an institutional and direct one between private and public control that allows for selective adjustments that ensure a regeneration of the system (Ranga & Etzkowitz, 2013, p. 240). Especially the move from Triple to Quadruple and Quintuple Helix is very much contingent on the view that

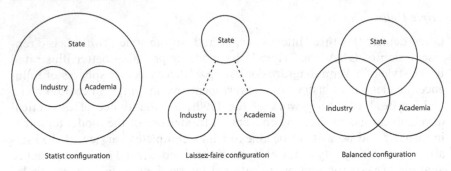

Figure 7.2 The three different configurations of the way the three spheres of influence interoperate in the institutional approach.

stakeholders must operate within a social system of change through individual actors that are able to set up lines of communications and work together in order to adjust and reshape the system.

Ranga and Etzkowitz (2013, pp. 241–254) break down the conventional understanding of the Triple Helix model using the innovation systems concept or, as it was later known; national innovation systems (NIS) (Ranga & Etzkowitz, 2013, pp. 240–241) in order to analyse it at a more granular scale. Their Triple Helix systems approach breaks down the Triple Helix model into three elements: First, system components (and their boundaries), these are the institutional spheres of university, industry, and government—consisting of the institutional actors themselves and any individual actors, and where these networks overlap. While the institutional actors can, as such, be argued to interact—the interest here is rather on the individual actors involved; innovational initiators, entrepreneurial researchers, R&D staff, the management or policy writers. Second, relationships between the system components, meaning direct collaboration, conflict moderation, collaborative leadership, and so on. It is through meaningful and carefully moderated relationships that actors are able to work together, transfer knowledge, network and replicate the system. Third, the functions of the system, relate to the generation, diffusion, and use of knowledge and innovation as an absolute main ideal. This is the end product of the model, the coming together of different stakeholders in order to produce and disseminate new knowledge that can be utilised to benefit a greater need.

Ranga and Etzkowitz (2013, pp. 247–250), however, argue that in order to get to this end product, it will need to be realised through what they call Triple Helix spaces: the Knowledge Space, the Innovation Space, and the Consensus Space. Very much relevant to our operational framework, they can, in an Arctic low-resource context, be understood as follows: The Knowledge Space is where an aggregation of local and regional research and knowledge exists. This could take the form of a local database managed by local research councils, tracking the output and needs of universities, industry, and government in order to ensure that research is not fragmented or needlessly duplicated.[2] Up to date data on the knowledge being produced and governmental or municipal projects needing attention could help streamline collaboration. The Knowledge Space should also aim to attract funding and leading research through networking and dialogue. The aim of the Innovation Space is to provide a common platform for stakeholders to manage the development of new and innovative firms and industries. This can be done through the creation of integrated environments where academia, research, and the needs of industry, governments, and municipalities can come together in order to find solutions. The Consensus Space is a coming together of stakeholders into a forum where they are interdependent and can begin to see themselves as part of something greater than their department, company, town, or country. This can take the form of brainstorming meetings and dialogue sessions that aim to solve issues that are not solvable by any one sector of society, such as housing shortages, unemployment, pollution etc.

Quadruple Helix

If we reconsider the earlier steps in this proposed framework, the academic mission statement of co-creation for sustainability is predicated on the collaboration between academia, industry, government—and also civil society

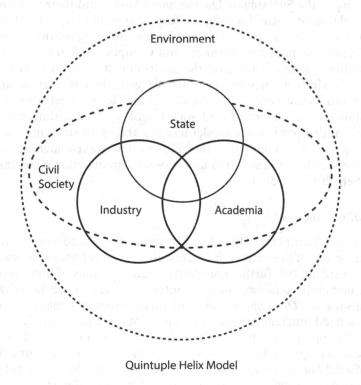

Quintuple Helix Model

Figure 7.3 Quintuple-Helix model.

and the media. This broad and inclusive approach is very much in line with the arguments made by Carayannis and Campbell (2009, pp. 206–207) for a fourth dimension to the existing Triple Helix model, and it is similarly very close to the evolutionary perspective of the triadic arrangement of academia-industry-government needing to exist within a malleable and changing social system and not a theoretical vacuum (Ranga & Etzkowitz, 2013, p. 240). For Carayannis and Campbell (2009, pp. 206–207), a Quadruple Helix model considers the necessity of the media as a means to transport public discourse and playing a vital role in the communication of social realities, norms, and values. Adding to that, a Quadruple Helix model also incorporates the actions of agents of culture, civics, and arts within the community. As such, the Quadruple Helix model is easily able to accommodate the Helix spaces mentioned above, and there are no practical barriers to the inclusion of NGOs, interest groups and cultural movements. This inclusiveness has the possibility of, as Carayannis and Campbell (2009, p. 207) argue, result in a "democracy of knowledge, driven by a pluralism of knowledge and innovation [...]."

Quintuple Helix

Returning to the Sustainable Development Goals and their focus on sustainable development and ecological stewardship, there is yet another Helix Model that concerns itself with these issues. Building on their argument for a Quadruple model, Carayannis and Campbell (2010) go on to argue for the fifth sphere of influence; the environment. While the Triple- and Quadruple Helix models concern themselves mainly with economic and cultural innovations and gains, the Quintuple Helix model adds elements of sustainable development and social ecology, and ultimately argues that this added level of the model has the ability to structure the Helix model approach in such a way that it can allow stakeholders to concern themselves with eco-innovation and eco-entrepreneurship (Carayannis & Campbell, 2010, pp. 58–63).

Innovation frameworks

While the three different Helix Models outlined above concern themselves with variations of theoretical and practical takes on collaboration and innovation, there are two further conceptual manifestations of innovation that can be favorably considered in this context. At one end, we have the profit maximisation of *Open Innovation*—and on the other end, the maximisation of social good inherent in *Social Innovation*. While these initially seem diametrically opposed, I hope to make a convincing argument that by combining these two approaches to innovation at the tail end of a framework an *Open Social Innovation* approach to innovation might have a useful role to play in making sustainable development viable within the Arctic.

Open innovation

The initial ideas relating to open innovation date back to the 1960s onset of the Information Age and the spread of computers, but the formalised term used in reference to industry opening up their silos of research and development to external researchers and developers in order to licence, spin out, and divest products was conceptualised and coined by Henry Chesbrough in 2003. While Chesbrough's (2003) initial conceptualisation was heavily slanted toward accelerated development of new technologies and goods by making an organisation's boundaries more permeable and hopefully more profitable, later refinements of the term argue that the process can also be a useful way of managing the flow of knowledge across organisational boundaries both internally and externally simultaneously using "pecuniary and non-pecuniary mechanisms in line with the organization's business model" (Chesbrough & Bogers, 2014)—that is, measured in *both* monetary and non-monetary terms.

In short, Chesbrough and Bogers (2014) argue that Open Innovation is based on the premise that the drive for innovation and the sources of knowledge that can drive innovation are dispersed widely in society and the economy. They invoke Bill Joy's (co-founder of Sun Microsystems) Law that states that *"most of the smartest people work for someone else"* (Chesbrough & Bogers, 2014). Open Innovation, briefly, is a way to scale the boundaries of otherwise closed-off organisations in order to allow for novel approaches to stagnant problems by outside expertise. In practice, this involves industry establishing an Open Business Model (Vanhaverbeke & Chesbrough, 2014) that is inclusive to external stakeholders and innovators. This type of model sets up a "division of innovation labour" where one party might research and develop a new idea—and another party, in turn, carries it to market (Vanhaverbeke & Chesbrough, 2014, pp. 52–53). Vanhaverbeke and Chesbrough (2014, p. 54) give a number of examples of how this approach can be useful; inside-out modes that result in licensing agreements and spin-offs, outside-in modes that draw in external expertise and combine it with an existing business model and modes where an industry makes use of external or internal knowledge to develop entirely new business models. While Chesbrough's definition of the term is generally applied to a commercial setting—the notion that very capable and innovative actors are to be found outside the boundaries of an organisation is just as likely to hold true to non-commercial organisations such as a university, a municipality or an NGO. While the literature on Open Innovation is too broad and varied to cover here, this main takeaway of being able to look beyond institutional and organisational walls for expertise and inspiration is the key. While the more typical combination of Open Innovation and academia is fairly commonplace at many technical universities and business schools around the world—applying the same mindset of openness that a business might employ to further commercial interest to an academic reality allows for even more and more diverse collaborative arrangements when combined with the tenets of Social Innovation. Especially for small non-technical Arctic universities, when combined with the fourth mission statement of co-creation for sustainability and a strong multi-helix approach, the concept of Open Innovation is easily transformed from a profit-making industry strategy into an academic collaborative approach based on openness and stakeholder inclusion. By utilising Open Innovation as a core strategy, the operational framework allows the university and its stakeholders to move forward both in a conventionally technical innovation track, or it can use the same tactics to focus its energy on maximising social good through Social Innovation. Further, Open Innovation when employed from the perspective of a university can even facilitate technical and social innovation in parallel—facilitating the development of a viable new product or service based on social issues. The majority of literature dealing with Open Innovation deals with how it benefits Technical Innovation, so I will not be covering it here. Its exclusion should, however, not be seen as it being of lesser interest or importance to the framework. For the sake of understanding how small institutions can better target social issues, we will concern ourselves with the value of Social Innovation and how Open Innovation can be used as an important tool to realise social innovation-based projects.

Social innovation

Open Innovation is, at least from a theoretical standpoint, very much tied to business innovations which in turn are motivated by profit maximisation. Social Innovation, on the other hand, is predominantly motivated by the goal of meeting social needs (Mulgan, 2006, p. 146). Social Innovation can be defined as "innovations that are social both in their ends and their means. [They are] new ideas (products, services, and models) that simultaneously meet social needs and create new social relationships or collaborations. In other words, they are innovations that are both good for society *and* enhance society's capacity to act" (Murray and Mulgan, 2010, p. 3).

A very different aspect to Social Innovation, as opposed to innovation in a more broad sense, is that Social Innovations need not be wholly new ideas—but can be new to those benefiting from them. Further, Social Innovations tend not to be inherently new in and of themselves, but rather a combination of already available technologies or solutions (Mulgan, 2006, p. 151). Boelman and Davies (2015, p. 6) also argue for this newness criteria along with the requirements that Social Innovations should meet a social need in positive ways, they should be put into practice, they should engage and mobilise beneficiaries through governance and they should transform social relations through greater access to power and resources. They further argue that Social Innovations (Boelman & Davies, 2015, p. 7) can take many different forms; new services, products, practices, processes, rules, regulations, and organisational forms.

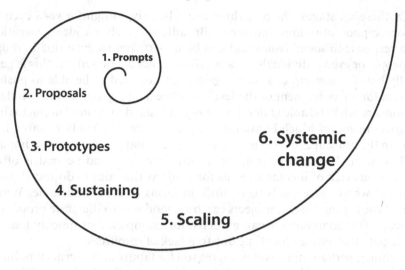

Social Innovation stages

Figure 7.4 Social innovation stages.

A core distinction between Open Innovation and Social Innovation, is that Social Innovation aims to solve issues not generally solved by commercial interests (Nicholls, Simon, Gabriel, & Whelan, 2015, p. 3). They can be issues such as human rights, environmental concerns, healthcare, or education—often in developing or remote regions of the world. According to Murray and Mulgan (2010) and Rayna and Striukova (2019) the Social Innovation process can be broken down into six different stages:

1 *Prompts, inspirations, and diagnoses:* Initial stage during which various factors trigger the need for innovation, after which a diagnostic of the problem and the framing of the question ensue.
2 *Proposals and ideas:* Idea generation stage using a variety of methods based on insight and experience.
3 *Prototyping and pilots:* Stage during which ideas are put into practice to be tested and, subsequently, refined.
4 *Sustaining:* Stage at which the idea is adopted for everyday use and is, as a result, streamlined. Income streams are identified at this stage.
5 *Scaling and diffusion:* Stage at which there is an attempt to scale up and diffuse the innovation beyond its original test bed.
6 *Systemic change:* This stage is the ultimate goal of social innovation, but also the most difficult to achieve due to its wide-scale, the large number of stakeholders it involves and the multiple barriers to change that exist.

Of these six stages, the first three are relatively straightforward even in resource-poor situations and areas. Broadly they rely on idea generation as a response to social issues that can be undertaken by even small groups of people or even individuals. In a small-scale academic setting, this is generally how far a group of students would be expected to be able to push a project for an assignment or the level of refinement a group of independent volunteers might be able to develop a project funded by a small municipality or government (of which I have taken part in more than my fair share). It is only in the last stages of the innovation process that any meaningful impact will be seen. The main reason for failure at steps 4 and beyond is often the obvious lack of income streams for projects that aim to do social good in areas where there tends to be little in terms of economic value. While these types of small-scale projects can be a good way to illustrate proofs-of-concept (free communal urban gardens, for example), they quickly tend to fizzle out after only a short time due to a lack of resources.

Another, perhaps not so obvious, reason for failure at this crucial point in the process is often the absence of supportive networks. Mulgan (2007) and Rayna and Striukova (2019) point out that failures to connect with networks that can provide expertise and experience are a common pitfall for social innovation processes. Similar to how the industry might employ Open Innovation as a means to entice external expertise into the fold—they argue that this is a tactic that could be extremely applicable in terms of bringing social innovation to market, so to speak. Rayna and Striukova (2019, p. 385) argue that "[b]y enabling access to a larger pool of resources and skills, as well as diffusion paths, applying open innovation paradigm to social innovation could enable to overcome critical challenges at all six stages of the social innovation process."

Open social innovation

I have argued that Open Innovation and Social Innovation are two rather different strands of innovation frameworks that serve very different needs. Yet the similarities of *bringing products to market* tie the two processes together in more ways than one. In both cases, maximising profits and maximising social good, the processes hinge on networking, external expertise, and novel business models. While commercial enterprises make use of business models in order to turn a profit, social innovations will ultimately have to rely on some form of operational resources (likely in the form of monetary funding) in order to sustain themselves and in turn scale, diffuse and enact systemic changes.

The concept of Open Social Innovation seems to be first proposed by Dominic Chalmers (2013). In his article, he argues for similarities between the Open and Social Innovation similar to the ones noted above. As argued above, some main reason for failures to launch social innovation processes are lack of long-term operation resources and a lack of expert networking.

Chalmers expands upon the knowledge searching and networking issues by arguing that social innovators would do well to focus more on their users as well as initiating boundary-spanning knowledge searches that look for expertise and innovation that exist beyond their organisation's traditional domains to combat "industry blindness" (Chalmers, 2013, p. 25).

Chalmers (2013) goes on to identify a third stumbling block (related to resources) that is common for social innovation processes; risk aversion. He argues that a common barrier "to the adoption of social innovation lies in the risk associated with disruptive innovation [in that] those offering support to social innovators in the form of capital are institutionally conditioned to favour incremental 'safer' forms of innovation" (Chalmers, 2013, p. 26). In order to combat risk aversion, Chalmers argues for five different propositions for social and community-based organisations (Chalmers, 2013, pp. 27–28):

1 Adopting a more "open" approach will mitigate risks associated with introducing new innovations.
2 Adopting problem solutions from different domains will reduce the risk of new innovations failing.
3 Incorporating user knowledge into the innovation process will increase their chances of success.
4 Participating in some form of open, networked innovation will be more effective at developing innovations addressing the root causes of social problems.
5 Engaging in "open source" collaboration will be more effective in tackling vested interests and dominant competitors.

Here we see the beginnings of merging of Open and Social Innovation. In applying Open Innovation mechanisms to the issues facing Social Innovation, social innovators should adhere to a networking paradigm that involves multiple boundary-spanning stakeholders—which in turn would help mitigate risk aversion from investors and funders.

Martins & De Souza Bermejo (2014) follow up on Chalmers work and argue as the main point that "when Social Innovation is seen from a collaborative point of view, organizations become more porous structures that make it possible to overcome the barriers that prevent communities from innovation from the bottom up. Thus, when organizations are open they strengthen localism and provide a means for civil society to become involved in finding solutions" (Martins & De Souza Bermejo, 2014). They go further to attempt a conceptual model of the relationship between Open Innovation and Social Innovation seen in Figure 7.5.

They argue that the collaborative interplay between Open and Social Innovation not only has the capacity to produce new solutions to social problems and changes to social practices, but at the same time is also able to facilitate and stimulate new technical innovations in the form of new

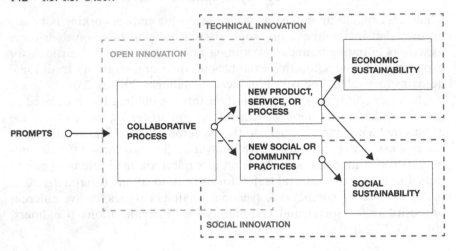

Figure 7.5 Conceptual model of the relationship between open innovation and social innovation.

products, services or processes that in turn result in economic development. At the same time, Chesbrough and Di Minin (2014, p. 170) venture a definition of the concept. They defined Open Social Innovation (OSI) to be "the application of either inbound or outbound open innovation strategies, along with innovations in the associated business model of the organisation, to social challenges" (Chesbrough & Di Minin, 2014, p. 170). They argue that Open Social Innovation is especially apt at solving stages 3–5 of the Social Innovation process (Chesbrough & Di Minin, 2014).

Regardless of where the prompt originates; industry, government, academia, or civil society—social innovation and socially innovative projects more broadly will mostly follow the six steps outlined above. They will need to pass through a phase of idea generation, prototyping, and testing phase, and eventually a process of sustaining itself so that it can scale and hopefully change society for the better. Most of these socially conscious community-based efforts at social change will likely struggle to move beyond the initial prototyping and proof of concept stages due to the lack of operation resources. And as Chalmers (2013) argues, it can be hard to find funding for risky disruptive innovation aimed at a segment of society that might not be seen as very profitable, politically incorrect or otherwise unpalatable for whatever reason.

Before moving on the how this might look from an academic perspective, I hope that (albeit brief) the arguments for utilising Open Innovation tactics in conjunction with Social Innovation processes will seem compelling. How they can be made to push Social Innovation initiatives beyond stages of initial prototyping and proof-of-concept, how they can mitigate risk and how they can break down institutional borders and attract external expertise. None of the examples used so far have concerned themselves with academia. They have all focused on industry and NGOs. What is wholly absent from the literature is a perspective from within academia. In the following concluding section, I will attempt to outline why I think the framework presented is such a natural fit for small Arctic universities.

An academic perspective

As I have noted above, Open and Social Innovation tactics are generally considered to belong in the domains of commercial enterprises and action-oriented civic groups such as NGOs or governmental agencies. Only in rare instances does the literature attempt to shine a light on how these tactics could be employed by academia in an effort to pursue issues of sustainability, innovation, entrepreneurship, or social change. The University of the Faroe Islands is comparable to a number of Arctic universities. It is a small, publicly funded university servicing around 800 students in five different departments. Apart from offering courses in social sciences, economy, law, natural science, and Faroese, it also houses a teacher's college and a degree in nursing. In relation to mainland Europe and the Arctic region broadly, it is a geographically isolated university with a relatively low level of contact with other institutions of higher learning in the region.

At the time of writing, the University is facing a number of issues. It is seeking to modernise, systematize, and professionalise a wide array of operational aspects of the organisation; this includes the development of a new university strategy and updated mission statements. It is also attempting to broaden its local, regional, and international reach in terms of collaborations, research networking, and funding—all with very limited resources. Further, the university is also reacting to the emerging trends within the Arctic academic environment described above (Blaxekjær et al., 2018). The small size of the University of the Faroe Islands and other similar universities in the Arctic that often exist to service the local population and seldom profile themselves in a manner that would attract an international student body or sizeable amounts of external funding, must often make do with local networks and draw on local sources of funding for research—or relegate themselves to taking on smaller and often non-technical roles within larger European or Arctic research projects.

The current lack of much international research presence facing many of the smaller Arctic universities does, however, allow them to occupy academic niche positions relating to local knowledge and development. This hyper-specialisation and localisation of knowledge has, especially in recent years, become a growing trend within the Arctic and often sees small Arctic universities being able to punch above their weight, participating in a much more level footing on their home turf due to valuable local insights and expertise. Coming to terms with how valuable these smaller universities are to their local communities and how much social good they are able to initiate through socially conscious activities would not only maximise organisational output, but it would also have the added benefit of building up local competences and credibility with regards to regional and international collaborations.

In order to operate optimally, an operational framework such as the one presented here would need to be functional on a number of levels in order to deliver practical lasting results. If the output is to be sustainable social innovation, it must cover the three general aspects of sustainability - economic, environmental and social. To reach that level, it must, therefore, encompass some form of technological innovation that creates local jobs or adds positively to the economy in the form of new products or services and it must also contain an element of social innovation that results in new sustainable practices or developments. However, in order to have commercial and social innovations complement each other in a sustainable way, university lead initiatives will need to rely on a collaborative process with industry on one side and civil society and government on the other. This process, the Innovation Unit argues, fits perfectly with the mechanisms of Open Innovation. It is here that university-led Open Innovation is able to facilitate openness, dialogue, networking, risk and stakeholder management, and a collaborative effort to push projects beyond stages four and five of the social innovation track.

To get to a point where this facilitation of sustainable development between academia, industry, and civil society becomes possible, the university must progress through a preparatory phase of internal auditing that ensures the inclusiveness of stakeholders, the strategic value for the university and a strict adherence to sustainability throughout the process. Before the Open Innovation tactics can be utilised in an open, collaborative process, the university and key stakeholders should conduct a Helix-model audit. Following a Triple Helix model stakeholders would only include academia, government, and industry (Etzkowitz, 1993; Etzkowitz & Leydesdorff, 1995). Using a Quadruple Helix model would include civil society in the process (Caryannis & Campbell, 2009). Even better, by utilising a Quintuple Helix model the university is able to include a fifth sphere of influence: the environment (Caryannis & Campbell, 2010). A prerequisite for successful Helix model collaborations led by the university is of course that the university takes on an active and participatory role. In order for the university to do so, projects must have a strategic value that complement university missions statements of education, research, and dissemination. In order to boost impact, the Innovation Unit advocates that the university goes beyond these three common mission statements and initiates a fourth mission statement of sustainable co-creation as is put forward by Trencher et al. (2013). As noted above, there is a clear wish from a number of Arctic universities that the UN Sustainable Development Goals be implemented as a guiding framework for curricula and daily operations. If the output of this operational framework is to be sustainability in its different forms, then the SDGs are currently likely the best way to audit the sustainability of initiatives. The SDGs are simple, easy to understand, give participants a common language through their goals and targets and are at the same time not very confrontational as a tool. As a yardstick, simply asking whether a project or an initiative would be counter to the tenets found in the SDGs would be a good start. If so, it should be reconsidered or scrapped. If a project is not counter to the SDGs, an audit and analysis of how well the project is in line with the SDGs, how many goals and targets it will cover and a measurement of impact should be produced. Also, the university should consider how participation in the project or initiative and subsequent adherence to the SDGs might positively impact internal operations and governance of the organisation and the institution, how it might be used for educational purposes, for research purposes—and how it might be used as a way for the university, along with stakeholders, to illustrate leadership and commitment toward sustainability within the community.

At the time of writing, the Innovation Unit at the University of the Faroe Islands is further developing the framework and lobbies for it to be implemented as a way to assess upcoming projects and processes. It is our firm belief that a structured and theoretically sound approach to social issues will allow the university to present itself as an open, inclusive stakeholder and champion of sustainability in the Faroese Islands.

Notes

1. At the time of publishing the Innovation Unit (now formally the *Research and Enterprise Unit*) has chosen to initially focus on mission two (research).
2. www.isaaffik.org is one such initiative.

References

Anderson, R. D. (2004). *European Universities from the Enlightenment to 1914*. Oxford Scholarship Online.

Arbo, P., & Benneworth, P. (2007). Understanding the Regional Contribution of Higher Education Institutions: A Literature Review. *OECD Education Working Papers, No. 9*. https://doi.org/10.1787/161208155312

Arctic Council. (2016). *Arctic Resilience Report*. http://hdl.handle.net/11374/1838

Boelman, V., & Davies, A. (2015). *Growing Social Innovation: A Guide for Policy Makers*. http://youngfoundation.org/wp-content/uploads/2015/04/YOFJ2786_Growing_Social_Innovation_16.01.15_WEB.pdf

Bush, V. (1945). *Science: The Endless Frontier: a Report to the President by Vannevar Bush, Director of the Office of Scientific Research and Development, July 1945*. www.nsf.gov/od/lpa/nsf50/vbush1945.htm

Caryannis, E., & Campbell, D. (2009). "Mode 3" and "Quadruple Helix": Toward a 21st century fractal innovation ecosystem. *International Journal of Technology Management, 46*, 201–34.

Caryannis, E. G., & Campbell, D. (2010). Triple helix, quadruple helix and quintuple helix and how do knowledge, innovation and the environment relate to each other? A proposed framework for A trans-disciplinary analysis of sustainable development and social ecology. *International Journal of Social Ecology and Sustainable Development, 1*(1), 41–69.

Carayannis, E. G., & Campbell, D. (2012). *Mode 3 knowledge production in quadruple helix innovation systems.* Springer.

Carayannis, E. G., Barth, T. D., & Campbell, D. F. J. (2012). The quintuple helix innovation model: Global warming as a challenge and driver for innovation. *Journal of Innovation and Entrepreneurship, 1*(1), 2.

Chalmers, D. (2013). social innovation, an exploration of the barriers faced by innovating organizations in the social economy. *Local Economy, 28*(1), 17–34.

Chiu, R. (2012). Entrepreneurship education in the Nordic countries: Strategy implementation and good practices. *Nordic Innovation Report,* 24.

Chesbrough, H. (2003). *Open innovation: The new imperative for creating and profiting from technology.* Harvard Business School Press.

Chesbrough, H., & Bogers, M. (2014). Explicating open innovation: Clarifying an emerging paradigm for understanding innovation. In H. Chesbrough, W. Vanhaverbeke, & J. West (Eds.), *New frontiers in open innovation* (p. 3–28). Oxford University Press.

Chesbrough, H., & Di Minin, A. (2014). Open social innovation. In H. Chesbrough, W. Vanhaverbeke, & J. West (Eds.), *New frontiers in open innovation,* (p. 169–188). Oxford University Press.

Clark, B. R. (1998). *Creating entrepreneurial universities: Organization pathways of transformation.* Pergamon.

E3M Project. (2012). *Green Paper: Fostering and Measuring "Third Mission" in Higher Education Institutions.* www.esna.tv/files/div/GreenPaper_ThirdMission.pdf

Etzkowitz, H. (1993). Enterprises from science: The origins of science-based regional economic development. *Minerva, 31,* 326–60.

Etzkowitz, H. (1998). The norms of entrepreneurial science: Cognitive effects of the new university-industry linkages. *Research Policy, 27,* 823–33.

Etzkowitz, H. (2002). *MIT and the rise of entrepreneurial science.* Routledge.

Etzkowitz, H., Webster, A., Gebhardt, C., & Terrad, B. R. C. (2000). The future of the university and the university of the future: Evolution of ivory tower to entrepreneurial paradigm. *Research Policy, 29*(2), 313–330.

Etzkowitz, H., & Leydesdorff, L. (1995). The triple helix, university-government relations a laboratory for knowledge based economic development. *EASST Review, 14*(1), 14–19.

Etzkowitz, H., & Leydesdorff, L. (1998). The triple helix as a model for innovation studies. *Science and Public Policy, 25*(3), 195–203.

Etzkowitz, H., & Leydesdorff, L. (2000). The dynamics of innovation: From national systems and "Mode 2" to a triple helix of university-industry-government relations. *Research Policy, 29*(2), 109–23.

Ferrer-Balas, D., Buckland, H., & de Mingo, M. (2008). Explorations on the university's role in society for sustainable development through a systems transition approach: Case study of the technical university of Catalonia (UPC). *Journal of Cleaner Production, 17,* 1075–1085.

Godemann, J., Bebbington, J., Herzig, C., & Moon, J. (2014). Higher education and sustainable development: Exploring possibilities for organisational change. *Accounting, Auditing & Accountability Journal, 27*(2), 218–33.

Government of the Faroe Islands. (2018). *United Nations Sustainable Development Goals.* The Government of the Faroe Islands. https://www.government. fo/en/the-government/ministries/prime-ministers-office/united-nations-sustainable-development-goals/

Laurent, A. et al. (2019). *UN SDG Assessment Methodology and Guideline: Version 1.01.* Quantitative Sustainability Assessment Group, Technical University of Denmark, Kgs.

Blaxekjær, L. O., Olsen, M. M., Thomasen, H., Gleerup, M. T., Lauritsen, S. N., Kappel, A. L. ... Friedlander, J. (2018). The sustainable development goals and student entrepreneurship in the Arctic. In L. Heininen, H. Exner-Pirot, & J. Plouffe (Eds.), *Arctic yearbook 2018 - Arctic development: in theory and in practice.* Northern Research Forum.

Leydesdorff, L., & Etzkowitz, H. (1998). The triple helix as a model for innovation studies. Science and public policy, 25(3), 195–203.

Lukman, R., & Glavič, P. (2007). Review of sustainability terms and their definitions. *Journal of Cleaner Production, 15*(18), 75–85.

Martins, T. C., & De Souza Bermejo, P. H. (2014). *Open social innovation.* In Ć. Dolićanin et al. (Eds.), *Handbook of research on democratic strategies and citizen-centered e-government services,* p. 144–163. IGI Global.

Mulgan, G. (2006). The process of social innovation. *Innovations, 1*(2).

Mulgan, G. (2007). *Social Innovation: What It Is, Why It Matters and How It Can Be Accelerated.* https://www.youngfoundation.org/publications/social-innovation-what-it-is-why-it-matters-how-it-can-be-accelerated/

Murray, C., & Mulgan, G. (2010). *The open book of innovation.* The Young Foundation.

Nicholls, A., Simon, J., Gabriel, M., & Whelan, C. (Eds.). (2015). *New frontiers in social innovation research.* Palgrave Macmillan.

Reffstrup, T., & Kærn Christiansen, S. (2017). *Nordic Entrepreneurship Islands: Status and Potential – Mapping and Forecasting Entrepreneurship Education on Seven Selected Nordic Islands.* https://eng.ffe-ye.dk/media/785770/nordic-entrepreneurship-islands-status-and-potential.pdf

Östling, J. (2018). *Humboldt and the modern German university: An intellectual history.* Lund University Press.

Ranga, M., & Etzkowitz, H. (2013). Triple helix systems: An analytical framework for innovation policy and practice in the knowledge society. *Industry and Higher Education, 27*(3), 237–62.

Rayna, T., & Striukova, L. (2019). Open social innovation dynamics and impact: Exploratory study of a fab lab network. *R&D Management, 49,* 3.

Rosenlund, J., Rosell, E., & Hogland, W. (2017). Overcoming the triple helix boundaries in an environmental research collaboration. *Science and Public Policy, 44*(2), 153–162.

SDSN Australia/Pacific. (2017). *Getting Started with the SDGs in Universities: A Guide for Universities, Higher Education Institutions, and the Academic Sector.* http://ap-unsdsn.org/wp-content/uploads/University-SDG-Guide_web.pdf

Sutherland, J. (2014). *SCRUM: The art of doing twice the work in half the time.* Random House.

Trencher, Gregory, et al. "Beyond the third mission: Exploring the emerging university function of co-creation for sustainability." Science and Public Policy 41.2 (2014): 151–179.

Trencher, G., Yarime, M., McCormick, K., Doll, C. N. H., & Kraines, S. (2013). Beyond the third mission: Exploring the emerging university function of co-creation for sustainability. *Science and Public Policy, 41*, 151–197.

Tushman, M. (1977). Special boundary roles in the innovation process. *Administrative Science Quarterly, 22*(4), 587–605.

Vanhaverbeke, W., & Chesbrough, H. W. (2014). A classification of open innovation and open business models. In H. W. Chesbrough, W. Vanhaverbeke, & J. West (Eds.), *New frontiers in open innovation*, p. 50–68. Oxford University Press.

Vanhaverbeke, W., Chesbrough, H. W., & West, J. (2014). Surfing the new wave of open innovation research. In H. W. Chesbrough, W. Vanhaverbeke, & J. West (Eds.), *New frontiers in open innovation*, p. 281–294. Oxford University Press.

Wittrock, B. (1993). The modern university: The three transformations. In S. Rothblatt & B. Wittrock (Eds.), *The European and American university since 1800: Historical and sociological essays*, p. 303–362. CUP.

Williams, P. (2002). The competent boundary spanner. *Public Administration, 80*(1), 103–24.

Wright, T. S. A. (2002). Definitions and frameworks for environmental sustainability in higher education. *Higher Education Policy, 15*, 105–20.

Zomer, A., & Benneworth, P. (2011). The rise of the University's third mission. In J. Enders, H. F. de Boer, & D. F. Westerheijden (Eds.), *Reform of higher education in Europe*, p. 81–101. Sense Publishers.

8 Sustaining Indigenous knowledge as renewable "resources"

Norma Shorty

Introduction

The objective of this chapter is to contribute to an understanding of Indigenous-led research processes which embrace Indigenous research agendas for the express purpose of sustaining Indigenous knowledge through Indigenous research, Indigenous philosophies, Indigenous methods, and Indigenous healing. As a curriculum developer, a scholar, and teacher of Tlingit constructs at Yukon University and privately, the researcher is Tlingit. The researcher's affiliation is to Tlingit people; this affiliation assists with international work which seeks to define Indigenous engagements and practices. Burdened with trauma from contact and colonisation, the researcher grew up believing that Tlingit stories were myth, that there was little or no value in understanding Tlingit history or language. These kinds of ideas are remnants of Canada's Indian Act policies; these policies shaped the researcher's educational pathway to the point of near amnesia on who are the Tlingit of Alaska, British Columbia, and the Yukon. Teaching Tlingit constructs in institutions of higher learning has led the researcher to realise that Tlingit people, like many other Indigenous Peoples in the Arctic and across the world, are healing and decolonising from contact and historical discrimination. For the Tlingit, the act of decolonisation is achieved through the Indigenization of their worldviews: speaking and thinking in Tlingit to each other; by hosting ceremonies and groups where there is clan balance so that speaking protocols are adhered to; by ensuring there are Tlingit-led enquiries and processes for Tlingit-led answers; by holding up Tlingit-led participation and assessments; and more. Tlingit Elders who are fluent in their Tlingit languages are the key to ensuring cultural ethics and protocols are adhered to (Dauenhauer N. & Dauenhauer R. 1987, 1990, 1994). Looking back, it is extraordinary what the Tlingit have endured: Boarding Schools, Indian Ceremonies Removal Acts, Indian Land Removal Acts, Indian Languages Removal Acts, the installation of International Borders, ad nauseum. Amongst many Tlingit Elders, the impacts of past government policies have laid an undeniable trajectory

DOI: 10.4324/9781003172406-8

which is struck with grief, sorrow, and general social unease and unrest due to loss of land, language, family, and culture.

Background

The day is done (November 5, 2017). The lead instructor/researcher closes the chapter on the boarding school curriculum she was asked to write by her employer in 2015. What resulted was pages and pages of notes, a draft curriculum framework, over one hundred hours of videotape, place-based curricula for Kindergarten to Grade 12 Social Studies for Juneau School District, and a YouTube (Sharing Our Knowledge Conference, 2016) Elders' presentation on boarding school at a Sharing Our Knowledge Tlingit Clan Conference. At the beginning of the Tlingit Elders' workshops, English was the dominant language being spoken and written. As time progressed, Tlingit became the dominant language, spoken and written. Together, over a two-year period, Tlingit Elders sat and discussed the impact boarding school and other Bureau of Indian Affairs' policies continue to have on Tlingit people. The lead instructor/researcher led the Elders through a series of focused and grant-supported workshops, which resulted in the application of Tlingit thinking, Tlingit language, Tlingit philosophies, Tlingit fluency, the articulation of Tlingit learning and teaching methods, and so much more. The method of data collection included the lead instructor/researcher being a learner, a teacher, and a facilitator. The focused discussion topics were for the express purpose of revitalising Indigenous languages, literacies, knowledge, and histories amongst Tlingit peoples.

Tlingit Elders and the lead instructor/researcher were paid for their participation by the employer, Goldbelt Heritage Foundation. Goldbelt Heritage Foundation is a non-profit organisation in Juneau, Alaska, and is dedicated to the revitalisation of the Tlingit language and culture. From the onset, the grant recipients and participants of the circle were the providers of traditional knowledge and traditional foods. The circle worked towards obtaining and recording Tlingit knowledge; the circle welcomed dynamic and responsive agendas; the circle encouraged the use of microphones and speakers; the circle enjoyed comfortable chairs, tables, and good light; the circle mandated an open-door policy for their families and youth; the circle made room for young and emerging Tlingit culture and language experts, for non-native Tlingit cultural experts, and literature on the Tlingit.

As part of a larger team of educators, Alaska and Yukon Elders and Knowledge Bearers worked on Indigenous knowledge inclusion in several meaningful and significant ways. The Elders recognised that the Alaska and Yukon border is a colonial construct. From the onset, the grant recipients and Elders and Knowledge Bearers were part of the planning team, laying foundations, planning language focus and culture activities, ensuring jurisdiction over the Tlingit place and Tlingit values, and cultural appropriateness. As the Elders embraced ideas, programs, and implementation concerning language and culture, the Elders focused on life before contact as a way to articulate and define foundational aspects to life in the past, today, and life in the future. To the Tlingit, perseverance is the key: "Before the great floods, we had already been through a lot" (Goldbelt Heritage Foundation Elders, Boarding School discussion, 2017). In rebuilding who are the Tlingit peoples of Alaska, British Columbia, and the Yukon, participating Tlingit Elders, parents, and youth determined what is healing to them. In developing place-based scholarship, Tlingit constructs and languages must be intact. Researchers and scholars of Tlingit knowledge must adhere to Tlingit ethics and protocols. This is especially true if the goal is to sustain Tlingit knowledge processes for the purpose of living and sustaining Tlingit knowledge in the context of social justice and other ideas such as renewable resource economies.

Sustaining Tlingit knowledge

Given the issues arising from contact and colonisation, how do Tlingit sustain their knowledge in the area of renewable economies in the Arctic? What do Tlingit think about solar energy, wind energy, geothermal pressure, wood, water, oxygen, fish, and precious metals? Tlingit concepts about solar energy are found in ancient stories, such as Raven giving birth to the sun, moon, and stars. That is, Raven's demand for the sun, moon, and stars produced the sun, moon, and stars, and provided for an ever-increasing population. Once Raven freed the sun, moon, and stars, Raven fixed man and woman. The Raven story also tells listeners how long the Tlingit have been in the area—since the beginning of time. In applying a Tlingit lens to precious metals, Tlingit value copper and historically went to any length to obtain it. Copper is dáanaa (currency, status, wealth) and receiving copper at a Ku.éex is a high honor. One of the trail networks inland is named by scholars and researchers as the Kohklux trail. This trail is known by Athabaskans and Tlingit as the "Grease Trail". This trail is one of many trade trails between Tlingit and Yukon First Nations. One route of this trail extends from Haines, Alaska to Skagway, Alaska then onto the Yukon and another route of this trail extends from Klukwan, Alaska to Haines Junction then onto Yakutat, Alaska and/or White River, Yukon for copper (de Laguna, 1972, p. 88 [citing Davidson 1901 who published the trail that Kohklux and his wives drew out]). After Alaska was sold to the USA, portions of the Kohklux trail on the Chilkoot (Skagway, Alaska) side were documented by US Army Lieutenant Schwatka in the summer of 1883. In 2020, the Grease Trails inland form the basis of our Alaska, Klondike, and Yukon highways. At the time of the Gold Rush in the Yukon, 1896, many Tlingit were hired as packers because of their strength and expertise. In our institutions of higher learning and in history, the focus when we talk about Yukon and precious metals is gold. Today, in the time of COVID-19, studies are suggesting that germs have a shorter life span on copper. Since the time of written history, Tlingit languages, histories, and methods have been overshadowed and diminished by "settler" histories and ideas. In the not-so-distant past, Tlingit knowledge had to fit into western frameworks in order to be understood.

An Alaska, British Columbia, and Yukon research example which highlights local Indigenous knowledge sustainability, local agreements, and treaties, and international research conventions is the Kwäday Dän Ts'ìnchi archeological find. The finding of the Long-Ago Man in the Ice, also known as Kwäday Dän Ts'ìnchi led to comanaged science and culture tables (Royal British Columbia Museum et al., 2017). Kwäday Dän Ts'ìnchi was found in ice near the Ruby Mountain Range (Champagne and Aishihik First Nations, 2009). Shortly after Kwäday Dän Ts'ìnchi was found, he not only received a proper burial, but he also received a forty-day smoking party (Klukwan) and a one-year memorial party (Champagne and Aishihik), also known as Ku.éex. Kwäday Dän Ts'ìnchi was cremated. This reveals Athabaskan and Tlingit law for the treatment of human remains. Based on Kwäday Dän Ts'ìnchi's clothing and stomach contents, he was traveling inland from the coast of Alaska. Kwäday Dän Ts'ìnchi was probably travelling at the same time that Columbus "discovered" the Americas. Athabaskan, Tlingit, and Western researchers determined that Kwäday Dän Ts'ìnchi was on one of the well-used Grease Trails between Klukwan, Alaska and/or Yakutat, Alaska and/or Copper River, Alaska (Royal British Columbia Museum et al., 2017).

The coproduction and comanagement of Southern Tutchone, Tlingit, and Western research revealed pathways to understanding Indigenous clan identity with respect to mitochondrial DNA markers (Monsalve, 2017). Most Alaska and Yukon First Nations communities are matrilineal. This means that Indigenous identity is set through the female line. Indigenous knowledge systems, including heritage language, thinking, objects, properties, families, ancestors, histories, songs, and values, are transmitted through matrilineal lines. Modern science has found that sperm carries its DNA markers in their tails and necks and is left behind during fertilisation (Monsalve, 2017). Kwäday Dan Ts'ìnchi DNA supports what Southern Tutchone and Tlingit ancestors inherently knew about DNA markers (S. Adamson, personal communication, 2019).

Today, many Alaska and Yukon First Nations people and Elders work with appropriate agencies to develop meaningful and relevant engagement connections regarding Indigenous finds and collections (Royal British Columbia Museum et al., 2017). Tlingit investigation and science is tied to balance, to ownership, to spirit, to the seen, to the unseen, to the known, and to the unknown (Goldbelt Heritage Foundation Elders, Tlingit Science Framework discussions, 2017). As seen in recent Kwäday Dan Ts'ìnchi publications, comanaged research tables can reveal Indigenous and western ways of knowing (Royal British Columbia Museum et al., 2017).

To coproduce and learn or teach Indigenous languages and knowledge is to address systemic discrimination in Alaska and Yukon public institutions. Discussions that embrace culture, language, and policy will bring up dialogue and memories concerning past and current treatment of Indigenous Peoples and equity and equality in public institutions, communities, and societies. Sustaining Indigenous knowledge means that modern research processes must allow for time and resources for Indigenous Peoples to reconcile their histories and languages amongst themselves. The coproduction of knowledge with other knowledge occurs after Indigenous Peoples have had time to go away and discuss matters at hand. Through this process of internal dialogue, Indigenous learning frameworks will form the basis for further work in articulating what is Indigenous knowledge for the purpose of influencing Indigenous lives, cultural endeavors, public and post-secondary schooling, and further research (Goldbelt Heritage Foundation Elders, Tlingit Science Framework discussions, 2016). Indigenous knowledge are collective and are often clan-based. At the practitioner stage, the researcher is a facilitator of knowledge as opposed to a gatekeeper of knowledge.

Indigenous knowledge-gathering activities and research must be seasonal to honor the various harvesting events. This was heard at various Social Economic and Culture Expert Group events in Finland and virtually.

Using Indigenous ways of recording events with history, art forms, face paints, and oratory, most Indigenous Peoples know that Indigenous stories are true (Shorty, 2015). Systemic discrimination and powerful English constructs surrounding the word "story" tend to ignore the history in the story and many believe that Indigenous stories are myth. Unpacking Indigenous stories brings forward ancestors' knowledge and wisdom, the strength of Elders, ideas of land, and working together (Sealaska Heritage Foundation, Core Cultural Values Poster). Dialogue and research amongst Indigenous Peoples must allow time for healing (Goldbelt Heritage Foundation Elders, Native American Graves Repatriation Act, 1990, discussions, 2017 for Native American Graves Repatriation Act 1990 due to the unearthing of unmarked graves in 2012 at an elementary school in Douglas, Alaska).

To engage with Indigenous Elders at the local level will produce new knowledge, research tools, and processes, on what is Indigenous thinking on Indigenous learning frameworks about the ancestors, the values, the land, the environment, the foods, the non-tangible, the tangible, and the history.

For many Indigenous Elders, Indigenous Peoples need to teach Indigenous constructs, especially if Indigenous Peoples are reconciling and sustaining Indigenous knowledge and languages which are based on the past and for the purpose of future children and grandchildren (Tlingit Elders, Boarding School discussions, 2015-2017). Protocols and laws on the handling of Indigenous knowledge and languages for public institution use by Indigenous and non-Indigenous educators is an area that will need exploration and definition. Perhaps Indigenous Peoples prefer to keep this area grey, perhaps due to ideas about power in the unknown.

Haa Kusteeyí[1] (Way of Life) and Dooli[2] (Law) sum up the ideas which govern Alaska and Yukon philosophies and views on sovereignty, spirituality, and stewardship of land, air, water, and the unseen. The idea is that Alaska and Yukon First Nations enjoy specific environments and territories, have clan crests, have ancestors, have a collective good, have history, sacred places, stories, songs, dances, symbols, law, respect for the land and her resources, traditional practices, knowledge, and an identity. Alaska and Yukon Elders say that the way they pass on Haa Kusteeyí and Dooli must be the foundation to the development and implementation of traditional/ contemporary practices and knowledge (Sealaska language proficiency meeting, September 24, 2010).

Using Indigenous constructs, the highest level of authority is reserved for Haa Shageinyaa (Our Heavenly Father). This reverence for a higher order of things is the first and most important idea of all (Goldbelt Heritage Foundation Elders, Tlingit Science Framework discussions, 2016). Everything has a spirit yet, very little research time is devoted to discussions on sacred spaces for sacred learning (Goldbelt Heritage Foundation Elders, Tlingit Science Framework discussions, March 2016). Indigenous Peoples in Alaska and Yukon are discussing and implementing "reconciliation" so there is a real need for Arctic research projects to embody Indigenous research agendas. Without prejudice, Indigenous Peoples need research support and space to explore and teach Indigenous constructs amongst themselves in order to collectively reconnect and/or reconcile with their heritage ancestors, land, languages, methods, and philosophies. Research agendas need to include space for Indigenous identity building, Indigenous culture sharing, Indigenous language renewal, Indigenous methods renewal, and healing amongst themselves. Throughout the teaching, learning, and research processes, Indigenous Peoples follow their protocols and their laws while demonstrating care for the continuation of language and culture by speaking and thinking in their languages together and with their children, adults, and youth.

At Arctic Council meetings, Arctic Indigenous Peoples still report that some researchers come to the villages and are sometimes disruptive and want to conduct research on the knowledge that is specific to "science" or to projects outside of what the villages want or feel is important. In other words, Indigenous Peoples are asked to participate and to fit their thinking into a western research paradigm. Not only is this "fitting" unethical, it is morally wrong considering concepts of cultural genocide through Indian Acts, policies, relocations, nationality, and land reassignments. To be clear, *Arctic Council is a high-level interagency that facilitates "soft law" arrangements for participating members on issues pertaining to Arctic issues. While the deliberations of the Arctic Council are not legally binding, the conversations of the Arctic Council do force consideration to Indigenous matters* (Amanda Graham, personal communication, August 26, 2020). Indigenous Peoples are Permanent Participants of Arctic Council. In this light, Arctic Council facilitates and promotes full consultation rights on negotiations and decisions which impact Arctic Indigenous Peoples through participation from the Permanent Participants. It is important to note that, most often, the results of Arctic Council deliberations do serve as a "moral guide" (term coined by Amanda Graham, personal communication, August 26, 2020) for participating countries on research issues impacting Arctic regions. How does the circumpolar research region engage with Indigenous Peoples and humans from the Alaska and Yukon region in areas such as sustainable development and environmental protections? Western research has provided many publications on Indigenous history, cosmology, epistemologies, and stories. With respect to decolonising the institution of Arctic Council towards the inclusion of Indigenous research methodologies, how do Indigenous Peoples define social, economic, and culture laws of engagement for issues of arising out of Arctic Council research?

As stated, Arctic Council does provide a permanent forum for Arctic Indigenous research concerns and, in this light, can advance Indigenous research for participating countries. Permanent Participants are part of the Arctic Council table and are given voice and time in Arctic Council roundtables and discussions. Arctic Council and Permanent Participant deliberations and declarations do influence local, regional, and federal research initiatives, policies, and programs such as the International Arctic Science Committee and the International Arctic Social Science Association (IASSA), as well as local and community-based research entities such as the independent Arctic Institute of Community-Based Research out of Whitehorse, Yukon Territory, Canada. In light of all this, the National Research Council of Canada calls for rigor regarding research standards and here is where the contact and colonisation history of Indigenous Peoples comes into play. Without repeating historical injustices such as land displacement, language loss, children and family connections severed, cultural ceremonies severed, recall that Goldbelt Heritage Foundation Elders collectively found their way through the tough conversations as they spoke to each other in their languages, using their ceremonies, and living their culture. Through this process, Tlingit knowledge was expressed in the Tlingit language, using Tlingit methods towards sustaining Tlingit knowledge. Collectively, and in their heritage languages, the Tlingit found their way back to their ancestors and were able to state that: "Before the great floods they'd already been through a lot" (Goldbelt Heritage Foundation Elders, Boarding School deliberations, 2016). In moments of despair, the Elders recalled what their ancestors endured over ten thousand years ago. Tlingit had been written out of history and what replaced Tlingit history was dominant culture history (English). Historically, Tlingit have been in this place since Raven released the sun, moon, and stars because Raven wanted the sun, moon, and stars. Tlingit use repetition as a way to teach values; Tlingit worldview is circular; advancing, going back, ebbing, flowing, dynamically, and in spirals. *The difference is between inductive and deductive reasoning. Indigenous thought is inductive, from the earth, sky, the unseen, the unknown* (Bernie Johnson, personal communication, August 16, 2020). Bernie Johnson is 75 and a deep Southern Tutchone thinker—Bernie has chosen not to live conventionally— he is a drifter. Bernie also noted that western thinking is in blocks, squares— Indigenous thought is in circles, spirals—the circle can fit in the square, but there are still the unknown parts. The square cannot fit in the circle (when it comes to translating Indigenous knowledge for science, research, writing). As declared in Canada, Indigenous research engagement protocols *are intended to support, not replace a direct relationship with Indigenous Peoples* (Assembly of First Nations, 2009).

Hence, what does this mean with respect to how public and research institutions display and use Indigenous knowledge? Often outside authors are celebrated for what they publish on local Indigenous experiences, history, peoples, and politics. At the public institution level, it should be the local history that is displayed, and the resources developed around those local experiences should be viewed as supports, not as replacements of local Indigenous knowledge. Using modern research methods, the author and/ or researcher is celebrated, and the multitudes of Indigenous Elders and knowledge bearers remain unnamed in author bylines.

In the application of an Indigenous research paradigm, the research question must involve the community right from the start. Research questions must place equal research value on Indigenous knowledge, languages, histories, and philosophies and engage with local Indigenous Peoples, especially when it comes to the formation and plans towards the research question itself. Research needs to be authenticated by Indigenous Peoples themselves (time to go away and discuss the research question, as well to define Indigenous enquiry and levels of research involvement). Most often, when there are research gaps with local Elders or traditional teachings, the research solution is to oversimplify or generalise specific and place-based knowledge. One of the current solutions to overcome this scenario is to ask a First Nation individual who is willing to volunteer time to come forward and address the issue. The term "volunteer" is used as the rate of honoraria for Indigenous knowledge inclusion is often not on par with modern expert inclusion.

Given new research on Indigenous Peoples by Indigenous Peoples what do Indigenous research instruments and research analysis look like today? There are now a growing number of Indigenous Peoples holding doctoral degrees, which could potentially level the research playing field as these doctoral degree holders can apply for research dollars and be Principal Investigators in matters which pertain to Indigenous Peoples. Past research has really paved the way to help develop and review how Indigenous researchers think today. In 1999, Indigenous researcher and author, Linda Smith recommended that Indigenous Peoples decolonise their methodologies.

Against today's social climate, Indigenous Peoples are still finding themselves being *othered* (Smith, 1999) in constructs and trade fields which clearly are the intellectual and cultural property of Indigenous Peoples. In 2000, Shawn Wilson said that Indigenous research paradigms include building respectful relationships, being responsible, and giving back. In 2005, Ray Barnhardt and Oscar Kawagley brought our attention to ideas of deep culture. In 2007, Beth Leonard wrote that the Indigenous research paradigms need to incorporate those elements not seen (intangible power); and Leonard further states, that the English language does not do an adequate job of translating intangible power as expressed through Indigenous histories, stories, oratories, medicine, and more. From 2015 to 2017, coastal Tlingit Elders wanted to set the stage on what is important to them; and to research and analyse elements of culture that will help boarding and mission school survivors and those that have been impacted by elements of contact and colonisation to relearn Tlingit dialects and language, know the Tlingit culture and history stories, and know deeply who are the Tlingit people. Tlingit people want to invest in research projects which build on their individual and collective identities while articulating and using cultural constructs that are based on their deep cultural values (Shorty, 2016). Tlingit people are seeking to heal themselves with their ancestors, lands, histories, heritage languages, cultures, worldviews, politics, learning and teaching methods, research, and data collection (Goldbelt Heritage Foundation Elders, personal communications, 2015-2017).

Sustainable Indigenous methods that plan forward by looking back include heritage language, include Youth and Elders to ensure that the "words of the Elders do not fall upon the ground" (G. Davis in Dauenhauer & Dauenhauer, 1981), and a time for healing (Goldbelt Heritage Foundation, NAGPRA discussions, 2017). Sustaining Indigenous knowledge methodologies allows for time on the land or sea with Indigenous adults, youth, Elders, and knowledge bearers. As demonstrated through Indigenous Peoples histories, acceptance is key to moving on. When Indigenous Peoples speak their language with one another, they are sustaining Indigenous knowledge and are working collaboratively and in their traditional ways. Land and food sovereignty reveal place-based research, ceremonies, philosophies, methodologies, and subject matter experts. In order to heal the impact of colonisation, Indigenous Peoples must be able to speak and think in their heritage languages and to make a life as an expert in their fields of study. As exemplified by the researcher's primary Indigenous research experiences, reconciliation of Indigenous Peoples to their culture, language, and traditions through public institutions can occur with careful, mindful, inclusive planning and dialogue lasting for more than two years.

When public school teachers of science sat with Tlingit Elders to coproduce knowledge on Kindergarten to Grade Twelve science, the center of the dialogue was worldview. Western or modern science at the public-school level addressed learning points that are not typically included in Tlingit "scientific" learning situations. One example was with hot rocks. Tlingit used rocks to cook their foods, when they cooked in wood or willow vessels.

The lesson was a chemistry lesson on specific-heat capacity: focusing on the types of woods or barks needed to make food containers, the kinds of weave to make the vessels watertight, the types of hot rocks needed to withstand cold to gradually cooking water. The science lesson asked the learners to make deductions on how the cooking rocks are formed, the heat transfer, the heat density (S.T., 2017, rough notes for Food Tools of Our Ancestors). Indigenous science brought forward clan-owned stories which were related to "hot rocks" and demonstrated how long this particular Tlingit clan have been in this particular area. As can be seen from the hot rock exercise, the coproduction of knowledge between schoolteachers and Elders can provide full circle, progressive, thinking on trends, solutions or recommendations with respect to Indigenous knowledge sustainability in a public-school context. Elders are utilised for sharing their knowledge on cooking rocks and food sovereignty and for their proficiency with philosophies, relevant cultural activities, language, and long-ago land connections. What is not written are the spaces for this lesson to occur. This lesson would take place outside with parents, children, youth, Elders, and teachers, including community specialists, all working together to co-produce knowledge in the area of fire rock cooking.

The coproduction of knowledge in curriculum writing processes can provide vital cultural and scientific knowledge which can inform contemporary Indigenous challenges such as wind energy, food sustainability, and mineral extraction.

Discussion

A glimpse at Canadian policies on Indigenous Peoples reveals laws that greatly disrupted Indigenous connections to family, heritage languages, history, land, oratory, place names, philosophies, and values. Contact and colonisation have negatively impacted many Indigenous Peoples through loss of heritage languages, through the lack of public space for cultural practices, through the loss of families, and identities. Because Indigenous knowledge, culture, and language have been lost through laws, boarding schools, and genocide, Indigenous philosophies, and worldviews may be incomplete: if one believes that culture is in understanding the context and details of Indigenous languages and thoughts.

Indigenous culture specialists, Elders, and non-Indigenous subject experts have tried to Indigenise science constructs and have found that through this process Indigenous constructs tend to remain subordinate to modern subject ideals and constructs (Goldbelt Heritage Foundation Elders, Hero's Journey with public schoolteachers' discussions, 2017). The attempt to make Indigenous knowledge fit into western research constructs for the purpose of the scholarship process often highlights what is present in literature and what is absent in the process. For instance, some course objectives have learning outcomes which include Indigenous focuses, and these outcomes remain silent on the laws and protocols that Indigenous Peoples follow when they are working with their culture. This lack of direction for working with culture, laws, protocols, and language processes in a modern context is quite concerning, especially when researchers come to Arctic communities with their research guidelines already set out.

Tlingit people, through their own research process, ensure that there are clan-balanced Elder and knowledge bearer tables formed specifically to "unpack" the task at hand. Collectively, Tlingit informs Tlingit thought, meaning no one person speaks for all. Tlingit speaking protocols ensure validation of what is being said. Clans speak to each other, weaving in history, land, oratory, and sacredness in their speeches. Research tables and discussions lasting more than two years often lead to Indigenous languages and thought application, deeper personal and professional understandings, values application, ancient and modern oratory, history, and social justice (Goldbelt Heritage Foundation Elders, Tlingit Boarding School discussions, 2015-2017). Dedicated research time and funds ensure that Indigenous and western research processes, which are often in opposition to one another, co-exist. Indigenous culture is oral. Written and non-written forms of Alaska and Yukon First Nations do record history through clan emblems, totem poles, personal names, place names. Without fluent Elder leadership or place-based significance, it is easy to teach Indigenous subject matter out of sequence or as myth. Indigenous Peoples pass on culture and knowledge through their clans and tribes, languages and lived experiences, place, and now through schools, both public and higher learning. Indigenous knowledge are collective and are often clan-owned and operated. Indigenous Peoples are actively teaching themselves about who they are on their own terms, with their own timelines. Alaska and Yukon Peoples continue to trace the development of their worldviews through their communication with each other, about past, present, and future times. In each of these contexts, Indigenous history and oratory, which have been constructed since time immemorial, are drawn upon in relation to the cultural context in which history and oratory is situated.

Those Elders who are fluent in their Indigenous languages and thoughts and who have been raised by their grandparents carry with them an authentic worldview which is based on the authentic teachings of their ancestors in their languages and with its culture. This authentic worldview embedded in Tlingit methods does contribute to an understanding of Indigenous-led research processes which embrace Indigenous research agendas for the express purpose of sustaining Indigenous knowledge through Indigenous research, Indigenous philosophies, Indigenous methods, and Indigenous healing. Indigenous Elders know that Indigenous knowledge must stand on its own because everything is interconnected. Indigenous knowledge is tied to Indigenous languages and is place-based and eco-specific (Barnhardt & Kawagley, 2005; Leonard, 2007; Topkok, 2015). Indigenous Peoples want time and resources to reconcile with their language, culture, and philosophies (Elders Boarding School Dialogue, Tlingit Clan Conference, September 2019). Public dedicated space and time for Elders and Indigenous languages will assist Indigenous Peoples methodologies on healing and reconciling who are Indigenous Peoples in light of what has happened to them (Goldbelt Heritage Foundation, Youth and Elders dialogue, 2017). Using an Indigenous research lens, Indigenous women recognise that gendered discussions include girls, women; men and boys; babies, and Elders (Empowering Indigenous Women Workshop, March 2018). Indigenous women want to have their voices include men and boys as Indigenous constructs are balanced. Co-constructed research agendas, including the preparation for the collaboration itself, will ensure the growth and sustainability of Indigenous knowledge and heritage languages at all research and implementation levels. Based on the presentation of Indigenous research practice, it is noted that Indigenous research methods deviate from western research practices because most often, Indigenous knowledge databases are formed at the time of needing the Indigenous data. Modern research processes uphold the literature and peer review processes. The Indigenous scholarship process requires local and place-based rigor with regard to how Indigenous knowledge is presented (Empowering Indigenous Women Workshop, March 2018). Researchers need to meet on Indigenous homelands where the topics being researched are located. Researchers must respect the interplay between Indigenous Peoples, lands, and Indigenous foods (Snowchange Cooperative, 2017).

With respect to engagement and co-production of Indigenous knowledge, there is a need for co-operation and a deeper analysis of what collaboration and coproduction looks like. In the area of food, plants, medicines, arts, culture, peoples, and knowledge, local Indigenous knowledge cannot be replaced with secondary resources.

Many Arctic Indigenous Peoples are recognising that knowledge and product appropriation continues to occur at alarming rates. For instance, Indigenous medicines such as willow bark are being marketed for profit by health and food scientists. These marketed items are copyrighted to the marketer, not to the Indigenous knowledge keepers. Ethically, any Indigenous research should ensure healing from contact and colonisation by embracing Indigenous language, methods, sovereignty, and worldviews. At the practitioner stage, the researcher or teacher of Indigenous knowledge is a facilitator of Indigenous and place-based knowledge as opposed to a gatekeeper of knowledge. Indigenous knowledge activities and research must be culturally relevant as well as seasonal to honor the various harvesting events (eulachan, herring eggs, seaweed, berries, salmon). The facilitator ensures meaningful Indigenous engagement (no tokenism) by ensuring the speakers are clan balanced. The facilitator ensures that Indigenous spirituality is incorporated by ensuring cultural laws and protocols are followed. The facilitator ensures that all contribute to the understanding of Indigenous-led research processes which embrace Indigenous research agendas for the express purpose of sustaining Indigenous knowledge through Indigenous research, Indigenous philosophies, Indigenous methods, and Indigenous healing. To coproduce, learn, and teach Indigenous languages and knowledge is to address systemic discrimination in public institutions. Discussions that embrace culture, language, and policy will bring up dialogue on, and memories of, land, language, and family losses. Sustaining Indigenous knowledge means that modern research processes must allow for time and resources for Indigenous Peoples to reconcile their histories, lands, and languages amongst themselves. The coproduction of knowledge with other knowledge occurs after Indigenous Peoples have had time to go away and discuss matters at hand. Through this process of internal dialogue, Indigenous learning frameworks will form the basis for further work in articulating what is Indigenous knowledge for the purpose of influencing Indigenous lives, cultural endeavors, public and post-secondary schooling, and further research (Goldbelt Heritage Foundation Elders, Tlingit Science Framework discussions, March 2016). Unpacking Indigenous stories brings forward ancestors' knowledge and wisdom, strength of Elders, ideas of Land, working together (Sealaska Heritage Foundation, Core Cultural Values Poster). Dialogue and research amongst Indigenous Peoples must allow time for healing (Goldbelt Heritage Foundation Elders, Native American Graves Repatriation Act discussions, 2017). To engage with Indigenous Elders at the local level will produce new knowledge, research tools, and processes, on what is Indigenous thinking on Indigenous learning frameworks about the ancestors, the values, the land, the environment, the foods, the non-tangible, the tangible, and the history.

For many Indigenous Elders, Indigenous Peoples need to teach Indigenous constructs, especially if Indigenous Peoples are reconciling and sustaining Indigenous knowledge and languages which are based on the past and for the purpose of future children and grandchildren (Tlingit Elders, Boarding School discussions, 2015–2017). Protocols and laws on the handling of Indigenous knowledge and languages for public institution use by Indigenous and non-Indigenous educators is an area that will need exploration and definition. Perhaps Indigenous Peoples prefer to keep this area grey due to ideas about power in the unknown.

Conclusion

Ancestral knowledge provides the background, the research, the scholarship to knowing who the Indigenous Peoples of Alaska and Yukon are. As evidenced in coproduced research tables, when Indigenous Peoples work in their culture and speak their languages, their philosophies, values, lands, and identities are revealed. Sustaining Indigenous knowledge means that Indigenous Peoples speak and think in their languages and acknowledge that their languages connect them to a deeper understanding of who they are, in context to themselves, society, and the Arctic. Arctic research tables must continue to host workshops to explore the question of how to be more inclusive of Arctic Indigenous research at discussions, conferences, workshops, meetings. As Indigenous Peoples rebuild and reconfirm their languages and societies, there is acute awareness about gaping social and land wounds resulting from colonisation (Truth and Reconciliation Commission of Canada, 2015). In spite of this tumultuous past, this chapter suggests how scholarship and research can engage with Indigenous Elders and knowledge bearers about renewable Arctic economies, including Indigenous knowledge sharing and research processes. Indigenous Peoples want to invest in research projects which build on their individual and collective identities while articulating and using cultural constructs that are based on the deep cultural values of their respective cultures. Research needs to be authenticated by Indigenous Peoples themselves (time to go away and collectively discuss the research question, as well to define Indigenous levels of involvement, including supports to bridge levels of involvement). With respect to traditional knowledge and its intended purpose, how does the circumpolar region engage with the human dimension in the Arctic? What will our research instruments and research analysis look like? What will our time together grow? How is our research ethic of "do no harm" implemented to concepts like traditional knowledge? Traditional knowledge is an oral construct and is exact (don't add anything in and don't take anything away[3]). Using Indigenous constructs, traditional knowledge is based on oral history and there are rules and laws for using traditional knowledge. Within Indigenous knowledge there are concepts of ownership, jurisdiction, and processes for working among the people. A basic philosophy about Indigenous knowledge is that the accumulation of Indigenous knowledge is for the benefit of future generations. Philosophies, methods, and heritage languages contribute to an understanding of Indigenous-led research processes, which embrace Indigenous research agendas for the express purpose of sustaining Indigenous knowledge about who are the Indigenous Peoples of this land. In conclusion, why did the ancestors leave migration and glacial stories behind? One answer might be, so that their great-grandchildren would know where to go and would know what to do when the ice and floods come again.

Acknowledgments

I fully acknowledge the many Elders and Scholars who have shown the way! A very special thank you to the late Mrs. Lillian Austin, Mrs. Irene Cadiente, Mrs. Della Cheney, the late Mr. George Davis, the late Mrs. Marge Dunston, the late Mrs. Diane Church, the late Mrs. Selina Everson, the late Mr. William Fawcett, Ms. Genevieve Gaunzon, Mrs. Flora Huntington, the late David Katzeek, Mr. Ed Kunz, the late Mrs. Percy Kunz, Mrs. Mary Lekanof, Mrs. Caroline Martin, the late John Martin, Mr. John Morris, Ms. Charlotte McConnell, Mrs. Lillian Perdue, Mrs. Florence Sheakley, the late Mr. Jimmy Johnston, the late Mrs. Auntie Pearl Keenan, the late Mrs. Percy Kunz, and Mom—the late Mrs. Emma Joanne Shorty.

Ray Barnhardt, the late Oscar Kawagley, the late Nora Marks Dauenhauer, the late Richard Dauenhauer for your wise and thoughtful contributions to how I think about Indigenous knowledge. For those of you un-named, thank you for your quiet wisdom and guidance.

Thank you to the Council of Yukon First Nations, Goldbelt Heritage Foundation, Sealaska Heritage Foundation, University of Regina, Yukon University for allowing me to facilitate learning frameworks with our precious peoples. A very special thank you to the Arctic Athabaskan Council and the International Association of Social Science for making my global Arctic experiences real.

Notes

1. Tlingit
2. Southern/Northern Tutchone
3. P. Keenan (personal communication)

References

Assembly of First Nations. (2009). *Ethics in First Nations Research*. Environmental Stewardship Unit, Assembly of First Nations. https://www.afn.ca/uploads/files/rp-research_ethics_final.pdf

Barnhardt, R., & Kawagley, A. O. (2005). Indigenous knowledge systems and Alaska native ways of knowing. *Anthropology and Education Quarterly*, 36(1), pp. 8–23. http://ankn.uaf.edu/Curriculum/Articles/BarnhardtKawagley/Indigenous_Knowledge.html

Champagne and Aishihik First Nations. (2009). Kwäday Dän Ts'ìnchi: long ago person found. *A Champagne and Aishihik First Nations Special Report*. http://cafn.ca/wp-content/uploads/2015/04/Kwaday_Dan_Tsinchi_Newsletter_March_2009.pdf

Dauenhauer, N., & Dauenhauer, R. (1994). *Haa kusteeyí, our culture: Tlingit life stories*. (Classics of Tlingit Oral Literature; Vol. 3). University of Washington Press.

Dauenhauer, N., & Dauenhauer, R. (1987). *Haa Shuká, our ancestors: Tlingit Oral narratives*. University of Washington Press & Sealaska Heritage Foundation. https://tlingitlanguage.com/wp-content/uploads/2015/01/Dauenhauer-1987-Haa-Shuká.pdf

Dauenhauer, N., & Dauenhauer, R. (1990). *Haa Tuwunáagu Yís, for healing our spirit, Tlingit oratory*. University of Washington Press & Sealaska Heritage Foundation. https://tlingitlanguage.com/wp-content/uploads/2015/01/Dauenhauer-1990-HTY.pdf

de Laguna, F. (1972). *Under Mount St. Elias: The History and Culture of the Yakutat Tlingit*. Part One: https://doi.org/10.5479/si.00810223.7.1, Part Two: https://doi.org/10.5479/si.00810223.7.2

Leonard, B. (2007). Deg Xinag Oral Traditions: Reconnecting Indigenous Language and Education through Traditional Narratives [Doctoral Dissertation, University of Alaska Fairbanks]. www.ankn.uaf.edu/curriculum/PhD.../Leonard Dissertation.pdf

Monsalve, M. (2017). *Origins of the Kwäday Dän Ts'inchi man inferred through mitochondrial DNA analysis in Kwäday Dän Ts'inchi: Teachings from long ago person found*. Royal BC Museum and Champagne and Aishihik First Nations.

Royal BC Museum & Champagne and Aishihik First Nations (2017). *Kwäday Dän Ts'inchi: Teachings from long ago person found* (R.J. Hebda, S. Greer, A.P. Mackie, Eds.). Royal BC Museum.

Sealaska Elders. (1981). *"Because we cherish you—"; Sealaska elders speak to the future* (N. Dauenhauer & R. Dauenhauer, Trans.). Sealaska Heritage Foundation Press.

Sharing Our Knowledge Clan Conference. (2016, May 1). *The Boarding School Curriculum Elders' Panel: Presented by Norma Shorty* [YouTube]. YouTube. https://www.youtube.com/watch?v=OkB9mxiZZTI

Shorty, N. (2016). Holding onto Tlingit culture through research and education. *Special Issue on Indigenous Knowledge for the Journal of Knowledge Cultures, 4*(3). Central and Eastern European Online Library GmbH.

Shorty, N. (2015). Inland Tlingit of Teslin Yukon: G̲aanax̲.ádi and Kóokhíttaan stories for the immediate and clan family of Emma Joanne Shorty [nee Sidney]. ProQuest 3723087.

Smith, L. T. (1999) *Decolonizing methodologies: Research and indigenous peoples.* Zed Books Ltd. https://nycstandswithstandingrock.files.wordpress.com/2016/10/linda-tuhiwai-smith-decolonizing-methodologies-research-and-indigenous-peoples.pdf

Snowchange Cooperative. (2017, June 9). Indigenous Knowledge Roundtable Discussion [Declaration on Indigenous Knowledge from ICASS IX]. Umeå, Sweden. http://www.snowchange.org/2017/06/declaration-on-indigenous-knowledge-from-icass-ix/

Topkok, C. S. A. (2015). *Iñupiat Ilitqusiat: Inner Views of Our Iñupiaq Values.* https://www.semanticscholar.org/paper/In%CC%83upiat-Ilitqusiat%3A-inner-views-of-our-In%CC%83upiaq-Topkok/73a45535db0bb81dfbeca2f6ab2748d74 74b0877

Truth and Reconciliation Commission of Canada. (2015). *Honouring the Truth, Reconciling for the Future: Summary of the Final Report of the Truth and Reconciliation Commission of Canada.* Truth and Reconciliation Commission of Canada. https://publications.gc.ca/collections/collection_2015/trc/IR4-7-2015-eng.pdf

Wilson, S. (2009). *research is ceremony: Indigenous research methods.* Fernwood Publishing.

9 Toward socially sustainable renewable energy projects through involvement of local communities

Normative aspects and practices on the ground

Karin Buhmann, Paul Bowles, Dorothée Cambou, Anna-Sofie Hurup Skjervedal, & Mark Stoddart

Introduction

While scientists' warnings about climate change and their calls for urgent action are gradually coming to be accepted by politicians and regulators across the globe, the implications of a warmer climate are particularly fast and acute for the Arctic. This has spurred projects in Arctic countries to shift to low-carbon energy sources, in particular wind, hydro and solar power (Business Index North, 2017; McCauley, Heffron, Pavlenko, Rehner, & Holmes, 2016). While these forms of energy are renewable and therefore environmentally sustainable in a narrow sense, decisions on their locations have caused a range of protests by local communities including Indigenous groups. The protests are typically fueled by concerns over the social and/ or environmental implications of the projects. On the one hand, the projects may offer jobs and economic development. On the other hand, they are seen as posing new risks for people living in the Arctic. Just because energy sources are renewable, this does not mean they are free from adverse social impacts or that they are regarded as socially acceptable. Developing and storing energy from renewable sources like wind and the sun depends on technical solutions, some of which in turn depend upon minerals, including rare earth elements (REE), copper or cobalt. In many countries, but particularly in the global South, issues of labor conditions and the environmental impacts of mining and the processing of minerals are well-documented. As deposits of such minerals in the global South are becoming depleted, and the Arctic increasingly more accessible due to the changing climate, Arctic countries are emerging as potent sources of minerals for the global production of wind and solar energy technologies and batteries for storing renewable energy. Scaling up renewable energy infrastructure is in line with Sustainable Development Goal (SDG) 13 on urgent action to combat climate change and its impacts, as well as SDG 7 on access to affordable, reliable, sustainable and modern energy for all (UN, 2015a). However, the texts of the Paris Climate Change Accord and the Sustainable Development

DOI: 10.4324/9781003172406-9

Goals recognise that the transition to a low-carbon economy should be implemented with respect for human rights (UN, 2015a, 2015b).

There is often a close connection between harmful environmental and social impacts. This is recognised by the inclusion of social aspects in environmental impact assessment processes (Esteves and al., 2012; Nenasheva, Bickford, Lesser, Koivurova, & Kankaanpaa, 2015). Likewise, the Aarhus Convention (UN, 1998) on public participation in environmental decision-making includes health aspects. The connection carries over into debates on injustice and disproportional burdens carried by certain groups, including Indigenous peoples, in the interest of others (see also the chapter by Cambou and Polzer in this volume). Environmental injustice is a situation where specific social groups disproportionately bear the risks or negative impacts of development projects (Mohai, Pellow, & Roberts, 2009). Similarly, an emerging energy justice literature defines energy injustice as a situation where the benefits and negative impacts of energy projects are disproportionately distributed and where those who are impacted by energy development are not able to fully participate in the decision making and planning that affects their communities (McCauley et al., 2016). Social expectations as well as national and international legal norms require that renewable energy projects include meaningful engagement with local communities that are or can be affected by a project. With variations across the Arctic, local law and regulations require companies to assess the environmental and other societal impacts of their planned projects and address adverse impacts, while local governments are required to organise consultation processes (Nenasheva et al., 2015). International soft law guidance issued by the United Nations (UN) and OECD require impact assessments and measures to address adverse impacts to be undertaken with meaningful engagement of potential or actual victims of harmful impacts (Buhmann, 2016; Ruggie, 2013). Referring to those whose human rights are or can be adversely affected by projects, the term "affected stakeholders" applied by the UN and OECD instruments includes a strong focus on rights-holders. Public demands to become part of the planning and decision-making processes increase as advances in technology enable access to information on projects and enable those affected to effectively organise to respond to such projects (Buxton & Wilson, 2013). Such demands reflect an expectation that impact assessments involve stakeholder engagement that is meaningful from the perspective of those affected by the project. Theory and practice recognise that processes for engaging local communities and other affected stakeholders in decision-making are conditions for making good decisions (Forester, 1989; Parenteau, 1988; Pearce, Edwards, & Beuret, 1979; Tauxe, 1995; Webler, Kastenholz, & Renn, 1995). As exemplified below in the section on stakeholder engagement, experience across the Arctic testifies to the need for stronger engagement by companies and governments with consultation processes in order to identify and address concerns from the local perspective.

The urgency of mitigating climate change and meeting the thresholds of the Paris Climate Change Agreement are legitimate needs recognised by a global society. However, when actions taken in response to that urgency are pitched against equally legitimate interests of communities and Indigenous groups related to their traditions and foundations for their own sustainable economies, then complexity increases. Further complexity is added by impacts on employment and rights to participate in decision-making, for example through consultations and other forms of impact assessment involving meaningful stakeholder engagement. Some Arctic countries have recognised the right of Indigenous groups to free, prior and informed consent (FPIC), but others have not. Handling climate change exacerbates moral dilemmas as communities in one area are being placed under social or environmental pressures in the larger global interest. The essence of the moral dilemma is that opposition to a project (for example by an Indigenous community in one place due to risks to their traditional lifestyles) may place other communities in far-away places under increased risk of climate-change-related harm (such as flooding). As one Sápmi leader said in June 2019 to the lead author of this chapter, it is a fundamental moral question that challenges conventional perceptions of balancing interests and rights. This emergent challenge is currently under-researched. This chapter contributes to addressing this gap by exploring the issue of stakeholder involvement in regard to renewable energy projects, in the recognition that involvement is often the key to solving difficult dilemmas.

In a global perspective, research-based knowledge on what makes stakeholder engagement meaningful is limited (Maher & Buhmann, 2019; Skjervedal, 2018; Zoomers & Otsuki, 2017). In turn this raises several sub-questions, including what makes a process "good" and how involved organisations can turn formal requirements and top-down approaches into meaningful engagement from the perspective of local communities. We contribute by addressing the issue in an Arctic context, drawing on a series of cases of stakeholder involvement that illustrate varieties of perceptions of processes on the ground, and analysing these against normative foundations. Selected cases from Arctic countries serve to identify what constitutes meaningful stakeholder engagement through "stories" of what works well and what does not work well or is perceived by Arctic communities as inadequate.

The chapter applies an interdisciplinary approach. Grounded within the social sciences, the chapter relies on legal, sociological and general social science and communication studies methods for analysis of political and normative foundations through document-based studies and fieldwork.

Background

The normative interface between climate change mitigation and socially sustainable economic activities

On a global scale, a connection between social sustainability and renewable energy is recognised in theory as well as in political agreements on sustainable development. While climate change and adequate responses are pressing challenges for all societies, responses involve navigating conflicting social priorities (Hoffmann, 2011; Hulme, 2009). The introductions to the declarations that embody the SDGs and the Paris Climate Change Agreement explicitly note that the transition to a more holistically sustainable and low-carbon economy must be accomplished with respect for social impacts, in particular human rights (UN, 2015a, 2015b). SDG 17 on partnerships explicitly notes that SDG implementation must not cause harm, and the Paris Agreement refers to responsibilities for human rights, including those of affected local communities and vulnerable groups.

The SDG implementation provisions explicitly refer to the United Nations Guiding Principles (UNGPs) on Business and Human Rights (UN, 2011). These are a set of globally applicable guidelines for how states and companies should act to avoid human rights harm related to business operations. Due to the comprehensiveness of human rights, this includes social and many environmental risks or impacts caused by economic activities. Jointly with the "Protect, Respect and Remedy" Framework (UN, 2008) (a UN study which provides a theory-based foundation for business respect for human rights), the UNGPs are considered current state of the art concerning business respect for human rights (Wettstein, 2012). They advance "human rights due diligence" as a management approach for companies to identify and manage their adverse impacts on society. It is clearly stated that impact assessment undertaken as part of the due diligence process should build on meaningful stakeholder consultation, especially with "affected stakeholders," including communities and individuals affected by proposed or actual projects. Under the term "risk-based due diligence," this approach has been adopted by several transnational business governance instruments, including the OECD Guidelines for Multinational Enterprises (OECD, 2011) and the IFC's performance standards which inform the Equator Principles. Applied by more than 30 export credit agencies and more than 90 banks and financial institutions, the Equator Principles have the capacity to influence decisions to fund renewable energy projects in many places, including the Arctic. The OECD's Guidelines apply to companies operating *in or out of* the currently 48 adhering states. As all Arctic countries except Russia are OECD members, and because the Guidelines' definition of "multinational" is inclusive, the Guidelines cover most companies and institutional investors involved in renewable energy across the Arctic. Testifying to the importance of involving stakeholders as part of the due diligence process, the OECD in 2017 issued guidance for the implementation of the Guidelines for meaningful stakeholder engagement in the extractive sector (OECD, 2017). Underscoring the Arctic relevance, translation into the Sámi language was published in 2019.

Impact assessment and meaningful stakeholder engagement

Theory and practice on social impact assessment is evolving in response to the confluence of established theory-supported best practice on social impact assessment, and emergent normative standards on human rights impact assessment (Vanclay & Esteves, 2011; Harrison, 2013).

Impact assessment is a process that involves scoping, assessing and mitigating impacts, often implemented through a permit conditional on the impact assessment being undertaken (Esteves, Franks, & Vanclay, 2012; Vanclay, 2003). Risk-based due diligence has come to be associated with impact assessments because this due diligence approach aims at preventing harm, mitigating harm that is inevitable (typically already occurring), and accounting for processes to do so (UN, 2011; Buhmann, 2018a). Meaningful stakeholder engagement is an integral element in a sound impact assessment process as well as in activities to monitor, follow up and adjust action to prevent or mitigate adverse impacts. As the OECD Guidance (2017:18) notes, "Stakeholders themselves can contribute important knowledge to help identify potential or actual impacts on themselves or their surroundings. The values and priorities of impacted stakeholders are vital considerations in evaluating impacts and identifying appropriate avoidance or mitigation steps."

Indeed, stakeholder participation is an essential qualitative component of an impact assessment process (Nenasheva et al., 2015). Involving affected stakeholders in line with the normative foundations of the UNGPs and the OECD Guidelines means that communities affected by projects related to renewable energy should be consulted in a meaningful manner, and their concerns addressed. Explicating implications for specific contexts, implementation guidance for the IFC performance standards notes, i.a., that

> Nomadic peoples may have rights—whether legal or customary—to pass through client-controlled land periodically or seasonally, for subsistence and traditional activities. Their rights may be linked to certain natural resources such as [...] herds of migratory animals [...]. In its due diligence, the client should establish whether nomadic peoples have such rights, and, if possible, with the safeguards mentioned above, the client should allow them to exercise these rights on company-controlled land.
>
> *(IFC, 2012, GN63)*

Aiming to prevent social harm, risk-based due diligence differs from conventional financial or legal due diligence that firms have long performed with the aim of preventing harm to the firm (Buhmann, 2018a). Meaningful stakeholder engagement with affected communities can benefit not only the community but also the involved company and government agencies (Udofia, Noble, & Poelzer, 2017). If done well it can help companies retain a "social licence to operate" that facilitates current and future operations and expansions, contributes to early identification of risks of adverse impacts at the site of operations or supply chains, and helps avoid the costs of conflict arising from lost productivity due to temporary shutdowns and senior personnel time being diverted to manage grievances (OECD, 2017, p. 14; Ruggie, 2013; Kapoor, 2001).

In some Arctic countries environmental impact assessments include broader societal aspects, such as impacts on health, employment, traditions and business operations (Nenasheva et al., 2015). Explicitly granted for Indigenous populations by ILO Convention 169, stakeholder engagement may be considered an extension of the human rights to participation in public decision-making affecting one's life (Mestad, 2002). Applying to environmentally related social issues, the Aarhus Convention grants citizens a right to popular participation that includes access to information, access to dialogue with authorities granting permits, and access to remedy. Yet, despite the international and national normative framework for stakeholder engagement, several recent incidents across the Arctic demonstrate that stakeholders often do not perceive their involvement in processes related to renewable energy to be adequate or meaningful.

Proposed or implemented mining projects have led to local conflicts in Northern Scandinavia (Bjørst, 2016; Hansen, Vanclay, Croal, & Skjervedal, 2016; Lindahl, Johansson, Zachrisson, & Viklund, 2018). Studies from other regions show that such conflicts can be devastating to the local community and undermine support from other stakeholders (Bebbington et al., 2008; Rodríguez-Labajos & Özkaynak, 2017). The adequacy of involvement by local communities and Indigenous groups and peoples in Arctic economic development projects has been questioned (Abram, 2016; Cambou, 2018). In Eastern Canada, protests around large-scale hydro-power projects testify to concern with environmental health and safety risks to Indigenous communities in Labrador and Newfoundland. In British Columbia, First Nations and others have protested the consultation processes and impacts related to various renewable energy projects. This is also the case in northern Fenno-Scandinavia, where Sámi communities have challenged decisions concerning the establishment of local wind project. In Greenland, new ideas have surfaced for involving youths in decisions that will ultimately affect their futures. In the following sections, we provide more details on some of these conflicts and developments concerning meaningful stakeholder engagement.

Stakeholder engagement on the ground

Sápmi

In Norway and Sweden, Indigenous involvement in energy projects in the Arctic parts of those countries has been a question of controversy for many years. Historically, large-scale hydroelectric projects spurred the first open conflicts between the Nordic governments and the Sámi people (Cambou & Polzer, 2020). In this context, the involvement of Sámi communities was often overlooked and their traditional livelihoods seriously impaired by the development of energy projects promoted by the nation states.

In the twenty-first century, the situation of the Sámi people has significantly evolved as a result of the adoption of international and national laws recognising their rights as a minority and an Indigenous people (Allard & Skogvang, 2017; Bankes & Koivurova, 2013). Yet, the impacts of projects related to a low-carbon transition continue to jeopardise their traditional livelihoods. In Northern Norway, the Nussir copper mine epitomises the dilemma faced by the Government of Norway between protecting the pristine ecosystem and providing the country with a mineral required for batteries for electric cars. However, for local Sámi reindeer herders, the project also means the potential loss of reindeer herding grazing pastures and the disturbance of the migration of reindeer and salmon fishing grounds (Koivurova et al., 2015, p. 32). The Sámi Parliament in Norway has opposed the Nussir project due to the lack of consultation with Sámi communities at the local level (Storholm, 2016). According to the UN Rapporteur on the Rights of Indigenous Peoples, Norwegian law does not comply with international standards on the rights of Indigenous peoples concerning consultation on the basis of the principle of FPIC (Human Rights Council, 2016, para. 29).

The expansion of wind power also highlights contestation over the rights of the Sámi people and what is entailed by meaningful stakeholder engagement. The rise in wind energy projects to advance ambitious national climate goals has raised important concerns among Sámi reindeer herding communities. The establishment of several wind energy projects, often large-scale, has been contested due to their adverse effects on reindeer pasture and migration (Cambou, 2018). Sámi reindeer herding communities have criticised and lodged legal protests against such projects due to their lack of meaningful consultation or loss of income due to participation in consultations (Cambou, 2018; Buhmann, 2018b). They also oppose wind energy projects because they do not benefit their communities and paradoxically also threaten the sustainability of their traditional livelihoods. Decisions by courts note the need for a collaborative effort to overcome the potential difficulties associated with the co-existence of both activities. Related to a Norwegian wind turbine project affecting a Sámi village in Sweden, the complaints institution to the OECD Guidelines underlines the responsibility of the company to engage with Sámi reindeer herding communities as a means to prevent and mitigate the adverse impact of wind projects (OECD Watch, 2012; Buhmann, 2018a).

These considerations underscore the need for ensuring the meaningful involvement of Sámi communities within energy and industrial projects located on their traditional lands in order to guarantee that they also reap some of the benefices of the energy transitions.

Greenland

Public participation in Greenland displays a contrast between solid national regulation and institutions, on the one hand, and the lived experience of meaningful stakeholder involvement, on the other (Olsen and Hansen, 2014). The changing climate has renewed interest in exploration and exploitation of mineral resources, including hydrocarbons. The Government of Greenland is determined to make extractives a primary business sector, in close collaboration and dialogue with the country's population (Government of Greenland, 2016, 2019).

Part of the Danish Kingdom, since 2009 Greenland has self-government with full decision-making powers in policy areas that have been claimed. Whereas the Inuit population, which forms the country's majority, can be argued to be entitled to special protection under international law as an Indigenous people in regard to decisions made by the Danish Government, the same does not apply to decisions made by Greenlandic authorities.

While renewable energy projects in Greenland are emergent (Nukissiorfiit, 2019), the mining sector offers insights on differences between formal requirements on stakeholder involvement and community perceptions of what constitutes meaningful stakeholder engagement. The mining sector is re-emerging decades after aluminum and coal extraction closed down (Sejersen, 2014).

Greenland has not acceded to the Aarhus Convention but national policy and legislation sets high goals for public participation. In the Mineral Resources Act Greenland's self-government has introduced explicit social sustainability assessment requirements for certain raw-material extraction projects (Hansen et al., 2016). The Social Impact Assessment Guidelines issued by the government refer to stakeholder involvement as a prerequisite for a good impact assessment process to promote sustainable development in Greenlandic society (Government of Greenland, 2016). Whereas the political, scientific and public debate on mineral extraction in Greenland has been extensive in the past decade, only two mines are currently operational (a ruby mine by a settlement of 235 inhabitants, and an aluminium mine located at some distance from settled areas). However, proposals to establish an iron mine (Isua) in the Nuuk fjord close to Greenland's capital and a REE mine (Kuannersuit) by Narsaq town in Southern Greenland have sparked extensive public debate on stakeholder influence, which provides insights on what makes stakeholder involvement meaningful. The Isua project led to strong public mobilising to protect the fjord, led by organised civil society groups (Nuttall, 2012). The project was eventually called off for reasons that also included economic viability. The Kuannersuit project is still under consideration and closer to the granting of an exploration license than some nearby smaller REE deposits. Kuannersuit stands out by containing 10% uranium that will be an unavoidable by-product of exploration. This has generated concern with impacts on human and animal health among some locals, in particular sheep farmers, whereas others welcome prospects for employment. Although consultations have occurred according to formal requirements, frustration nevertheless resulted with some sheep-farmers and others in the local community having a sense that the process was not accessible because of their locations or due to work requirements. The frustration was aggravated by the insecurity caused by the risk of uranium dust and uncertainty of actual health impacts (Buhmann et al., 2019/2020). This example shows that conducting impact assessments by the book is not enough if consultation meetings are seen to be inaccessible, and that complex technical and health implications must be explained in a manner that makes sense from the perspective of the audience (i.e.Cunsolo & Ellis, 2018).

The Nordic Council has a policy of involving youth. However, studies show a lack of focus on engaging young people in Greenland. Such oversight may cause young people to miss out on having a say on project development that will affect their futures. It may also reduce their access to relevant information to plan for the future in terms of education and employment opportunities. Research demonstrates that the use of social media as a complementary communication channel enables young people to engage in project development in a manner that is meaningful to them (Skjervedal, 2018). Capturing their interests, values, fears, hopes and aspirations for the future along with their thoughts on future extractive development in Greenland, this approach engaged a wide range of youth across Greenland in a manner they perceived as meaningful. Providing a "safe forum" for active and collaborative engagement, social media allowed youths to engage in their own time in a manner aligned with how this age group normally communicate and share information on a daily basis (Skjervedal, 2018).

These observations highlight the need for applying methods that are appropriate and relevant to specific types of stakeholders and tailoring the participation form(s) to the specific project, local context, and target groups.

Canada

While the Eastern (Atlantic) Canadian province of Newfoundland and Labrador sits well below the Arctic Circle, Labrador is often defined as a northern place due its landscape and histories of inhabitation by Inuit, NunatuKavut, and Innu communities (Proctor, Felt, & Natcher, 2012). The province is the site of the 1990s cod fishery collapse, regarded by many as one of the worst ecological disasters in Canadian history (Bavington, 2010). In the decades since the cod fishery collapse, the provincial political economy reoriented around offshore oil extraction. Given the dominance of oil development in provincial politics and the public imaginary, it is unsurprising that the province tends to resist federal government moves toward stronger climate policy, downplaying the responsibility of the oil sector and other large industry (Sodero & Stoddart, 2015).

This sets the context for understanding renewable energy transitions in Newfoundland and Labrador. Public discourse and planning has been dominated by the large-scale Muskrat Falls hydro-electric project. This new dam is located on the lower Churchill River in Labrador and involves a link across the Labrador Straits to the island of Newfoundland where it will feed into the provincial energy system. Muskrat Falls will allow the province to meet its energy needs while decarbonising the provincial energy system and meeting its climate change goals (Nalcor Energy, 2020).

The project has generated considerable controversy and contention, especially from downstream Indigenous communities, including the grassroots movement known as the Labrador Land Protectors. The process of planning and approving the Muskrat Falls dam was handled by the provincial government in partnership with Nalcor, the public energy corporation. Many opponents of the project have argued that Indigenous communities were not appropriately consulted or engaged in the process of planning and implementing this project, especially as environmental health concerns were still being researched and documented as the project was approved (Allen, 2017).

The negative impacts of Muskrat Falls are not only financial, which have been the main focus of a public inquiry. There are significant downstream environmental health risks. There are concerns that methylmercury from rotting vegetation inside the dam reservoir will flow downstream and contaminate fish and wildlife populations. Fish and game remain integral to Indigenous community diets and cultural traditions. The ability to retain land-based food culture is essential in a region where high food costs and food security are serious issues (Cox, 2019b; Penney, 2018). There are also concerns around slope stability related to the dam infrastructure. If dam infrastructure fails and collapses, then downstream communities will be flooded, which is a source of fear and stress (Cox, 2019a; Philpott, 2018).

Opposition to the project and calls to "Make Muskrat Right" have taken the form of hunger strikes and grassroots protest by the Labrador Land Protectors and their allies at the Muskrat Falls site, as well as at the provincial legislature in St. John's and other locations around the province (Allen, 2017; Cox, 2019a, 2019b). On-site protests have used civil disobedience, which has been met with arrests and imprisonment for breaching court orders that restrict access to the work site. Land Protector Denise Cole has described protest as "an act of ceremony" and as a responsibility to Indigenous laws as it rejects a view of Labrador as a resource warehouse for Newfoundland and bears witness to the destruction of sacred spaces (Cole, 2018). Grassroots opposition has been amplified by formal representatives of Indigenous communities, though there were also divergent responses from various community leaders (Doherty, 2018).

The Muskrat Falls project is an example of environmental injustice and energy injustice, as downstream Indigenous communities are being asked to bear the health costs and safety risks of a mega-project that is being promoted as a provincial climate change solution and core part of a renewable energy transition (Municipal Affairs and Environment: Climate Change Branch, 2011).

In contrast to Newfoundland and Labrador, British Columbia (BC) in Western Canada is historically a province that relies heavily on renewable energy. The province gets approximately 95% of its power from renewables with hydroelectricity accounting for about 85% (National Energy Board, 2018) and so it is widely seen as a "green energy" province. The environmental movement has been strong in the province for many decades, indeed the "global environmental movement started in British Columbia" (Byers, 2012), and, in 2008, it became the first jurisdiction in North America to introduce a carbon tax.

Despite these contrasts with Newfoundland and Labrador there are also striking similarities. Just as the Muskrat Falls hydro project raised questions of environmental injustice and environmental racism in its treatment of Indigenous peoples, so the same issues are evident in BC's hydro projects. The WAC Bennet dam, constructed in the Peace river region of northern BC in the 1960s, is a major contributor to hydropower but was built without the local Indigenous population of Kwadacha being informed, let alone consulted. Their traditional territory was flooded when Williston reservoir, the world's seventh largest, was formed. The livelihoods of Indigenous individuals and groups were destroyed with the Kwadacha community forgotten in the push to develop the power source that would drive industrialisation and resource extraction in the northern part of the province (Loo, 2007; Stanley, 2010). The community was promised electricity as part of the project but this was never honored.

This sets the historical background for the construction of the Site C dam in the same geographical area today, designed to further increase hydro capacity. The approval of this project has also proved to be contentious. While consultation processes did take place and there have been significant improvements since the 1960s when the WAC Bennet dam was constructed, Site C was opposed by a coalition of Indigenous groups, farmers and environmentalists (Cox, 2018a). It is still subject to on-going legal challenge by the West Moberly and Prophet River First Nations with West Moberly First Nation's Chief declaring the project "cultural genocide" (Cox, 2018b). The various review processes have been widely criticised as politically driven and biased in the knowledge that they viewed as important (Bakker & Hendriks, 2019; Muir, 2018).

Again, similar to the Newfoundland and Labrador case, the project was initiated by one government and reluctantly approved by a new one on the grounds that sunk costs made cancellation too costly. Even when the economic case for the project and the impacts on Indigenous relations are recognised as problematic, once started, these mega-projects become difficult to halt. This reinforces the need for the initial consultation processes to be thorough, transparent and seen as legitimate by all stakeholders and rights holders.

A further complicating factor with the Site C dam is that unspecified amounts of electricity from the dam will be used to develop BC's emerging liquefied natural gas (LNG) export industry. Ironically, the expansion of renewable energy will support the expansion of fossil fuel production, in the form of LNG, rather than to replace it. The development of renewables around the Arctic therefore has also to be seen in the context of what the renewable energy will be used *for*, a question which has also arisen in the case of the uranium that will be a by-product from the Kuannersuit mine in Greenland.

Hydro mega-projects have proven too often be contentious but there are smaller renewable energy projects which are much less so, including many involving First Nations. These include attempts by off-grid First Nations to replace diesel with renewables such as solar power as in the case of the T'Sou-Ke First Nation (Bhattacharya, 2017). Several local governments and Band Councils in Haida Gwaii, an island off British Columbia's northwest coast, have been active in plans to reduce reliance on diesel and move to renewable energy sources.[1] The key to these projects is that they have been initiated and led by local communities themselves. Higher level governments have often been facilitators by supplying needed financial support but the decision-making has been local community based. This provides a quite different model of community involvement and empowerment than occurs with renewable energy mega-projects.

Implications for planning and implementing socially sustainable renewable energy projects

These examples underscore the need to take social sustainability seriously in discussions and analyses of how to promote renewable energy in Arctic countries. This must be central in plans to expand renewal energy to both address climate change and provide new economic opportunities.

The Muskrat Falls controversy highlights how renewable energy transitions can produce new energy injustices for local communities, in particular in the Arctic context northern Indigenous communities. Some of these, including Indigenous communities in Labrador, are already experiencing the harmful impacts of climate change more acutely than elsewhere in the country.

The Canadian cases also show that scale is an important factor in the quest for genuine community participation, involvement and ownership of renewable energy projects. Community participation in a way that appears meaningful to those affected has proven to be more feasible for smaller projects. It remains a pressing issue whether and how this can be scaled up for larger projects which may contribute more to combat the urgency of the global climate crisis.

The case of the Sámi people also highlights how the development of industrial and energy projects promoting sustainable development can paradoxically jeopardise the sustainable livelihoods of Indigenous peoples. The lack of meaningful consultation of Sámi communities continues to loom large in the debate concerning the development of energy projects on their traditional lands and questions the adequacy of the legal framework of the Nordic countries to guarantee their rights as an Indigenous people. This case also demonstrates the need to integrate energy justice considerations in order to ensure a transition to renewable energy and economies that are socially sustainable and just for all.

One of the first states to sign the ILO Convention 169, Norway was early to recognise the principle of FPIC. Yet, despite important legislative changes, the lack of meaningful consultation with the Sámi people concerning the development of industrial and energy projects on their traditional lands remains problematic. The Nussir mine case illustrates the problem when stakeholder consultation is perceived as an empty process. On the one hand, the Sámi cases demonstrate the need to improve legislative frameworks that ensure that the rights of the Sámi people as an Indigenous people are protected and respected by states and companies, especially when decision and measures concerning energy projects affects them. On the other, the Sámi and Greenlandic cases demonstrate that for stakeholder engagement to be meaningful, formal legislation must be matched by implementation and planning of consultations that respect existing commitments of affected stakeholders.

The Greenland case also highlights how broadening the range of forms of public participation can increase the perception of consultations as meaningful for those involved. Opportunities for a broad representative voice among local communities can be provided through tailoring the participation form(s) to the specific project, local context, and target groups, with a focus on creating a "safe forum' for active and collaborative engagement.

The social legitimacy of renewable energy projects and the public and private organisations behind them require careful planning to avoid such injustices. This is a problem across the globe, but its acuteness hits strongly in the Arctic, exacerbated by past injustices between colonial settlers and Indigenous groups that with variations apply to all the areas discussed above. Moreover, the rise of conflicts in rule-of-law based Arctic states discomfortingly reminds us of conflicts that scholars and practitioners in the North often associate with countries with weak governments. Yet the Sápmi, Greenland and Canadian cases above all demonstrate that the transition from the ideals of meaningful stakeholder involvement, expressed in the globally applicable UNGP and elaborated through the OECD Guidelines and related guidance texts, is easier said than done. Moreover, the Greenland example illustrates that even where local law and regulations aim to ensure at a high level of stakeholder involvement, the experienced effects may be different. Jointly, the cases demonstrate that meaningful stakeholder engagement from the perspective of those affected is a core element in upholding energy justice and environmental justice in order to ensure that climate change solutions do not amplify other forms of social inequality. This also confirms findings from studies in a Global South context (for example, Maher & Buhmann, 2019; Zoomers & Otsuki, 2017) that more research is needed on how to transform formal—and often well-intended—rules into practical application that is truly meaningful and effective from the bottom-up perspective.

Conclusion

Renewable energy in the Arctic takes several forms in this chapter, including hydro and wind power. Renewables transitions are predominantly led by government and corporate actors and framed in terms of technological and economic responses to environmental sustainability. These projects are often presented in pro-environmental terms of responding to climate change, shifting away from fossil fuels, and fitting into a governmental self-image of a "green society"

Contestation around renewables transition takes multiple forms, including landscape impacts, concerns with infrastructure siting, and social impacts. Indigenous communities and environmental groups are actors that create friction and raise questions about the potential downsides of renewables transitions.

Our overview shows that renewable energy projects that are promoted as part of necessary climate change action can have perversely negative impacts on community environmental health and safety as well as the traditions and income-generating activities of Arctic Indigenous groups. The need for energy justice underscores the importance of approaching climate change responses and renewable energy transitions in ways that adequately address local concerns and needs in a manner that is meaningful to those who may be adversely affected.

The urgency of mitigating climate change means a surge in renewable energy projects. While in line with the SDGs, in particular SDG 7 and 13, the examples above demonstrate the risk that such projects may cause social harm to local communities, including—but not limited to—Indigenous groups. This accentuates fundamental issues of environmental, energy and human rights justice, with strong connections between the environmental and social (including human rights) aspects. In turn, this underscores the pertinence of scholarship and practice considering the practical realisation of the SDG's implementation provision No 67 on not causing harm and acting in accordance with the UNGP's provisions. In an Arctic context it highlights that compliance with rules must be complemented by practices that consider and respect local experiences and perceptions of impacts.

A key take-away argument of this chapter is that as circumpolar countries grapple with the necessities of decarbonising energy systems in response to climate change, we see that renewable transitions will also create new points of tension. Renewable does not necessarily equal socially sustainable or just transitions. To address this actual or potential conflict with commitments under the SDG Declaration and the Paris Agreement, there is a need to move beyond governmental and corporate-led models of renewable transitions to more participatory, deliberative processes for ensuring a just renewables transition. Involvement of communities and other affected stakeholders including rights-holders may help uncover alternative placements and local benefits. Involvement of affected stakeholders can contribute to understanding and acceptance of unwelcome social impacts that are necessary for the green transition in the larger interest of responding to climate change. It may therefore contribute to transitions that are perceived and accepted as socially fair under the given circumstances. Although it does not eliminate difficult challenges around adverse impacts, it can help alleviate some of the social and moral dilemmas related to urgent action to mitigate climate change.

Acknowledgements

This chapter has benefitted from the following projects (funding agencies in brackets): NOS-HS project *"Best practice for Impact Assessment of infrastructure projects in the Nordic Arctic: Popular participation and local needs, concerns and benefits"* (Nordic Research Councils); University of the Arctic (UArctic) Thematic Network on Arctic Sustainable Resources and Social Responsibility (UArctic with the Danish Agency for Science and Higher Education); and the Insight Grant project *The Oil-Tourism Interface and Social-Ecological Change in the North Atlantic* (SSHRC).

Note

1. see Swiilawiid.org

References

Abram, S. (2016). Jokkmokk: Rapacity and resistance in Sápmi. In G. Huggan & L. Jensen (Eds.), *Postcolonial perspectives on the European high North*, pp. 67–92. Palgrave.

Allard, C., & Skogvang, S. F. (Eds.). (2017). *Indigenous rights in Scandinavia: Autonomous Sámi law.* Routledge.

Allen, V. (2017). Muskrat falls. In A. Marland and L. Moore (Eds.), *The democracy cookbook: Recipes to renew governance in newfoundland and Labrador*, pp. 318–320. ISER Books.

Bankes, N., & Koivurova, T. (2013). *The proposed Nordic Saami convention: National and international dimensions of indigenous property rights.* Hart Pub.

Bakker, K., & Hendriks, R. (2019). Contested knowledges in hydroeletric project assessment: The case of Canada's site c project. *Water, 11*(3). https://doi.org/10.3390/w11030406

Bavington, D. (2010). *Managed annihilation: An unnatural history of the newfoundland cod collapse.* UBC Press.

Bebbington, A., Bebbington, D. H., Bury, J., Lingan, J., Muñoz, J. P., & Scurrah, M. (2008). Mining and social movements: Struggles over livelihood and rural territorial development in the Andes. *World development, 36*(12), 2888–2905.

Bhattacharya, A. (2017). *Building A Network of Clean Energy Systems: A Case Study of T'Sou-Ke First Nation Solar Project* [MA Thesis, University of Northern British Columbia]. Canadian Centre for Policy Alternatives.

Bjørst, R. (2016). Saving or destroying the local community? Conflicting spatial storylines in the Greenlandic debate on uranium. *The Extractive Industries and Society, 3*(1), 34–40.

Buhmann, K., Sanne, V. L., & Anna-Sofie, H. S. (2019/2020). Why consultations on large projects matter to citizens as well as companies, entry No 1 in blog-post series 'Consultations, Public Participation, and Meaningful Stakeholder Engagement' (Center of Business and Development, Copenhagen Business School), https://www.cbds.center/post/blogpost-series-1-consultations-public-participation-and-meaningful-stakeholder-engagement, republished in 2020 under the Business of Society BOS (CBS Center for Sustainability) blog series (http://www.bos-cbscsr.dk/2020/02/07/why-consultations-on-large-projects-should-matter-to-citizens-as-well-as-companies/).

Buhmann, K. (2018a). Neglecting the proactive aspect of human rights due diligence? A critical appraisal of the EU's non-financial reporting directive as a pillar one avenue for promoting pillar two action. *Business and Human Rights Journal, 3*(1), 23–45.

Buhmann, K. (2018b). Analyzing OECD national contact point statements for guidance on human rights due diligence: Method, findings and outlook. *Nordic Journal of Human Rights, 36*(4), 390–410.

Buhmann, K. (2016). Public regulators and CSR: The "Social licence to Operate" in recent United nations instruments on business and human rights and the juridification of CSR. *Journal of Business Ethics, 136*(4), 699–714.

Business Index North. (2017). *Renewable energy in the North, business index North* (Issue 1). Business Index North. https://businessindexnorth.com/sites/b/business-indexnorth.com/files/bin2017_5_renewable_energy_in_the_north_web.pdf

Buxton, A., & Wilson, E. (2013). *FPIC and the extractive industries: A guide to applying the spirit of free, prior and informed consent in industrial projects.* International Institute for Environment and Development.

Byers, M. (2012, March 9). My Cage Fight in Fort McMurray. *The Tyee.* https://thetyee.ca/Opinion/2012/03/09/Fight-In-Fort-McMurray/

Cambou, D. (2018). Renewable energy in the Arctic and the human rights of indigenous peoples: Past, present and future experiences of the Sámi people. In G. Xue, & L. He (Eds.), *Law and governance: Emerging issues of the polar regions* (p. 291). Shanghai: China University of Political Science and Law Press.

Cambou, D., & Polzer, G. (2020). Energy justice in the Arctic: The limits and challenges of the transition to renewable energy for Arctic indigenous peoples. In D. Natcher & T. Koivurova (Eds.), *Renewable economies in the Arctic: A state of knowledge.* Routledge.

Cole, D. (2018). *Muskrat Falls Symposium.* Memorial University.

Cox, S. (2018a). *Breaching the peace: The site c dam and a Valley's stand against big hydro.* On Point Press

Cox, S. (2018b, October 30). Be Prepared to Be Surprised: What's Next for Site C Dam?. *The Narwhal.* https://thenarwhal.ca/be-prepared-to-be-surprised-whats-next-for-the-site-c-dam/

Cox, S. (2019a). A Reckoning for Muskrat Falls. *The Narwhal.* https://thenarwhal.ca/a-reckoning-for-muskrat-falls/

Cox, S. (2019b). Mercury rising: How the Muskrat Falls dam threatens Inuit way of life. *The Narwhal.* Retrieved from https://thenarwhal.ca/mercury-rising-muskrat-falls-dam-threatens-inuit-way-of-life/

Cunsolo, A., & Ellis, N. R. (2018). Ecological grief as a mental health response to climate change-related loss. *Nature Climate Change*, *8*(4), 275–281.

Doherty, T. (2018). *Muskrat Falls Symposium*. Memorial University.

Esteves, A. M., Franks, D., & Vanclay, F. (2012). Social impact assessment: The state of The art. *Impact Assessment and Project Appraisal*, *30*(1), 34–42.

Forester, J. (1989). *Planning in the face of power*. University of California Press.

Government of Greenland. (2016). *Social Impact Assessment: Guidelines on the process and preparation of the SIA report for mineral projects*. Ministry of Industry, Labour and Trade.

Government of Greenland. (2019). *Oliestrategi 2019-2023*. Government of Greenland. https://naalakkersuisut.gl/~/media/Nanoq/Files/Hearings/2019/Oliestrategi%20 for%202019_2023/Documents/Oliestrategi%202019-2023%20DK%204.pdf

Hansen, A. M., Vanclay, F., Croal, P., & Skjervedal, A. S. H. (2016). Managing the social impacts of the rapidly-expanding extractive industries in Greenland. *The Extractive Industries and Society*, *3*(1), 25–33.

Harrison, J. (2013). Establishing a meaningful human rights due diligence process for corporations: Learning from experience of human rights impact assessment. *Impact Assessment and Project Appraisal*, *31*(2), 107–117.

Hoffmann, A. J. (2011). Talking past each other? Cultural framing of skeptical and convinced logics in the climate change debate. *Organization & Environment*, *24*, 3–33.

Hulme, M. (2009). *Why we disagree about climate change: Understanding controversy, inaction and opportunity*. Cambridge University Press

Human Rights Council. (2016). *Report on the human rights situation of the Sámi people in the Sápmi region*. United Nations Special Rapporteur on the Rights of Indigenous Peoples. http://unsr.vtaulicorpuz.org/site/index.php/en/documents/country-reports/155-report-sapmi-2016

IFC. (2012). *Guidance note no 5: Land acquisition and involuntary settlement*. International Finance Corporation.

Kapoor, I. (2001). Towards participatory environmental management? *Journal of Environmental Management*, *63*, 269–279.

Koivurova, T., Masloboev, V., Hossain, K., Nygaard, V., Petrétei, A., & Vinogradova, S. (2015). Legal protection of Sámi traditional livelihoods from the adverse impacts of mining: A comparison of the level of protection enjoyed by Sámi in their four home States. *Arctic Review on Law and Politics*, *6*(1), 11–51.

Lindahl, K. B., Johansson, A., Zachrisson, A., & Viklund, R. (2018). Competing pathways to sustainability? Exploring conflicts over mine establishment in the Swedish mountain region, *Journal of Environmental Management*, *218*.

Loo, T. (2007). Disturbing the peace: Environmental change and the scales of justice on a Northern River. *Environmental History*, *12*(4), 895–919.

Maher, R., & Buhmann, K. (2019). Meaningful stakeholder engagement: Bottom-Up initiatives within global governance frameworks. *GeoForum*. doi: https://doi.org/10.1016/j.geoforum.2019.06.013.

McCauley, D., Heffron, R., Pavlenko, M., Rehner, R., & Holmes, R. (2016). Energy justice in the Arctic: Implications for energy infrastructural development in the Arctic. *Energy Research & Social Science*, *16*, 141–146.

Mestad, O. (2002). Rights to public participation in Norwegian mining, energy, and Resource development. In D. N. Zillman, R.L. Alastair, & G. Pring (Eds.), *Human rights in natural Resource development. Public participation in the sustainable development of mining and energy Resources* (pp. 382–400). Oxford UP.

Mohai, P., Pellow, D., & Roberts, T. (2009). Environmental justice. *Annual Review of Environment and Resources, 34,* 405–430.

Municipal Affairs and Environment: Climate Change Branch. (2011). *The Way Forward on Climate Change in Newfoundland and Labrador.* Government of Newfoundland and Labrador. https://www.exec.gov.nl.ca/exec/occ/action_plans. html

Muir, B. (2018). Effectiveness of the EIA for the site c hydroelectric dam reconsidered: Nature of indigenous cultures, rights, and engagement. *Journal of Environmental Assessment Policy and Management, 20*(4). https://doi.org/10.1142/S146433321850014X

Nalcor Energy. (2020). Muskrat Falls. *Nalcor Energy.* http://muskratfalls.nalcorenergy.com/

National Energy Board. (2018). Canada's Renewable Power Landscape 2016 – Energy Market Analysis. *Canada Regulator Regulator.* https://www.neb-one. gc.ca/nrg/sttstc/lctrct/rprt/2016cndrnwblpwr/prvnc/bc-eng.html

Nenasheva, M., Bickford, S., Lesser, P., Koivurova, T., & Kankaanpaa, P. (2015). Legal tools of public participation in the environmental impact assessment process and their application in the countries of the Barents Euro-Arctic region. *Barents Studies, 1*(3), 13–35.

Nukissiorfiit. (2019). Vindenergi. *Nukissiorfiit.* https://www.nukissiorfiit.gl/vedvarende-energi/vindenergi/

Nuttall, M. (2012). The Isukasia iron ore mine controversy: Extractive industries and public consultation in Greenland. *Nordica Geographical Publications, 41*(5), 23–34.

OECD. (2011). *OECD guidelines for multinational enterprises* [rev. May 2011]. Organisation of Economic Cooperation and Development.

OECD. (2017). *Due diligence guidance for meaningful stakeholder engagement in the extractive sector.* OECD.

OECD Watch. (2012). *Jijnjevaerie Saami village vs Statkraft — OECD watch case database.* OECD. https://complaints.oecdwatch.org/cases/Case_280

Olsen, A. S. H., & Hansen, A. M. (2014). Perceptions of public participation in impact assessment: A study of offshore oil exploration in Greenland. *Journal of Impact Assessment and Project Appraisal, 32*(1), 72–80.

Parenteau, R. (1988). *Public participation in environmental decision-making.* Minister of Supply and Services, Government of Canada.

Pearce, D. W., Edwards, L., & Beuret, G. (1979). *Decision-making for energy futures: A case study of the windscale inquiry.* Macmillan, in association with the SSRC.

Penney, J. (2018). *Muskrat Falls Symposium.* Memorial University.

Philpott, E. (2018). *Muskrat Falls Symposium.* Memorial University.

Proctor, A., Felt, L., & Natcher, D. C. (2012). Introduction. In D. C. Natcher, L. Felt, & A. Proctor (Eds.), *Settlement, subsistence, and change among the Labrador Inuit* (pp. 3–13). University of Manitoba Press.

Rodríguez-Labajos, B., & Özkaynak, B. (2017). Environmental justice through the lens of mining conflicts. *Geoforum, 84,* 245–250.

Ruggie, J. (2013). *Just business.* Norton.

Sejersen, F. (2014). *Efterforskning og udnyttelse af råstoffer i Grønland i historisk perspektiv. Baggrundspapir. Udvalget for samfundsgavnlig udnyttelse af Grønlands naturresourcer.* Ilisimatusarfik & University of Copenhagen.

Skjervedal, A. (2018). *Towards Meaningful Youth Engagement: Breaking the Frame of the Current Public Participation Practice in Greenland* [Ph.D. dissertation, Ilisimatusarfik – The University of Greenland and Aalborg University].

Sodero, S., & Stoddart, M. C. (2015). A typology of diversion: legitimating discourses of tourism attraction, oil extraction and climate action in newfoundland and Labrador. *Environmental Sociology, 1*(1), 59–68.

Stanley, M. (2010). *Voices from two rivers: Harnessing the power of the peace and Columbia.* Douglas & McIntyre.

Storholm, L. (2016). Sámi Parliament Council refers to Nussir as "the new Alta case". *High North News.* https://www.highnorthnews.com/en/Sámi-parliament-council-refers-nussir-new-alta-case

Tauxe, C. S. (1995). Marginalizing public participation in local planning: An ethnographic account. *Journal of the American Planning Association, 61*(4), 471–481.

Udofia, A., Noble, B., & Poelzer, G. (2017). Meaningful and efficient? Enduring challenges to aboriginal participation in environmental impact assessment. *Environmental Impact Assessment Review, 65*, 164–174.

UN. (1998). *Convention on Access to Information, Public Participation in Decision-Making and Access to Justice in Environmental Matters* (Aarhus Convention). United Nations.

UN. (2008). *Protect, respect and remedy: A framework for business and human rights, Report of the Special Representative of the Secretary-General on the issue of human rights and transnational corporations and other business enterprises, John Ruggie* (UN Doc. A/HRC/8/5). United Nations Human Rights Council.

UN. (2011). *Guiding Principles on Business and Human Rights: Implementing the United Nations "Protect, Respect and Remedy" Framework* (UN Doc. A/HRC/17/31). United Nations Human Rights Council.

UN. (2015a). *Transforming our world: the 2030 Agenda for Sustainable Development* (UN Doc. A/Res/70/1). United Nations.

UN. (2015b). *Framework Convention on Climate Change – Conference of the Parties: Adoption of the Paris Agreement* (UN Doc. FCCC/CP/2015/L.9/Rev.1). United Nations.

Vanclay, F. (2003). International principles for social impact assessment. *Journal of Impact Assessment and Project Appraisal, 21*(1), 5–11.

Vanclay, F., Esteves, A.M., 2011. New Directions in Social Impact Assessment: Conceptual and Methodological Advances. Edward Elgar Publishing, Cheltenham, UK.

Webler, T., Kastenholz, H., & Renn, O. (1995). Public participation in impact assessment: A social learning perspective. *Environmental Impact Assessment Review, 15*, 443–463.

Wettstein, F. (2012). Human rights as a critique of instrumental CSR: Corporate responsibility beyond the business case. *Notizie de POLITEIA, 116*, 18–33.

Zoomers, E. A., & Otsuki, K. (2017). Addressing the impacts of large-scale land investments: Re-engaging with livelihood research. *Geoforum, 83*, 164–171.

10 Enhancing energy justice in the Arctic

An appraisal of the participation of Arctic indigenous peoples in the transition to renewable energy

Dorothée Cambou & Greg Poelzer

Introduction

Renewable energy is a key to the transition toward a more sustainable development. The importance of renewable energy both for mitigating the impact of climate change and for helping to alleviate poverty on a global scale is recognised widely. Policy makers also recognise its especial importance in the Arctic region, where temperature is warming twice as fast as the global average. In this context, Arctic states have placed increasing policy emphasis on the development of renewable energy in order to ensure a sustainable development of the region and recognises the importance of Arctic communities' access to energy that is not only cleaner, but also more affordable, accessible, and reliable (SDWG, 2019).

Yet, the full impacts—both positive and negative—of the development of renewable energy for Arctic Indigenous communities require much greater study. At the community level, scholars need to focus more attention to the needs and interests of Arctic communities, who remain heavily dependent on fossil fuels and build a program of inquiry that is systematic and comparative to better understand energy needs and usage within Indigenous communities. At regional and national levels, utility-scale renewable energy projects often pose a conundrum: large-scale hydro, wind, and solar projects serve to meet national objectives to reduce greenhouse gas emissions; however, the siting of these projects are often on the traditional land and territories of Arctic Indigenous peoples, almost invariably infringing on Indigenous well-being, if not also, Indigenous rights (Lempinen & Cambou, 2018; Cambou,2020). Thus, this situation raises clear concerns about the principles and considerations of basic fairness of the transition to renewable energy as it related to the implications for Indigenous peoples in the Arctic.

The purpose of this contribution is to examine in what extent Indigenous communities currently share the benefits and costs of the energy transition in the Arctic region. While all Arctic states have committed to ensure that "no one will be left behind in the sustainable transition" (UN General Assembly, 2015), this contribution questions the extent of this commitment

DOI: 10.4324/9781003172406-10

and its practical effects in relation with Arctic Indigenous peoples. In this regard, the analysis takes a multidisciplinary approach to the question while building upon the concept of energy justice with a specific focus on the rights of Indigenous peoples. Energy justice seeks to identify when and where injustices occur and just opportunities can be realised in order to ensure policy and legal response that contribute to a more representative and inclusive energy decision-making (Sovacool, Burke, Baker, Kotikalapudi, & Wlokas, 2017). "It calls on academics and practitioners to critically evaluate the implications of energy policies" and "begins with questioning the ways in which benefits and ills are distributed, remediated and victims are recognized" (Heffron, McCauley, & Sovacool, 2015; McCauley, Heffron, Stephan, & Jenkins, 2013). Using the concept of energy justice, this contribution relies on legal, policy, and empirical arguments to demonstrate the need to take into account the rights of Indigenous peoples in the transition to renewable energy. In addition, the empirical study of different cases aims to provide a more nuanced analysis of the specific situation and needs of Arctic Indigenous communities to ensure their fair inclusion in the transition to renewable energy in the Arctic. For this purpose, the analysis draws upon different case studies and relies on various sources, including legal texts as well as personal observations and fieldwork experiences. Ultimately, the interpretation of the case studies contributes to improve knowledge about the progress and challenges raised by the energy transition to renewable energy in the Arctic context and further theorise how to ensure a transition that is sustainable and just for all.

In what follows, the chapter first sets out to consider the legal and policy background that links the transition to renewable energy to the social dimension of sustainability and the rights of Indigenous peoples. In this regard, Section I highlights the central tenets, rights and duties attached to the social commitments made by states and business to promote renewable energy as a means to ensure a sustainable and just transition at the international level and highlights the specific status and rights of Indigenous peoples in this context. Next, Section II outlines the broader contexts and corresponding patterns of renewable energy development in the Arctic. Against this backdrop, Section III examines the actual state of play of the energy transition and its impact on Indigenous peoples in the Arctic based on illustrative examples. For this purpose, the contribution examines more particularly the situation of Arctic communities living in Canada, Alaska, Russia, and in the Nordic countries of Norway and Sweden.

A just energy transition for Indigenous peoples: Policy and legal background

With a few exceptions, a global consensus now acknowledges that the transition from fossil fuel to renewable energy is urgent and critical to address climate change, ensure sustainable development and fulfil human rights.

Indeed, access to affordable and reliable energy is central to the fulfilment of basic human rights including that of life, food, shelter, health, and education (Tully, 2006; Walker, 2015). In addition, the importance of providing clean energy is increasingly recognised as a baseline to ensure sustainable development. In this regard, the United Nations Sustainable Development Goals (SDG), which provides a set of policy commitments to ensure sustainable development, recognises the central importance of affordable and clean energy "to nearly every major challenge and opportunity the world faces today" (United Nations, 2019). This is more particularly articulated in the implementation of SDG 7, which highlight the need to ensure "access to affordable, reliable, and sustainable energy," and SDG 13, on urgent action to combat climate change that equally requires a transition to a new energy paradigm in which "flow" (renewable) energies are used and consumption is lower and more efficient (UN General Assembly, 2015; Council of Europe, 2011). With the specific mention of sustainable energy in the Agenda for sustainable development, the UN member states therefore recognise the global and urgent need for a transition to renewable energy, in order to achieve sustainable development and address the world's major twenty-first century challenges (Calzadilla & Mauger, 2018, p. 237).

Paradoxically, however, the transition to renewable energy can also undermine the achievement of the SDGs and adversely affect human rights. For example, research has evidenced the negative impacts of large-scale renewable projects such as hydropower projects and biofuels production on local communities and Indigenous peoples as well as the environment (Zehner, 2012; Cambou,2020). Yet, the issues concerning the social risks and justice concerns of renewable energy continue to be overlooked in the debate concerning the sustainable transition (Calzadilla & Mauger, 2018, p. 235). In this context research also evidences "how major national energy policy and planning documents concentrate almost exclusively on energy technologies, while social considerations tend to be narrowly economic, focusing on energy prices, jobs and, to some extent, energy access" (Miller & Richter, 2014, p. 76) and therefore "rarely incorporates justice dimensions" (Healy & Barry, 2017, p. 452). Thus, although it is argued that "'social sustainability' has emerged as a theme in its own right" (Dempsey, Bramley, Power, & Brown, 2011), the fact remains that the environmental and economic dimensions of sustainability often override social and justice concerns in practice. As a response, an emerging transdisciplinary scholarship on the concept of energy justice is currently being shaped, which highlights the need to integrate "justice" as a meta concept in the narrative of the energy transition (Heffron et al., 2015; Sovacool, Burke, Baker, Kotikalapudi, & Wlokas, 2017; McCauley et al., 2013).

Beyond its academic foundation, the call for energy justice is also grounded in policy and legal discourses concerning sustainable development and the increasing importance attached to the fact that sustainable development relies upon the golden thread that connects economic development with environmental sustainability and social equity. Furthermore, with the adoption of the 2030 Agenda for sustainable development, it is now unequivocally recognised that sustainable development is anchored in human rights: The new Agenda is explicitly aimed to "realize the human rights of all" (preamble) and emphasises "the responsibilities of all States... to respect, protect and promote human rights and fundamental freedoms for all" (UN General Assembly, 2015, para. 19). Additionally, UN Member States have also pledged to ensure "no one will be left behind" and to "endeavour to reach the furthest behind first" (UN General Assembly, 2015, para. 4). In practice, these pledges reflect the fundamental human rights principles of non-discrimination and equality. Applied to the context of the energy transition, these pledges also mean that the transition to renewable energy "require the consideration of social justice in terms of fairness in access and allocation of resources and technologies" (McCauley et al., 2019). They require that the transition to renewable energy, which is grounded in the principle of equity, must be accessible and beneficial for all. In this context, it is also recognised that "addressing the critical needs of indigenous peoples, the elderly, people with disabilities and other marginalized groups is the best way of ensuring that no one is left behind by the 2030 Agenda for Sustainable Development" (UN Department of Economic and Social Affairs, 2017).

Since Indigenous peoples are often among the poorest and most marginalised of the world's population, the UN Agenda for sustainable development calls for specific attention to their situation (UN General Assembly, 2015, para. 23). In this respect, although Indigenous peoples have faced challenges to seeing their interests reflected in the Agenda, the framework highlights that "People who are vulnerable must be empowered" and reflects the needs of Indigenous peoples in several extents in the framework (i.e., four indictors specifically mention Indigenous peoples) (UN General Assembly, 2015). In addition, the UN Permanent Forum on Indigenous Issues has advised that the "the 2030 Agenda must be implemented in full accordance with the United Nations Declaration on the Rights of Indigenous Peoples and other relevant international agreements" (Permanent Forum on Indigenous Issues, 2016). In the context of the energy transition, this means that states must ensure the rights of Indigenous peoples, including their rights to self-determination, to land and natural resources and to maintain their traditional culture. The importance of the right of Indigenous peoples to participate in the energy transition is also enshrined in numerous international instruments, which highlights the duty of states to consult and cooperate in good faith with Indigenous peoples in order to obtain their free and informed consent prior to the approval of any project affecting their traditional lands or territories and other resources (UNDRIP, 2007, art. 32; ILO Convention 169, arts. 6–7). The right of Indigenous peoples to participation based on the principle of FPIC is a core element of their right to self-determination, which must be respected and protected by all states and respected by corporate actors, especially when energy projects affect their lands and resources and the maintenance of their culture (OHCHR, 2011).

The need to pay attention to the issues of energy justice and to the right of Indigenous peoples to participate in the energy transition is a global issue that finds specific relevance across the Arctic. There are about 500,000 Indigenous peoples living in the Arctic, which represents 10% of the total population in the region. While there is great diversity of cultural, histori-cal economic and political situations, many communities share legacies of colonisation, the impacts of new industrial processes, and exposure to the effects of pollution and climate change. In addition, many Indigenous com-munities in the Arctic still lack access to clean and affordable energy or more singularly face the adverse impacts of renewable energy projects. For example, while many Inuit communities struggle to get affordable and clean energy in Alaska, Canada and Greenland, the Sámi reindeer communities who live in Norway, Sweden and Finland face the adverse impacts of renew-able energy projects on their traditional territory. Against this background, the following sections explore more closely the situation of Indigenous peo-ples living in Canada, Alaska, Russia and the Nordic countries in order to identify how Indigenous peoples living in the Arctic participate and benefit from the energy transition.

Renewable energy in the Arctic: Context and patterns

Before we discuss the respective case studies, it is important first to outline the broader contexts and corresponding patterns of renewable energy devel-opment. The Canadian North (territorial and provincial North), Alaska, and the Russian North and Siberia share a significant number of challenges and historical parallels, especially compared to the Nordic countries. First, size and population density matter: Russia and Canada are the two largest countries in the world and, Alaska is the largest state in the United States. To put things into perspective, the state of Alaska is about 1.5 times larger than the territory of Norway, Sweden, and Finland combined, but has less than 5% of the population of those countries combined. Canada and Russia have comparable territorial size to small population densities. This has a significant bearing on energy infrastructure, particularly electrical grid build-out. Whereas the Nordic countries are almost entirely grid con-nected on the mainland, most of the territory of Alaska, eastern Siberia, the Canadian territories of Nunavut and the Northwest Territories, and much the provincial North of Canada is not. The cost of erecting and maintain-ing transmission lines between communities of less than a thousand people that are separated by many hundreds of kilometers is prohibitively expen-sive. In this regard, it is not surprising that outside the Nordic world, the Circumpolar North has a very high number of off-grid, diesel communities.

Second, legal recognition of Indigenous rights, especially land rights, have significant bearing on Indigenous participation in the electricity sector. In Canada, Alaska, and Russia, Indigenous peoples enjoy formally recognised Indigenous land rights not found in the Nordic countries, as well as forms of self-government. Importantly, all three countries are federal countries in contrast to their Nordic Circumpolar neighbors, which on the one hand complicates energy decision-making as a result of jurisdictional complexities, but on the other hand fosters a political culture of multilevel decision-making. The latter creates political space for Indigenous participation in energy decision-making and ownership. Finally, socio-economic gap between Indigenous peoples and the broader society in Canada, Alaska, and Russia is significantly wider that of the Nordic countries. This is a result of a mix of historic factors beyond the scope of this paper. However, what is important to understand is that many Indigenous communities in Canada, Alaska, and Russia confront much higher unemployment rates, lower household incomes, high social pathologies, and lower formal educational attainment than the larger societies. In all three countries, many remote Indigenous communities confront the "heat-or-eat" dilemma as energy costs are often significantly higher than urban centers, even though household incomes of Indigenous communities is often less than half of their urban counterparts. These factors need to be kept in mind as they shape the discourse of energy transition and the role of and impact on Indigenous peoples, especially compared to their Nordic counterparts.

Notwithstanding important commonalities among Canada, Alaska, and Russia, these three cases also have important differences that shape and constrain Indigenous opportunities to achieve greater energy justice. In Canada and Russia, for instance, large-scale public utilities play significant roles compared to Alaska, which has a mix of private, cooperative, and parastatal and no state-wide public utility. In Canada and Russia, there are significant subsidies and cross-subsidies in many provincial and territorial jurisdictions, but Alaska is almost a pure market system in its electrical utility sector, with subsidies limited to consumers downstream and essentially none available for independent power producers. In Alaska, Native Alaskan communities own and operate their own utilities through parastatal or cooperative models; in Canada and Russia, there are no Indigenous owned, generation and distribution utilities, although there are many Indigenous community-owned renewable energy generation projects, operating effectively as small-scale independent or community-owned producers who sell energy to a larger utility. In Alaska, in the absence of a state-wide grid, there is little opportunity for Indigenous communities to participate as utility scale, independent power producers, with the exception of the Railbelt, which has a transmission line that connects Anchorage to Fairbanks. In Canada, however, there are significant opportunities for Indigenous communities to participate in utility-scale power projects, as many of these communities are connected to large-regional or the North American grid.

On the surface, the Nordic welfare state model would appear to apply to the electricity sector, including a shared Nordic grid; however, there are important differences between Norway and Sweden. In Norway, as a result of late modern state-building foreign capital and local municipalities played a much larger role in the development of the electricity sector. To this day, there are many more, small municipal and cooperative electricity utilities in Norway compared to Sweden, and a number of these small communities have large Indigenous populations, the Sámi. While the question appears clear when it comes to the conflict between reindeer herding communities and large-scale regional and national energy projects, it is not so clear to what extent the Sámi people in Norway exert control and decision-making power through local parastatal and cooperative electricity institutions. Questions of energy justice are never as clear-cut in the real world as they are in the theory.

The participation of Indigenous peoples in the Arctic energy transition

The next section of the paper demonstrates how these contexts fundamentally shape patterns of participation by Indigenous communities in current energy transitions. We suggest that four distinct patterns have emerged in the contemporary Circumpolar North: (a) utility-scale projects on Indigenous traditional territory that are opposed by local communities and/or present a threat to indigenous rights ; (b) utility-scale projects (typically projects in the order of 10MW or larger and is sold to a large-scale grid-connected utility) in which Indigenous peoples are owners or co-owners of the project; (c) community-scale projects driven by national and regional governments or public utilities; (d) community-scale projects driven by communities (typically less than 1 MW and intended to serve the immediate needs of a small community). Canada has examples of all four patterns; Alaska (b) and (d); Russia (a) and (c); and the Nordic countries mainly a), but with some qualifications as it relates to (b) and (d). It should be apparent, that pattern a) does meets least, if at all, any notions of energy justice for Indigenous peoples; pattern c) meets better energy justice goals; but, clearly patterns (b) and (d) are the most robust in achieving energy justice, as Indigenous peoples are able to participate in energy decision-making on a much more level playing field and accrue the benefits that come from those projects to improve the well-being of their communities. Below, this paper outlines specific examples in each of Canada, Alaska, Russia, and the Nordic countries.

Canada

Canada may have the most complex story among these cases. For most of the twenty-first century, the legacy of renewable energy development was dominated by utility-scale hydro-electric projects, for industrialisation and resource development, domestic and international exports markets, and residential consumers. Indeed, hydro-electricity accounts for 60% of Canada's total electrical production. The Kemano Hydro facility in the northern British Columbia, for instance, was built for the sole purpose of producing electricity for an aluminium smelter in the town of Kitimat. Island Falls Hydro facility was built in the northern Saskatchewan to power mining and smelting in northern Manitoba. The James Bay project in northern Quebec was built not only to serve domestic needs in the province, but also as an export, primarily to New York State and the Eastern seaboard of the United States. All of the facilities generated and, in some cases, continue to generate significant controversies with Indigenous peoples. The large-scale projects in Manitoba and Quebec, in particular, flood tens of thousands of hectares of Indigenous traditional lands, changing hunting and fishing patterns, transportation, among other things. Although many of these legacy hydro projects have produced impact benefit agreements and, in the historic case of James Bay, land claims settlements and Indigenous self-government, there are several notable cases still under litigation.

The policy environment in Canada today is radically different from the preceding century, with three new developments shaping the energy transition toward a low-carbon future: climate change and Canada's commitments to the Paris Agreement; the United Nations sustainable development goals; and, seeking Reconciliation between the Crown and Canada's Indigenous peoples. The developments may and often reinforce positively energy transitions and energy justice for Indigenous people; they can also be at odds. This becomes evident when we contrast national and provincial policy imperatives with those of local Indigenous communities.

At the national level, Canada already has a high level of electricity produced by non-emitting sources at approximately 80% (65% from renewable energy, most of which is hydro-electricity, and 15% from nuclear energy). Nevertheless, Canada intends to meet the goals of reaching 90% by 2030 and eventually 100% of electricity from non-emitting sources. These goals align with both Paris commitments, as well as the UN Sustainable Development Goals. Meeting these goals will require adding tens of thousands of MW of capacity through large, utility-scale energy projects. This has the potential to conflict with Indigenous land rights, if not done properly. Done properly, meeting this goal opens up new avenues for reconciliation with Indigenous peoples through "steel in the ground" by creating equity ownership opportunities and long-term, sustainable revenue and employment streams.

At the local level of Indigenous communities, a very different picture emerges: clean energy generation and its benefits are distributed highly unevenly; so, too, are the costs. Canada has more than 280 off-grid diesel communities representing 200,000 Canadians, 144 of these communities are Indigenous; in other words, Indigenous people constitute about 5% of the Canadian population, but represent 50% of the off-grid diesel communities. Remote communities typically experience the highest energy costs, even though they are also the poorest in Canada. In Nunavut, for instance, communities have "electricity costs subsidized at 28.4 cents per kilowatt hour (kW.h) for the first 1,000 kW.h per month in the winter or the first 700 kW.h per month in the summer. If electricity consumption exceeds these amounts, the consumer is charged the unsubsidized rate" (National Energy Board, 2018). The unsubsidised rates vary by community; for instance, the rate is 56.7 cents per kWh in Iqaluit and 112.3 cents per kWh in Kugaaruk (National Energy Board, 2018). By comparison, the rate for residents in Saskatchewan in 14.5 cents per kWh. If the energy justice is to be achieved for Indigenous communities, then meeting clean energy sustainable development goals means addressing not only the "clean" part of renewable energy, but also increasing energy access and energy security. Renewable energy almost by definition means local energy, potentially reducing both transportation costs and risks of interrupted supply chains. Renewable energy holds the promise of the reducing energy costs for communities with the lowest household incomes. Locally produced energy also provides opportunities for increased, sustainable employment, thus raising household incomes, if only modestly across a community as a whole.

Given the above context, it is no surprise that all four patterns outlined in the previous section of the paper are found in Canada. We briefly provide examples of each below.

Meeting Paris commitments and the Clean Energy SDG continues to lead to contestation between utility-scale renewable energy and Indigenous rights. From a greenhouse emissions perspective, hydro-electric power is one the best options for renewable energy development. Upfront capital costs are generally very high, but operational costs per kWh are among the lowest over the long term. Importantly, in contrast to wind and solar, hydro-electricity has the huge advantage of providing dispatchable generation and with capacity factors typically five times greater than solar and at least double that of wind, even with advances in wind technology. Hydro-electricity also has the advantage of providing flood control to help mitigate due to extreme weather events, which are becoming increasingly common due to climate change.

Notwithstanding the advantages, in Canada, the conventional wisdom is that mega-projects and even large-scale, conventional hydro-electric dam development is largely a relic of the past. There are many excellent sites for hydro development in Canada that are technically accessible but politically inaccessible. Environmental requirements, Indigenous rights claims, and public social licence requirements, among other things, make such projects very difficult to advance. However, two notable cases prove the exception to the rule: Site C Dam in northern British Columbia and Muskrat Falls in Labrador. Both of the cases have resulted in political and legal challenges from Indigenous communities affected by the development of these projects. (For details on these cases, please read Buhmann et al., Chapter 9.)

Utility-scale electricity projects, however, are also points of reconciliation rather than conflict. Increasingly, Indigenous peoples in Canada are seeking opportunities to be independent power producers. Two innovative developments, among many to date, deserve special note: First Nations Power

Authority and Wataynikaneyap Power. The First Nations Power Authority was created through agreement of the Government of Saskatchewan and the Federation of Sovereign Indigenous Nations in the province of Saskatchewan to facilitate the participation of Indigenous communities in the electricity sector and to provide a formalised pathway for working the publicly-owned crown utility, SaskPower. In its own words, "First Nations Power Authority (FNPA) was established as a not-for-profit to create a landscape favorable to Indigenous inclusion in the power sector. Created in 2011, FNPA was mandated to facilitate the development of First Nations led power projects and promote Indigenous participation in procurement opportunities with the crown utility in Saskatchewan, SaskPower" (FNPA, 2019). The Province of Saskatchewan has a public goal of obtaining 50% of its electricity from renewable energy by 2030 from a starting point of 24%. SaskPower is providing set-asides to FNPA to procure electricity from greenfield renewable energy projects that are majority owned by First Nations communities. Although FNPA started with smaller pilot renewable energy projects, it has now successfully tendered two utility-scale solar projects, 10 MW that will be owned by Cowessess First Nation and another 10 MW jointly-owned by Starblanket Cree Nation and George Gordon First Nation. First Nation Power Authority also has 20 MW set aside for turning waste flare gas into recovered energy for electricity production. The goal for FNPA is to continue to increase the independent power producer opportunities for Indigenous communities not only in Saskatchewan, but across Canada, and FNPA has been very active with community energy planning with First Nation in the neighboring province of Alberta.

The construction of transmission lines is often a source of contention with Indigenous communities. Wataynikaneyap Power provides a counter example. In Northern Ontario, there are 32 electrically remote communities, 25 of which are First Nations communities. A provincially commissioned study determined that creating a transmission line to connect 16 of these communities would create a net savings in diesel of more than $1 billion over the next 40 years. In addition, it would provide an opportunity to upgrade an outdated transmission line connecting those services with 5 additional communities. Whereas other Northern Ontario grid connected communities experienced approximately 3 outages per year, those on the outdated line experienced 14 unplanned outages per year. This is particularly hard for households that rely on electricity for their primary or even supplemental heating during winter months that commonly go to 40 degree below zero. The solution was to invest in a $1.6 billion transmission line connecting 17 off-grid communities and, in the process, upgrading the transmission to the 5 grid-connected communities. The new transmission line be owned and operated as Wataynikaneyap Power LP with the 22 communities as the owners of Wataynikaneyap Power LP. The creating of this transmission utility achieves not only Clean Energy SD goals by taking 17 communities off diesel, but also achieves energy justice by providing reliable and affordable power and by providing Indigenous ownership and control over electricity transmission infrastructure.

Canada is not only making progress toward great energy justice in electricity generation and transmission, but also in local Indigenous clean energy generation. The federal government has launched to programs to eventually reduced or eliminate the reliance on diesel generators in Canada's more than 280 off-grid communities, half of which are Indigenous. To that end, the Government of Canada has committed more than $800 million for remote and Indigenous communities to develop local renewable energy generation projects, primarily for electricity, but also for biomass heating. Rather than the Government of Canada dictating to communities their energy choices, communities apply for funding for energy projects that they would like to see developed for and owned by their communities. Projects are assessed on the financial and technical feasibility by reviewers external to the Ministry. Critically, proposals must include a capacity-building plan to operate and maintain community energy projects. In addition, the Government of Canada has invested in training programs such as the 2020 Catalyst Program, which provides training to Indigenous participants so they have the capacity to lead community energy planning in their home communities. Although the program is in its beginning two years as of 2019, already dozens of projects have been funded by the federal government, including a $40 million wind turbine project to be sited near the Arctic community of Inuvik, Northwest Territories, lead and owned by the Gwich'in.

Alaska

Alaska is unique within the Circumpolar North. It is the story of more than a half a century of Indigenous owned and operated electric utilities. No part of the State of Alaska is connected to the North American electrical grid. And, outside of the Railbelt, the only major regional transmission grid connection Fairbanks to Anchorage and its local environs, and of the immediate Juneau area, no communities in Alaska are connected to an electrical grid. Moreover, there is not state-wide public or private electric utility. There are approximately 90 electric utilities spanning private, parastatal, and cooperative. In Alaska, about 45% of the electricity comes from natural gas, 25% from hydro, 15% from oil and the other 15% of the electric capacity comes from predominantly a mixture of coal, wind and biomass (EIA, 2018). Due to their dependence on fossil fuel, scarce transportation infrastructure and great distances between communities, Alaskans also pay nearly double the national average for energy, which cause or exacerbate social, economic and environmental issues at the local level.

However, there is also significant momentum in community-level renewable energy project in Alaska. Among the numerous success stories, two electricity stories exemplify Native Alaskan ownership and self-determination: the Alaska Village Electric Cooperative (AVEC) and Cook Inlet Region Incorporated (CIRI) Fire Island project.

AVEC was created in 1967 as Native Alaskan owned electricity cooperative. Since 1967, it has grown to include 57 Native Alaskan communities and is the largest electricity cooperative in the world by territory. The cooperative model has been highly successful and uses economies of scale and pooling of resources to operate an efficient and reliable utility in one of the most challenging environments in the world. One of the major challenges of remote communities is affording technicians who can operate generation systems, especially highly variable renewable energy systems. One of the solutions is a circuit rider program in which technicians are shared among communities and visit communities for repair and maintenance on a regular and scheduled basis. AVEC has been very successful in integrating renewable energy generation in small communities with diesel generation, especially wind power and it has been on the forefront of cutting-edge technologies such as the adoption of smart meters and prepaid card for using electricity in the home. These measures have resulted in greater energy literacy among many residents of AVEC served communities, certainly higher than that of most larger urban centers in North America or Europe. All of this has been achieved in the closest example market system with little operating subsidies, certainly compared to Canada, Russia, or the Nordic countries.

CIRI is the regional cooperation, formed out of the 1971 Alaska Native Claims Settlement Act. CIRI represents Native Alaskan beneficiaries in the Cook Inlet areas, including individuals of Athabascan, Southeast Indian, Inupiat, Yup'ik, Alutiiq/Sugpiaq and Aleut/Unangax descent. CIRI is an example of a global leader in Indigenous independent power production. CIRI constructed, owns, and operates a 17.6 MW wind farm on Fire Island just outside of Anchorage. It is the first independent power producer in the region and has a 25-year power purchase agreement with the Anchorage-based utility, Chugach Electric Association. The eleven turbines have the capacity to provide electricity to the equivalent of 7,000 homes. CIRI not only is the owner of this project, but also has equity positions in renewable energy projects in the continental United States, thousands of kilometres away. CIRI demonstrates that Indigenous peoples and communities can be owners and drivers in the global energy transition.

With its increasing number of renewable energy facilities, the transition to renewable energy is underway in Alaska. Yet, it remains to be seen whether this development will be sufficient to achieve the goal to generate 50% of Alaska's electricity from renewable energy before 2025 and enhance energy security and sustainability for all Indigenous communities in the region.

Russia

Indigenous communities in the Russian North and Siberia share many of the level energy security challenges with similar communities in Canada and Alaska; they also share the same challenges that large-scale renewable energy projects present to traditional ways of life, especially hydro-electric development, with Indigenous communities in the Nordic countries.

However, it may come as a surprise to many, but Russia is also home to important advances toward greater energy justice and meeting the Clean Energy SDG, notably in the Sakha Republic (Yakutia), located in eastern Siberia, which constitute about one-fifth of the entire territory of the Russian Federation. These advances include renewable projects and assessment processes, which are contributing to create greater energy security in Northern Indigenous communities and also accommodate Indigenous needs surrounding utility-scale greenfield energy development.

Among these advances, projects concerning Russian energy supply are notable. Although higher energy costs present significant challenges to Indigenous households, energy supply may be an even more threat to achieving energy security. Many remote communities in the Russian Arctic and Siberia depend on shipping supply lines to provide basic necessities, especially fuel for heating and for diesel electric generation. The supply lines are complex and require shipping along the Northern Sea Route and along major river systems such as the Lena. Weather conditions can result in significant delays and even missed shipping seasons, placing communities in highly vulnerable positions. To address both costs and vulnerabilities, Sakhaenergo (Sakhaenerg is a subsidiary of PJSC Yakutskenergo, part of the RAO Energy Systems of the East holding company) has taken the initiative to build and deploy significant solar arrays in some of the most remote communities that Sakhaenergo serves. In fact, the largest solar facility in the Arctic is located in Batagay, located in northern Sakha Republic (Yakutia) over 1 MW in installed capacity. Beyond the ground-breaking Batagay facility, Sakhaenergo has deployed solar arrays in 16 communities with an additional total installed capacity of 1.47 MW (Bellini, 2019). Although these efforts clearly address Indigenous energy security, the decisions to build and operate are top-down, not bottom-up, and there is no ownership by local communities. This case therefore fits pattern "c" of Indigenous participation, due to the lack of agency provided for Indigenous peoples in the development of the project.

The Sakha Republic (Yakutia) has also served as a pilot region within the Russian Federation for undertaking environmental impact assessment and innovative social impact assessments, the latter through a formal legally established process called Ethnological Expertise, which is overseen by an Expert Committee, whose composition includes appointments made by republic level Association of the Small-Numbered Peoples of Yakutia. To date, this process has reviewed applications for industrial resource development on approximately 50 projects (three of which include, in 2012, the proposed Cancun Hydro-electric station on the Timpton River in the southern part of the republic and, in 2013, two transmission lines, also in southern Yakutia). The Expert Committee found that RusHydro, the proponent of the Cancun HPS project, had not assessed the impact on traditional livelihoods and deemed the proponent's assessment as inadequate and grossly underestimated appropriate levels of compensation owed to Indigenous communities affected by the proposed project. It also placed as further requirement "a full environmental and social impact assessment be conducted during the construction of the Cancun HPS" (Sleptsov, 2015, p. 94). The proponent revised its submission; it was reassessed, and was approved by the Expert Committee. In the case of the transmission lines, the Expert Committee approved the applications. Although important challenges remain in the Russian Arctic to consolidate Indigenous peoples'rights, it is suggested that this level of Indigenous agency in approval processes represents a progress in institution building to ensure energy justice for Indigenous peoples (Sleptsov, 2015).

Norway and Sweden

In comparison with other Arctic states, the Nordic countries have already emerged to be leaders in renewable energy and efficiency, with all countries aiming to be virtually fossil free by 2050 (Sovacool, 2017). However the energy transition is not without its own cost and has raised a number of social issues, more specifically in relation to the rights of the Sámi Indigenous people, who live as a minority group across the territory of Norway, Sweden, Finland and the Kola Peninsula of Russia.

Paradoxically, the transition to renewable energy to promote sustainable development is among the issues that jeopardise the sustainability of Sámi livelihoods due to the adverse impacts of renewable energy projects on their traditional land and resources. Historically, it was the impact of large-scale hydroelectric projects that spurred the first opened conflicts between the Nordic governments and the Sámi people in the energy context. The transition to hydroelectric energy was grounded in the policy needs for increasing energy supply and ensures self-sufficiency in order to guarantee the industrialisation and modernisation of the nation states. However, the development of these projects had also for adverse effect to damage the traditional land and resources of Sámi communities. In the beginning of the twenty century, the construction of the first stage of the great reservoir at Suorva in northern Sweden, signed for example the beginning of an extensive encroachment on reindeer husbandry and Sámi culture (Össbo & Lantto, 2011, p. 330). In Norway, the conflict of the Alta-Kautokeino hydropower project, which began in the 1970s is also notorious and has served to expose the lack of recognition of the rights of the Sámi as an Indigenous people in the energy context and more generally in mainstream national politics (Howitt, 2002, p. 280; Paine, 1982).

Today, the rights of the Sámi people have also become an issue of controversy in the context of the transition to renewable energy. This is more particularly due to the development of wind energy projects and their adverse effects on reindeer herding. The establishment of wind turbines has multiple effects on reindeer herding (Skarin, 2016). It can affect the migration of reindeer, disturb grazing and calving areas and increase the workload and cost of activities for reindeer herders. With the multiplication of onshore wind energy projects, Sámi reindeer herding communities also face the cumulative effects of projects, which are magnified by the impacts of other industrial development processes. Several UN reports have criticised the impact of these projects on the traditional livelihoods of the Sámi (UN report & Anaya, 2011; UN Report & Tauli-Corpuz, 2016). Renewable projects have also led to an increasing number of court cases highlighting the responsibility of companies to mitigate the impact of wind energy in the purpose to allow co-existence between reindeer herding and the development of wind energy projects (Cambou, 2020).

Among the numerous projects that are currently being opposed by Sámi reindeer communities, two large renewable energy projects stand out due to their large-scale impacts at the local level: the Fossen project in Norway and the Markbygden project in Sweden.

With its 273 planned turbines, the Fossen project constitutes one of the largest onshore wind projects in Europe. The Norwegian Petroleum and Energy Ministry gave its approval for the construction of the project in 2017 despite important concerns, protests and court petitions from the southern reindeer herding communities, who have traditionally used the land where the project is being built. The project was also authorised in defiance to the request of the UN Committee on the Elimination of Racial Discrimination, which asked Norway to temporarily suspend the project in order to examine a complaint from Sámi communities (CERD, in Reuters, 2018). In the view of the Norwegian ministry, the fact that the project has successfully met all legal national requirements and also survived a number of court petition justified the authorisation of the project. The project is currently underway and should be completed by 2020.

In Sweden, the development of renewable energy has also raised important issue in relation to reindeer herding. Sweden aspires to be a leader in the transition to renewable energy and for this purpose has set the target to produce 100% renewable electricity by 2040. In this context, wind energy represents an important tool to achieve this goal, which has materialised with an increase of wind energy projects being established in the country. In the northern most counties, where the Sámi reindeer herders' territories is located, the amount of wind turbines has increased from approximately 1,000 to more than 3,000 turbines between 2008 and 2018. While this development is supported by a growing number of small and medium projects, the Markbygden project represents by itself 1001 turbines, which is approximately 7% of the share of total electricity generation in Sweden. Despite several court cases against the project and protests from Sámi reindeer communities, the project has been authorised by the government and is planned to be completed by 2021.

The Fossen and Markbygden cases epitomise pattern (a) of participation by Indigenous communities in renewable projects in so far as these projects create conflict and present a threat to Indigenous rights. In order to ensure a just transition, the current challenge for the Nordic governments therefore remains to ensure that the transition to renewable energy does not adversely affect the Sámi Indigenous people by undermining their right to maintain their traditional culture.

Conclusion

The global energy transition toward a low carbon future means significant investment in renewable energy. But green energy is only good energy if generated properly and for the benefits of all people. In this regard, Indigenous peoples confront two main challenges, one from new, large-scale renewable energy projects that have the potential to undermine Indigenous land rights and traditional livelihoods; and the struggle to secure affordable and reliable energy in their communities. At the very core is the question of how can energy justice be achieved for Indigenous peoples in this transition, a transition that fundamentally affects the rights and well-being of Indigenous peoples living in the Arctic.

In this analysis, four patterns of participation by Indigenous communities in the current energy transitions were identified: (a) utility-scale projects on Indigenous traditional territory that are opposed by local communities and present a threat to Indigenous rights; (b) utility-scale projects in which Indigenous people are owners or co-owners of the project; (c) community-scale projects driven by national and regional governments or public utilities; (d) community-scale projects driven by communities. At the outset, this analysis demonstrates that many Indigenous peoples continue to encounter pattern (a) in all Arctic states as a result of their opposition and/or lack of adequate participation in the development of large-scale renewable development projects, which do not meet their community demands in energy and encroach on their rights This situation is more particularly illustrated in the case of the Site C Dam and Muskrat Falls in Canada, the Fossen project in Norway and the Markbygden project in Sweden. However, we also have seen cases where projects meet indigenous energy security needs but where participation is not adequately accommodated (e.g. Sakha Yakutia) notably because these projects fall short of enabling Indigenous agency—ownership and control—over local energy generation.

In contrast, this contribution also identified patterns (b) and (d) of participation that are more adequate in achieving energy justice for Indigenous peoples. When Indigenous peoples act as owner or drivers of renewable energy projects and are therefore able to participate in energy decision-making and accrue the benefits that come from those projects for improving the well-being of their communities, there is more certainty to achieve the sustainable developments goals and ensure energy justice for Indigenous communities. Yet, the development of such participation pattern remains insufficient today. This paper recommends that Arctic states at both the national and regional levels proactively develop robust institutional arrangements and policy environments that promote solutions that follow patterns (b) and (d) in accordance with the rights of Indigenous peoples, in particular their right to maintain their traditional culture. As this paper has shown, patterns (b) and (d) are not merely possible but are already in practice. If the Arctic states strive to promote energy justice, they will in so doing provide a global model for other regions on our planet to emulate as countries seek appropriate pathways to achieve the Clean Energy Sustainable Development Goal and contribute to combat climate change and its impacts.

References

Bellini, E. (2019, February 26). Floating solar reaches new territories. *PV Magazine.* https://www.pv-magazine.com/2019/02/26/floating-solar-reaches-new-territories/

Buhmann, K., Bowles, P., Cambou, D., & Skjervedal, A. S. H. (2020). Towards socially sustainable renewable energy projects through involvement of local communities: Normative aspects and practices on the ground. In D. Natcher & T. Koivurova (Eds.), *Renewable economies in the Arctic: A state of knowledge.* Routledge.

Calzadilla, P. V., & Mauger, R. (2018). The UN's new sustainable development agenda and renewable energy: The challenge to reach SDG7 while achieving energy justice. *Journal of Energy & Natural Resources Law, 36*(2), 233–54. doi: https://doi.org/10.1080/02646811.2017.1377951.

Cambou, D. (2020). Uncovering Injustices in the Green Transition: Sámi rights in the development of wind energy in Sweden. *Arctic Review on Law and Politics, 11,* 310–333. doi: https://doir.org/10.23865/arctic.v11.2293.

Council of Europe. (2011). Energy supply and energy efficiency at local and regional level: promoting energy transition (S. Orlova, Rapp.). Council of Europe, 21st session, CG (21)11, Russian Federation.

Dempsey, N., Bramley, G., Power, S., & Brown, C. (2011). The social dimension of sustainable development: Defining urban social sustainability. *Sustainable Development, 19*(September), 289–300. doi: https://doi.org/10.1002/sd.417.

Energy Information Administration (EIA). (2018). "Alaska. State Profile and Energy Estimates". *US Energy Information Administration.* https://www.eia.gov/state/analysis.php?sid=AK#71

First Nations Power Authority (FNPA). (2019). About FNPA. *FNPA.* https://fnpa.ca/about-fnpa/

Healy, N., & Barry, J. (2017). Politicizing energy justice and energy system transitions: Fossil fuel divestment and a "Just Transition". *Energy Policy, 108*(September), 451–59. doi: https://doi.org/10.1016/j.enpol.2017.06.014.

Heffron, R. J., McCauley, D., & Sovacool, B. K. (2015). Resolving society's energy trilemma through the energy justice metric. *Energy Policy, 87*(December), 168–76. doi: https://doi.org/10.1016/j.enpol.2015.08.033.

Howitt, R. (2002). *Rethinking Resource management: Justice, sustainability and indigenous people.* Routledge.

Lempinen, H., & Cambou, D. (2018). Energy security in the barents region: A focus on societal perspectives. In K. Hossain & D. Cambou (Eds.), *Society, environment and human security in the Arctic barents region* (pp. 118–133).

McCauley, D. A., Heffron, R. J., Stephan, H., & Jenkins, K. (2013). Advancing energy justice: The triumvirate of tenets. *International Energy Law Review, 32*(3), 107–10.

McCauley, D., Ramasar, V., Heffron, R. J., Sovacool, B. K., Mebratu, D., & Mundaca, L. (2019). energy justice in the transition to low carbon energy systems: Exploring key themes in interdisciplinary research. *Applied Energy, 233–234*(January), 916–21. doi: https://doi.org/10.1016/j.apenergy.2018.10.005.

Miller, C. A., & Richter, J. (2014). Social planning for energy transitions. *Current Sustainable/Renewable Energy Reports, 1*(3), 77–84. doi: https://doi.org/10.1007/s40518-014-0010-9.

National Energy Board. (2018). *Market Snapshot: Overcoming the challenges of powering Canada's off-grid communities.* Canada Energy Regulator. https://www.neb-one.gc.ca/nrg/ntgrtd/mrkt/snpsht/2018/10-01-1cndffgrdcmmnts-eng.html?=undefined&wbdisable=true

OHCHR. (2011). *Guiding Principles on Business and Human Rights: Implementing the United Nations "Protect, Respect and Remedy" Framework* (HR/PUB/11/04). OHCHR. https://www.ohchr.org/Documents/Publications/GuidingPrinciplesBusinessHR_EN.pdf.

Össbo, Å, & Lantto, P. (2011). Colonial tutelage and industrial colonialism: Reindeer husbandry and early 20th-century hydroelectric development in Sweden. *Scandinavian Journal of History, 36*(3), 324–48. doi: https://doi.org/10.1080/03468755.2011.580077.

Paine, R. (1982). *Dam a River, damn a people?: Saami.* International Work Group for Indigenous Affairs.

Permanent Forum on Indigenous Issues. (2016). *Substantive Inputs to the High Level Political Forum on Indigenous Issues: Thematic Review of the 2030 Agenda for Sustainable Development.* UN Sustainable Development.

Reuters. (2018, December 21). Norway to build wind farm despite concerns of reindeer herders. *Reuters.* https://www.reuters.com/article/us-norway-wind-un-idUSKCN1OK1WS

SDWG. (2019). *Arctic Sustainable Energy Futures: The Arctic Community Energy Planning and Implementation (ACEPI) Toolkit Final Report.* Arctic Council. https://oaarchive.arctic-council.org/handle/11374/2301.

Sleptsov, A. (2015). Ethnological expertise in Yakutia: Regional experience of legal regulation and enforcement. *Northern Review, 39*, 88–97.

Sovacool, B. K. (2017). Contestation, contingency, and justice in the Nordic low-carbon energy transition. *Energy Policy, 102*(March), 569–82. doi: https://doi.org/10.1016/j.enpol.2016.12.045.

Sovacool, B. K., Burke, M., Baker, L., Kotikalapudi, C. K., & Wlokas, H. (2017). New frontiers and conceptual frameworks for energy justice. *Energy Policy, 105*(June), 677–91. doi: https://doi.org/10.1016/j.enpol.2017.03.005.

Skarin, A. et al. (2016). *Renar och Vindkraft II - Vindkraft i drift och effekter på renar och renskötsel.* Swedish University of Agricultural Sciences

Tully, S. (2006). The contribution of human rights to universal energy access. *Northwestern Journal of Human Rights, 4*(3), 518.

UN Department of Economic and Social Affairs. (2017). *Reaching the furthest behind first is the answer to leaving no one behind.* United Nations Department of Economic and Social Affairs (UN DESA). https://www.un.org/development/desa/en/news/sustainable/reaching-furthest-behind.html.

UN Report & Anaya, J. (2011a). *Report of the Special Rapporteur on the Rights of Indigenous Peoples Extractive Industries Operating within or near Indigenous Territories* (UN Doc. A/HRC/18/35). United Nations.

UN Report & Anaya, J. (2011b). *The situation of the Sami people in the Sápmi region of Norway, Sweden and Finland* (UN Doc. A/HRC/18/35/Add.2). United Nations.

UN Report & Tauli-Corpuz, V. (2016). *Special Rapporteur on the rights of indigenous peoples on the human rights situation of the Sámi people in the Sápmi region of Norway, Sweden and Finland* (UN Doc. A/HRC/33/42/Add.3). United Nations.

UN General Assembly. (2015). *Transforming our world: The 2030 agenda for sustainable development.* United Nations. https://doi.org/10.1891/9780826190123.ap02.

United Nations. (2019). *Sustainable development goal 7.* United Nations. https://www.un.org/sustainabledevelopment/energy/.

Walker, G. (2015). The right to energy: Meaning, specification and the politics of definition. *L'Europe En Formation, 378*(4), 26–38.

Zehner, O. (2012). *Green illusions: The dirty secrets of clean energy and the future of environmentalism.* University of Nebraska Press.

11 Adding value from marketing origin of food from the Arctic Norway

Bjørg Helen Nøstvold, Ingrid Kvalvik,
& Morten Heide

Introduction

The Norwegian Arctic is important as a food-producing region, producing food from both marine and terrestrial resources. Fisheries, aquaculture, agriculture (both meat and vegetables) and reindeer herding, and processing of these products are important for sustaining the economy, population, and culture in many local communities. Within Arctic Norway, there is considerable variation in products and markets. Fisheries and aquaculture are large-scale and export-oriented, but also include a high number of companies producing for the local and national market. Agriculture is quite marginal compared to farming in more favorable locations further south but produce both for the big national cooperatives and for smaller companies. All food sectors are producing both commodities and high-value niche products for local, regional, and national markets.

Natural conditions like harsh weather and cold climate, combined with a high general cost level in Norway makes it impossible to compete with low cost, high volume, mass-produced products from lower cost countries. Hence, agriculture is heavily subsidised, and it is a public debate that too much seafood is exported unprocessed as a result of high production costs in Norway (Iversen et al., 2016). At the same time, there is a growing market for authentic and niche food products. According to Luceri, Latusi, and Zerbini (2016), emphasising the geographic origin of a product is one way of ensuring authenticity, quality, and food safety to consumers, and hence, a potential marketing strategy for added value for producers. In an Arctic context then, a very relevant question is whether there is a potential for attaining added value by exploiting the image of the Arctic region in marketing of food products. Studies by van Ittersum, Candel, and Meulenberg (2003) and Kuznesof, Tregear, and Moxey (1997) show that a condition for region of origin branding to positively influence consumers' perception of food products is that the product and its qualities coincide with what the consumer associate and perceives as authentic for the region. Successful use of Arctic origin in the branding of food products from the Arctic requires knowledge about what consumers perceive as "authentic Arctic food products" and the

DOI: 10.4324/9781003172406-11

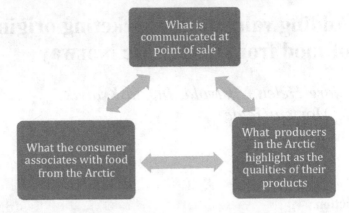

Figure 11.1 Consumer and producer perceptions, and the marketing of Arctic food.

characteristics of such products. In this study, we examine Norwegian consumers' perceptions of Arctic food because it is reasonable to assume that
the perception of consumers within the Arctic region differs from that of the
consumers outside the region. Consequently, the marketing strategies might
need to be different if you target a local/regional market or a market outside
the region. The consumers are therefore split into two groups: local consumers, that are consumers who live within the Arctic region themselves, and
consumers from the southeastern part of Norway around the capital where
the population density is the highest. Knowledge about consumer preferences can provide input about targeted marketing, which can increase sales
and added value of Arctic food products. To complement the analysis, we
also examine the producers' view of their own products and to what degree
and how they take advantage of "Arctic qualities" and the Arctic origin in
their marketing. By doing this, the study reveals to what degree producers'
marketing strategies correspond with the consumers' perception of Arctic
food, illustrated in Figure 11.1. Based on this, we discuss the potential for
adding value by exploiting the Arctic origin in the marketing of food from
Norwegian Arctic, in the region and outside the region.

Background: Artic Norway – geographical, cultural, and economic setting

With a total mainland area of 324,000 km² and 5.3 million inhabitants,
Norway is one of the least densely populated countries in Europe. The
built-up area, including roads, amounts to only about 2%, and a further
3% is cultivated agricultural land. Norway has a 2,500 km coastline, with
90,000 km² of sea within the baseline and an exclusive economic zone of
788,000 km² of the mainland, providing plenty of suitable space for aquaculture production and good fishing grounds. In addition, the zone outside
the Svalbard archipelago is important for fishing (see Figure 11.2).

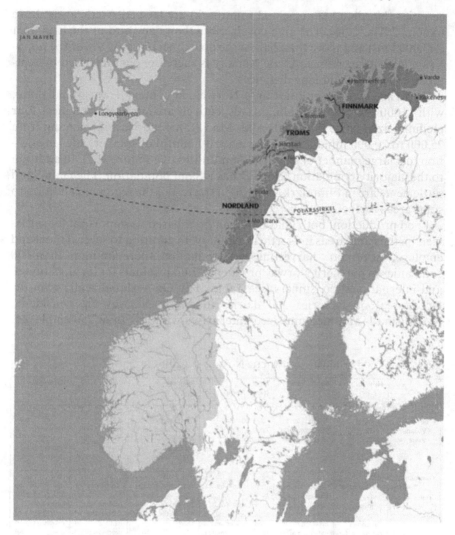

Figure 11.2 Arctic Norway.

Based on climatic parameters only a tiny fraction of mainland Norway is in the Arctic. If defined by the Arctic Circle (located at a latitude of 66°33′ North), about half of Norway is however considered the Arctic. This includes the two northernmost counties (Troms and Finnmark) and approximately half of the third northernmost county (Nordland). Politically, however, all the three counties are considered part of Arctic Norway or Northern Norway, and the Norwegian High North policy is directed at all three counties (in addition to Svalbard, the archipelago in the Barents Sea between 74° and 81° North, which have a small population involved in mining, research, and tourism). Similarly, food policies directed at the three northernmost counties (Nordland, Troms, and Finnmark) are termed Arctic agriculture (Anon., 2011a, 2016). The area covers more than one-third of the land area (113,000 km²) and more than half of the coastline, but only 9% of the population (486,000 persons). Arctic Norway is therefore characterised by scattered settlements and long distances.

There are two bigger cities (in a Norwegian context Tromsø and Bodø, with 75,000 and 51,000 inhabitants respectively are relatively big cities), four medium-sized cities (Harstad, Alta, Mo i Rana and Narvik, ranging from 25,000 to 20,000 inhabitants), and a good handful of smaller cities in addition to a large number of smaller communities (see Figure 11.2). In addition to the majority population, Norway has a Sami population. The main Sami settlements are in Finnmark, but also in the rest of the region, in addition to some areas further south.

Food production, both land-based and sea-based, is important for settlement in the rural areas in Northern Norway. In addition to several thousand employed in primary harvesting and production, there are more than 600 companies in northern Norway producing food products for local, national, and international consumers (see Table 11.1). On a global scale, even the largest seafood factories and dairies in Arctic Norway are considered SMEs (small- and medium-sized enterprises with less than 250 employees

Table 11.1 Food production in Arctic Norway (in 2016)

	Fisheries	Aquaculture	Meat (cattle, sheep, and pigs)	Reindeer	Dairy products	Horticulture
Annual Production - Volume - Value	-354 000 tons - 12 billion NOK	-511 000 tons -29 billion NOK (eksport value)	-186 000 sheep -28 000 cattle 104 000 pigs -2,15 billion NOK	-82 000 animals -1 800 tons meat -1 500 tons by-products -500 mill NOK	-151 000 mill litres cow milk -7 mill litres goat milk -1,43 billion NOK	-8 400 tons potatoes, and some vegetables -100 mill NOK
Companies - Primary 1roduction - Secondary production	-97 purchaser and producers -app. 50 companies producing for national market	-39 companies dominated by 4 -35 smaller companies producing for national market	-dominated by 1 large national cooperative (with 3 slaughterhouses), 1 medium sized and a few smaller slaughterhouses - in addition, 10 processing companies and 60 micro processors	-7 slaughterhouses -15 processing companies, 5 with 95% of the market share	-dominated by 1 large national cooperative (with 6 processing plants), 20 small scale producers	-dominated by 1 large national cooperative, and one national and one local company. Also several small companies (farm level)
Employment - Direct - Supporting inudstries	-4 500 fishers -3 000 to 7 000 in processing depending on season	-2 200 -5 500	-3000 farmers -1000 in processing	-950 reindeer owners -350 in slaughter and processing companies -with all processing companies, app 1000 employees	-1000 farmers -500 to 600 in the processing industry	-500 farmers -200 in production companies, and seasonal workers
Markets	95% exported	95% exported	regional and national market	Mainly sold in the north, some to southern Norway	Local, regional and national market	HoReCa important market for local and regional market, also national

and annual turnover up to EUR 50 million) (EC, 2016). Still, in a North Norwegian context they are quite large and contribute with significant revenue to the Norwegian national, regional, and local economy. Most companies are however more in the scale of micro to small enterprises (1–50 employees).

The government emphasises the need for increased and sustainable food production and the need to take advantage of national resources (Anon., 2016). It is a government aim to have food production in the whole country. As mentioned, both fisheries and aquaculture are big scale and export-oriented, while agriculture is rather small scale and heavily subsidised. Agricultural production in Norway, and Northern Norway, in particular, is taking place under challenging climatic and geographical conditions, specific programs are therefore created to support it. Increased food production in Arctic Norway is a part of the government's high north policy (Anon., 2011a), and Arctic food production has benefitted from the government's increasing interest in local food, amongst other voiced in several reports to the Parliament (Anon., 2011b, 2015). The government in June 2019 announced that it would double its economic support to "Arctic agriculture" from 2 to 4 mill NOK annually (Anon., 2019). As such, the government is taking action to promote food production in the Arctic. Even though food security is not a pressing matter in Arctic Norway compared to some other Arctic areas, these initiatives contribute to the Norwegian government's fulfillment of the obligation to realise the UN Sustainable Development Goal #2. In particular, providing knowledge, financial services and opportunities for value-adding of local food are important initiatives in a Norwegian Arctic context.

It is the Gulf Stream that allows agriculture production in Northern Norway. You will not find the same production on the same latitude in the other Arctic states. Even though the production is lower than further south (yield per acre), the unique climatic northern growth conditions with long days and low temperatures, provide some advantages for food and fodder products. For instance, broccoli contains more C-vitamin, while carrots and turnips get a sweeter taste, and the cloudberries get bigger (Johansen, Uleberg, & Mølmann, 2018). The cold climate also gives very little diseases, and hence the use of pesticides and medicines is very low and animal welfare high (Johansen, Uleberg, & Mølmann, 2018). Future climate changes also offer possibilities for northern horticulture production through prolonged growing seasons in Northern Norway. As for seafood, the most productive areas and biggest fishing grounds are in the north (in the Lofoten area and in the Barents Sea), and Nordland and Troms are the two biggest counties for aquaculture production and are still increasing. Primary processing is taking place in all three counties, while the number of secondary processing companies is decreasing as one moves northwards.

Primary processing includes slaughtering, sorting and cleaning, processing, and packing of a raw material product from primary production. It can also include adding some conservatives (salt and sugar), drying, freezing, etc. Secondary production is used to denote processing that significantly changes the product by adding other ingredients and produce a consumer-ready and value-added product (Elde et al., 2018). The differences in the number of secondary production companies are not surprising, given the much smaller population in the northernmost county, and hence also a smaller local market.

There are about 3,000 farmers in Northern Norway and 950 reindeer owners. In addition, about 2,000 persons are working in the processing industry. The agricultural processing industry in Norway is dominated by large cooperatives owned by the farmers, like TINE (dairy) and Nortura (meat). In addition, there are some other large companies and many small, local enterprises. The number of small companies in the agricultural sector has been increasing rapidly the last decades due to increasing demands for locally produced special food products (Norsk mat, 2018), and today there are about 500 local food producers in Northern Norway.[1]

As for fisheries, there are approximately 4,500 people directly employed in fishing and another 3–7,000 working in the fish processing industry throughout the year in Northern Norway. In aquaculture, 2,000 are employed in the main industry and 5,500 in supporting industries. Despite declining employment in the industry as a whole, production is higher than ever before. There are 100 landing stations/primary processing companies and 40 aquaculture companies. There are about 80 companies producing for the national market. The fisheries' catch varies considerably from year to year. In 2015, about 94,000 tons were landed in the north, which constitutes 42% of the total landings in Norway (SSB, 2016). About half of the production of farmed salmon is taking place in Northern Norway, with 517,000 tons in first-hand sales from Northern Norway in 2015. The economic significance of the farming industry, for Norway as a whole, by far exceeds the traditional fisheries. Only a small amount of Norwegian seafood goes to domestic consumption (<5%). Still, Norway is self-sufficient in seafood, as well as in dairy products.

Despite growing importance for the economy and employment, food production in the Arctic region is associated with some challenges. Not only are the food producers faced with challenging environmental conditions, the region also suffers from poor and/or costly infrastructure, limited entrepreneurial capacity, relatively small local markets and long-distance to bigger markets. Most food producers in the Norwegian Arctic region are as mentioned small and micro-sized companies, with less than 5 employees. It is a known challenge for such small companies to manage all aspects of running a healthy company (Dawar & Frost, 1999), and according to Statistics Norway approximately 70% of SMEs fail within five years (SSB, 2019). Those who survive though, have a four time increase in the number of employees, proving the importance of these SMEs to rural Arctic communities (SMB Norge, 2018).

Competence is the key to success. The government supports regional Centers of Expertise for local food production, where one center is covering the Arctic region. The aim is to build competence in small- and medium-sized local food producers and increase the value of the Arctic food. In addition, food producers have established various types of network corporations to overcome or mitigate some of the challenges of being a small producer. The networks are often partly publicly funded through Innovation Norway. Most networks are geographically linked (like *Nordlandsmat* or *Vesterålsmat)*, while others are limited to a product type like *Lofotlam* (sheep meat) and *Arktisk kje* (goat meat). In addition to knowledge exchange and support, the main areas of cooperation are distribution, sales, and marketing (Natcher et al., 2019). There has been an initiative to develop a regional food label, "Northern Norwegian food label" (Nordnorsk Matmerke) (Nordnorsk Landbruksråd, 2018). The aim was to increase value creation through better visibility of Arctic food products in stores and give the products an advantage in the market. Originally, only the agricultural sector was involved, but it was a stated goal to include seafood from the region. The idea was to develop an area of origin label for products based on raw material and processed in the Norwegian Arctic region. Despite the lack of realisation of the label, so far, this shows that many actors in the sector have a belief in the branding of Arctic origin. Still, it does not seem to exist any thorough analysis neither of the use of Arctic origin in branding of food products or of which attributes or characteristics has the highest potential in branding of Arctic food.

Research question and analytical approach

Internationally there is increasing consumer interest in the origin of food, traceability, and in supporting local companies (Feldmann & Hamm, 2015; Hingley, Boone, & Haley, 2010). Especially for high-quality food products the use of regional origin is often successfully incorporated in the branding (Luceri et al., 2016; Trognon, 1998), conveying a message of authenticity and tradition (Ilbery & Kneafsey, 1998) and health and sensor properties (Bryla, 2015). Displaying origin can, therefore, function as a risk reduction, reassuring consumers about where the products come from and how they have been produced (Luceri et al., 2016). In some cases, it can even replace the need for building a brand name (van Ittersum et al., 2003).

There is also an expressed desire to reduce the number of steps the food goes through from production to plate (Murtagh, 2015). It is in the context of this consumer interest and concern, and the governmental support one sees growth in the production of local food products. Several studies show consumers' preference for food from their own region, i.e. local food (see, for instance, Aprile, Caputo, & Nayga, 2016; Feldmann & Hamm, 2015; Ilbery, Morris, Buller, Maye, & Kneafsey, 2005). Local food is getting increasingly important also for the Norwegian consumer, reflected in increasing sales numbers in the retail sector. In 2013, local foods had twice the increase in turnover compared to other food products. Norwegians also show an increased willingness to pay for food from their own region (Ipsos MMI, 2014). Local and regional foods however also have potential for distribution toward national consumers. Extending the local or regional market to a national market offers a larger segment of high-end customers with a high willingness to pay. The marketing strategy might, however, need to be different. If local food producers aim to extend their market outside their region, should the products be marketed the same way as in the local market? And should they brand their product based on origin toward consumers outside the region? To answer this, one needs to know how consumers outside the region perceive food from this region, compared to the local consumer.

Within marketing, branding, i.e. the marketing practice of creating a name, symbol or design or a combination of these that identifies and differentiates a product from other products, is one way to achieve higher willingness to pay, increase market penetration and/or develop new markets (Keller, 1998; Murphy, 1998). A successful brand is known to give better prices, more loyal consumers, and strengthen the company's reputation (Keller, 1998; Murphy, 1988). Even established companies can increase turnover and/or profit through small adjustments in how and what they communicate about their products and to whom. Branding is, however, expensive, time-consuming, and a risky process, and exploiting existing positive associations toward the product, company or region of origin could be beneficial. A brand is more than a name, and to create a value of a brand name a set of associations need to follow (Aaker, 1991). As shown by Trognon (1998), if a region already is perceived positively based on subjective associations, these associations might be successfully used as part of the branding if they are transferable to the regional product. A regional product is defined as a product whose quality and/or fame can be attributed to its region of origin and which is marketed using the name of the region of origin (van Ittersum et al., 2003). Examples of products marketed as regional specialties are abundant, like Parma ham, Champagne, and Florida Oranges. The regional image can have an influence on how the product is perceived, and according to van Ittersum et al. (2003) the place of origin can also have very little or even a negative influence on the evaluation of the product. To avoid this, it is eminent that the particular product is perceived as authentic for the region and that the consumers should perceive the region to be suited for production of the product (Kuznesof et al., 1997; van Ittersum et al., 2003. The qualities associated with these products may or may not be scientifically documented, like more vitamins, sweeter, etc. (Johansen, Uleberg, & Mølmann, 2018).

The question raised here is therefore, whether and how can the Arctic origin be used to create a higher perceived value for consumers in the Arctic region and in a national market? From this follows two groups of questions: (1) What do consumers perceive as authentic "Arctic food"? What are the characteristics of Arctic food? And is there a difference in these perceptions between consumers living in or outside the Arctic region? (2) Based on the above, is there a potential for promoting food from the Arctic more effectively by utilising certain characteristics, in the region and outside the region, and for different food products? What are the producers doing today, and is there a potential for increased value creation by adapting marketing communication? Arctic food is defined as food based on raw materials from the Arctic which is also being processed in the Arctic region.

Material and methods

A survey was conducted of 472 consumers, 246 in Arctic Norway and 226 in Southern Norway, in the area around the capital. The consumers were all more than 18 years of age. The aim was to reveal what type of food consumers think of as Arctic food and what qualities they ascribe to Arctic food. By having respondents both within and outside the Arctic region, we would be able to test our assumption about different perceptions of Artic food between those living in the Arctic region and consumers in the south. Attaining knowledge about how consumers perceive food originating from their own region (local Arctic food) compared to consumers living outside of the region (Arctic food), will provide producers with important knowledge about Arctic origin as a marketing opportunity in the differed markets.

The survey first had a section aiming to capture association about species and characteristics. This part started with an open-ended question, followed by closed questions. This is to make sure also the consumers' own thoughts are captured without any influence from the survey's predetermined characteristics (Altintzoglou, Sone, Voldnes, Nøstvold, & Sogn-Grundvåg, 2018). In the open question, the consumers were asked to state what were the three first words that came to mind when thinking about Arctic food. In the closed questions the consumers were asked to state how much they associated predefined factors to food from the Arctic region. The scale was running from 1 to 7, where 1 was denoting "not at all" and 7 "very much." Brands of the Arctic food products for sale in the Tromsø groceries and specialist stores were used to identify relevant factors.

A similar survey was developed targeting food producers in Arctic Norway. The survey was sent to 86 companies, of which 29 companies responded. The aim of this survey was to gather information about how the producers market their products, what they think are their main selling points, and to see if these correspond to the consumers' perception of Arctic food. The producers were also asked an open question first; "What words would you use for marketing the Arctic origin of your food products?" They were then asked, "Do you use any of the following words in your marketing" and were asked to rate their answers from 1 "not at all" to 7 "to a large degree." The words were based on the same information as the consumer factors.

The rank order of importance of factors that describe and add value to Arctic food was tested using Friedman's related samples test (a Shapiro–Wilk test showed that normality was violated). Differences in importance between individual factors were determined using Wilcoxon related samples test for pairwise comparison.

Findings: Consumer perceptions, producer perceptions, and the Arctic food marketing

In the following first the results from the consumer study are presented followed by the results from the producer survey. The results will then be discussed in the next chapter.

What is the Arctic food – species and characteristics from the consumer perspective?

The survey started with an open question, asking the consumers to state three words that describe what they associate with Arctic food. Some chose to describe species, some wrote different characteristics, and some both. More than 900 different words were given. The words were sorted, categorised (species or characteristics), and translated. Some choices were made when sorting a word into "species" or "characteristics," for instance "reindeer meat," "reindeer steak," and "reindeer" were all translated into "reindeer" as a species. When a specific traditional dish was stated, for instance "boknafisk" (cured cod), it was translated and categorised as a characteristic of Arctic food and not categorised under the main ingredient species (in this case cod). The reason for this is that typical traditional dishes and food preparation is more related to culinary heritage and tradition than species.

The species "berries" consists of several types of berries. Cloudberries were mentioned the most. Also, lingonberries and blueberries were mentioned. As they all are wild berries, they were generalised into "berries." Two respondents mentioned strawberries, but these were not included as they are cultured. Words like flavor, flavorful, rich flavor, tastes good, jummy, and good taste have all been included in the characteristic "tasty." And as mentioned, specific traditional dishes and types of cooking like tørrfisk (dried cod or other whitefish), lutefisk (dried whitefish treated with lye), klippfish (salted and dried whitefish), biljo (a specific reindeer stew/soup) and the generalisation "traditional food" have all been included in the characteristic "tradition."

For the respondents in Northern Norway, we see that the species "reindeer" is most stated (100) when thinking about Arctic food. With the exception of "berries" (45), four types of seafood; "fish" (81), "cod" (49), "king crab" (32), and "salmon" (23) are the species most stated after "reindeer," as illustrated in the word cloud in Figure 11.3.

The respondents in the South of Norway clearly state "fish" (110) the most, while "reindeer" comes second (60). As in Northern Norway, "Cod"

Figure 11.3 Species suggested by Northern consumers.

(71) and "King crab" (60) are the seafood species most stated. "Berries" (all types of wild berries) (28) are the fifth most mentioned, while "salmon" (22) comes sixth, as illustrated in Figure 11.4.

The consumers who rather thought of quality characteristics when thinking about Arctic food gave a wide range of descriptions. More

Figure 11.4 Species suggested by Southern consumers.

Figure 11.5 Quality characteristics mentioned by Northern and Southern consumers.

than 130 different words were used. Sorted, categorised, and translated, Figure 11.5 shows the main characteristics identified by the northern and southern respondents, respectively.

The results show that consumers mainly think of different seafood species as Arctic, this range from cod, whale, and saith to seaweed and urchin (in total 600). Terrestrial species can be grouped into two; farmed and wild. Reindeer, game and wild berries are mentioned more often (a total of 266) than farmed vegetables (26) and animals (24). This applies to both northern and southern consumers.

For the northern consumers the characteristics "pure" (32), "local" (31) and "tasty" (30) were mentioned the most. "Tradition" (25), "quality" (22)

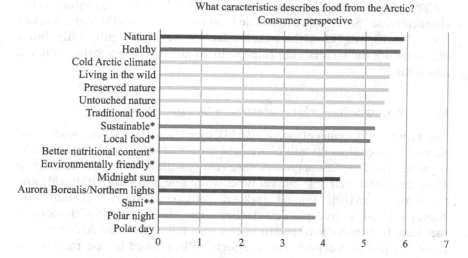

Figure 11.6 What characteristics describe food from the Arctic? Consumers' perspective.

and "fresh" (20) were also frequently used. For the southern consumers "tradition" (46) was by far the most mentioned characteristics, both the word tradition itself and description of traditional cooking and traditional dishes. "Healthy" (20) was the second most mentioned characteristic, followed by "fresh" (14) and "pure" (14).

In the next part of the survey, the consumers were asked to think about food from the Arctic and then rate different characteristics for how much they associate them with Arctic food (Figure 11.6). Different colors and nuance of color indicate significant differences between traits so that the dark green, lighter green and pale green are significantly different. The characteristics are then grouped into three groups; green, blue and, red where green is considered to be the most important factor, blue medium importance, and red considered a group of not important factors.

The analysis shows that the characteristics the consumers mostly use to describe food from the Arctic are "natural," followed by "healthy." "Cold Arctic climate," "living in the wild," "preserved nature," "untouched nature," and "traditional food" are all in the group of the third most associated characteristics of food from the Arctic. "Traditional food," the characteristics that scored the highest in the open question also scores high in the rating. Being pale green, it is in the third strongest group of associations with Arctic food.

The data was split between the responses from the Northern and Southern consumers. The consumers in Northern Norway gave a significantly higher score to "local food," "environmentally friendly," "sustainability," and "better nutritional qualities." These factors are marked with * in Figure 11.6. The respondents from Southern Norway associate the characteristic "Sami" more to Arctic food than do the northerners, marked with ** in Figure 11.6. Otherwise the factors were not significantly different, meaning the factors were rated with equal importance between north and south.

Producer perspective—what characterises the Arctic food

The producer respondents were all SMEs with less than 20 employees and turnover lower than 40 mill NOK, and 50% had less than 5 mill NOK in turnover. 19 of the 29 companies were selling within the Arctic region, and 10 of these had their sole market here, while one also sells nationwide and two moves from their regional market directly on to export. Only four companies sell both within- and outside the region and exports. 15 of the companies have their main market within Norway, but outside the Arctic region, three of these also export their products. When asked to rate their major challenges, market-related issues were chosen by 18 companies.

Parallel to the question to the consumer about what they perceive as characteristic of Arctic food, the producers were asked in an open question which characteristics they would use to promote their products. As shown in the word cloud (Figure 11.7), "pure" (25), "Arctic" (23), "flavor" (23), and "natural" (19) are the words used the most. Within "flavor" words like "flavorful," "strong flavor," "nice flavors," and "natural flavors" are all included. A wide variety of words describing the nature where the products are grown or raised are used by several producers, like "the fish lives in pure cold waters" or "luscious grassland with natural herbs and an eternal spring under snow-covered mountains."

Figure 11.7 Producers' choice of characteristics in communication of their products.

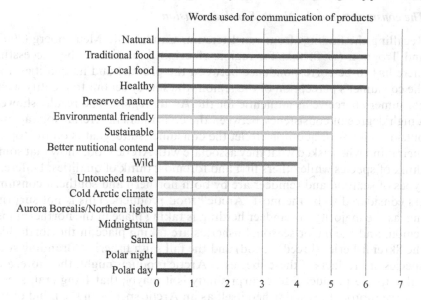

Figure 11.8 Words used for communication of products.

For the producers, the open question was followed by closed questions where they were asked to rate to what extent they used different words in communicating about their own products. The results are presented using a median because the producers are very polarised in their answers. The answers are not checked for significance due to the low number of respondents. Figure 11.8 presents the scores where the different colors represent the median scores for the factors. The group that is used the most includes the words "natural," "traditional," "local," and "healthy." It must be mentioned that the closed question in the questionnaire did not include "z," "Arctic origin," and "purity of taste." There was no difference in the choice of words between producers aiming for regional or national markets. Between the open and closed questions tradition and healthy are hardly mentioned in the open question but gains a high score in the closed question.

Discussion

The question raised in this study is whether there is a potential for adding value by exploiting the regional origin in the marketing of food products from the Arctic, within and outside the region. Where one expected to find many differences between perceptions of northern and southern consumers, results show a high degree of correspondence. Below the similarities and differences will be further explored, starting with the consumer dimension, followed by the consumer-producer dimension.

The consumers and the North-South dimension

Recalling the findings from van Ittersum, Candel, and Meulenberg (2003) and Trognon (1998) that for region of origin branding to be successful, there has to be correspondence between the product and its qualities and the consumer's perception of the region, the survey set out to identify what consumers perceive as authentic for the Arctic region. The results showed a high degree of consistency between the consumers in Arctic Norway and outside. The open questions to let the consumer state what is on the top of their mind when asked what they associate with Arctic food, show that some think of species while others first and foremost think of qualities. Different types of seafood and reindeer are by both northern and southern consumers considered to be the most "Arctic" food products. This is not surprising, as the majority of reindeer herding is taking place in the northernmost region, and the major seasonal fisheries are taking place in the north, like the Skrei fisheries (breeding cod) and the fall saith fisheries. Branding such species or products of these species as Arctic products might, therefore, add value to the product. More surprisingly is it maybe that King crab seems to have thoroughly established itself as an Arctic species in the mind of the Norwegian consumers, both in the north and in the south. King crab is an Arctic crab species, but it is a relatively new species in Norway, first being commercially harvested in the early 2000s. This illustrates that it is not only traditional species or products that can be marketed as Arctic and that it is possible for new species and products to be established as Arctic. For instance, the more exotic species like urchin and seaweed, which were mentioned by a few consumers, may have potential as "Arctic food products" in the future.

Besides reindeer and seafood, berries were among the top five species mentioned, both in the north and south. Thus, it seems that wild-caught, harvested and gathered products that are perceived to be "wild" is more authentic Arctic food than agricultural products like potatoes, vegetables, and farmed animals. This despite efforts being done within collective branding of the origin of such species, for instance Målselvnepe (turnip from Målselv), Lofotlam (lamb from Lofoten).

When it comes to qualities, the northern consumers have strong associations with "purity," "taste" (always in a positive sense), "high quality," and "health." The southern consumers to a large degree mention the same characteristics, but here "traditional food production or dishes" are mentioned the most. Most of the characteristics are corresponding with current food trends focusing on health-related qualities; less processing, purity of production, purity of flavor, and no additives (Asioli et al., 2017). For the producers in Arctic Norway, the bias toward traditional dishes or methods can be both positive and negative, depending on what type of products they want to promote. The study indicates that a highly modern and/or novel product should maybe not be branded as an Arctic product. The food trend of convenience is apparently not put in context with Arctic food, and if a producer should use Arctic origin in the branding of such a product it might only be successfully applied if the raw material used in the product is associated with Arctic food (i.e. some type of ready meal based on reindeer). On the other hand, this also might mean a higher acceptance for products that are not so convenient.

It is interesting that favorable taste (sweet berries, pure seafood, and strong and rich taste) are amongst the first thoughts in the mind of the consumer, both in the south and the north. Elements of taste were not included in the part where the consumer was to rate different set qualities because this was not identified in the marketing of the products in shop that were used as a basis for the survey. In the set qualities characteristics related to nature and the environment came out strong. It seems that food production in the Arctic is considered to take place on the terms of nature (being untouched, preserved, wild and natural). These aspects could also be connected to health, as a clean and natural production often is considered using fewer preservatives, fewer additives, etc. Within these most important characteristics, there is no difference between perception in the south and in the north. This implies that a stronger focus on healthy and natural aspect could be a useful marketing strategy, both regionally and nationally.

The survey shows that the consumers in the north to a much higher degree think food from their own region is more "sustainable," more "environmentally friendly" and has "better nutritional content." This could be used more actively by producers aiming for a regional consumer. The potential of exploiting documented better nutritional content or more favorable taste should be utilised. In general, the results indicate that it is not necessary to develop a different branding strategy for the southern and northern markets. Though, if you are including less important elements like better nutritional content or sustainability when communicating about your product this will resonate more with the northern consumers. Commonly used phrases like "grown under the midnight sun" or "the northern lights" are generally not considered to ascribe value neither for northern nor southern consumers. One could assume that this would resonate more with foreign tourists and such phrases are observed on many food products that seem to be targeting tourists in the region. Using such characteristics might however result in the product losing out on local customers. Whether it fits the expectations of tourists and to what degree national customers disregard these products remains to be studied.

Luceri et al.'s (2016) argument is that emphasising the geographic origin of a product has a positive effect on consumers because it ensures authenticity, quality, and food safety. The survey shows that consumers, both in northern and southern Norway have associations of especially quality and food safety to food from the Arctic. In the open-ended questions pure, "healthy" and "fresh" are dominating. In the closed-ended question "health" and "natural" are significantly higher rated than other characteristics. The results indicate that not only does the Arctic origin give the assurance about food safety (natural and fresh) as emphasised by Luceri, Latusi, and Zerbini (2016), it also incorporates several health-related elements (pure, clean, and fresh) that is important according to Bryla (2015). Finally, the Arctic origin includes authenticity and traditional elements (tradition and real), characteristics identified by Ilbery and Kneafsey (1998) to be important. This shows that food from the Arctic has high potential for benefitting on Arctic origin branding.

The consumer – producer dimension

In the open-ended question, the producers taking part in the survey list natural, flavor, pure and healthy as the characteristics they would use in communication of their own food products. The closed question also enhances the importance of tradition and environmental concerns, like preserving nature, environmentally friendly and sustainability. Natural, tradition and healthy corresponds with the most chosen criteria by the consumers. Traditional was specially mentioned in the south. This shows that the producers to a high degree understand and do take advantage of the Arctic origin of their products by using words that resonate with that of the consumers. It might, however, seem that the producers are less aware of the potential in exploiting the associations to the wild, cold Arctic climate and untouched nature in their branding. This might be because the producers in the survey do not think that their products are based on such "wild material." But it might also be that the producers are unaware of the strong association the consumer has between food from the Arctic and the nature associated with the Arctic region. For instance, reindeer producers might focus on their natural and traditional production rather than the association to a (more or less) wild animal, the wild nature where they live, and the cold Arctic climate creating this nature, and hence product. The producers and consumers give the same rating to the elements like midnight sun, northern lights and polar night. These are not very important when it comes either to association or communication of Arctic food.

Conclusion

The study has explored Arctic origin as a marketing opportunity for food producers in Arctic Norway. The assumption was that it is possible to achieve added value based on Arctic origin in strategic marketing, but that to do this, it is vital to know what consumers perceive as Arctic qualities. The study finds a high correspondence between northern and southern consumers on what they associate with the Arctic food products both in regard to species; reindeer, seafood, and game, and in regard of characteristics; natural, pure, healthy, tasty, and traditions. This means that a producer often can use the same strategy in their communication with their northern and southern customers. They can also extend their markets, from regional to national without extensive change in their branding, unless they have a strong focus on a local food image.

Another important finding is that many of the characteristics associated with Arctic food are in line with current international food trends, like health, natural, authenticity, and tradition. Within these associations, we, therefore, see many opportunities for branding products based on their Arctic origin. The associations of the Arctic as something wild, pure, and untouched, thus using "Made in the Arctic" in strategic marketing would correspond with these traits.

Acknowledgment

This work is part of the Arctic Council project "Arctic as a Food Producing Region" and the Norwegian part has been supported by the Norwegian Ministry of Foreign Affairs through the Arktis2030 programme, project number NOR15/0117.

Note

1. Numbers from The Centre of Expertise for local food production North (Kompetansenettverket for lokal mat Nord).

References

Aaker, D. (1991). The Negative Attraction Effect? a Study of the Attraction Effect Under Judgment and Choice. *NA - Advances in Consumer Research*, 18, eds. Rebecca H. Holman and Michael R. Solomon, Provo, UT: Association for Consumer Research, 462–469.

Altintzoglou, T., Sone, I., Voldnes, G., Nøstvold, B., & Sogn-Grundvåg, G. (2018). Hybrid surveys: A method for the effective use of open-ended questions in quantitative food choice surveys. *Journal of International Food & Agribusiness Marketing*, *30*(1), 49–60. doi: 10.1080/08974438.2017.1382422.

Anon. (2011a). Report to the Parliament no. 7 (2011–2012). *The High North* (Nordområdene), Oslo, November 2011.

Anon. (2011b). Report to the Parliament no. 9 (2011–2012) *Welcome to the table* (Velkommen til bords), Oslo, December 2011.

Anon. (2015). Report to the Parliament no. 31 (2014–2015) *The farm as a resource – the market as the goal* (Garden som ressurs – marknaden som mål), Oslo, June 2015.

Anon. (2016). Report to the Parliament no. 11 (2016–2017) *Change and Development. A future-oriented agricultural production* (Endring og utvikling. En fremtidsrettet jordbruksproduksjon), Oslo, December 2016.

Anon. (2019). Vil produsere mer mat i Nord-Norge https://www.regjeringen.no/no/aktuelt/vil-produsere-mer-mat-i-nord-norge/id2662274/. Dato: 28.06.2019

Aprile, M. C., Caputo, V., & Nayga, R. M. (2016). Consumers' preferences and attitudes toward local food products. *Journal of Food Products Marketing*, *22*(1), 19–42.

Asioli, D., Aschemann-Witzel, J., Caputo, V., Vecchio, R., Annunziata, A., Næs, T., & Varela, P. (2017). Making sense of the "clean label" trends: A review of consumer food choice behavior and discussion of industry implications. *Food Research International, 99*, 58–71.

Bryla, P. (2015). The role of appeals to tradition in origin food marketing: A survey among Polish consumers. *Appetite, 91*, 302–310.

Dawar, N., & Frost, T. (1999). Competing with giants: Survival strategies for local companies in emerging markets. *Harvard Business Review, 77*, 119–132.

EC. (2016). User Guide to the SME definition. European Commission (Ref. Ares (2016) 956541 – 24/02/2016). *European Union*. https://ec.europa.eu/regional_policy/sources/conferences/state-aid/sme/smedefinitionguide_en.pdf

Elde, S., Kvalvik, I., Nøstvold, B., Rødbotten, R., Dalmannsdottir, S., Halland, H. … Sidorova, D. (2018). *The Arctic as a food producing region. Phase 1: Current status in five Arctic countries* (Nofima Report 10/2018). Nofima AS.

Feldmann, C., & Hamm, U. (2015). Consumers' perceptions and preferences for local food: A review. *Food Quality and Preference, 40*, 152–164.

Hingley, M., Boone, J., & Haley, S. (2010). Local food marketing as a development opportunity for small UK agri-food businesses. *International Journal on Food System Dynamics, 1*(1012-2016-81180), 194–203.

Ilbery, B., & Kneafsey, M. (1998). Product and place: Promotion quality products and services in the lagging rural regions of the European Union. *European Urban and Regional Studies, (5)*, 329–341

Ilbery, B., Morris, C., Buller, H., Maye, D., & Kneafsey, M. (2005). Product, process and place: An examination of food marketing and labelling schemes in Europe and North America. *European Urban and Regional Studies, 12*(2), 116–132.

Iversen, A., Hermansen, Ø, Henriksen, E., Isaksen, J. R., Holm, P., Bendiksen, B. I. … Dreyer, B. (2016). *Fisken og folket* (A. Iversen, Ed.). Orkana.

Johansen, T. J., Uleberg, A. L., & Mølmann, E., J. (2018). *Arktisk kvalitet – En beskrivelse av nordlige natur- og klimaforhold og virkning på egenskaper hos nordnorske matprodukter*. NIBIO RAPPORT 4 (40).

Ipsos, M. M. I. (2014). *Norske Spisefakta 2014*.

Keller, K. L. (1998). *Strategic brand management, building, measuring and managing brand equity*. Prentice Hall, Inc.

Kuznesof, S., Tregear, A., & Moxey, A. (1997). Regional foods: A consumer perspective. *British Food Journal, 99*(6), 199–206.

Luceri, B., Latusi, S., & Zerbini, C. (2016). Product versus region of origin: Which wins in consumer persuasion? *British Food Journal, 118*(9), 2157–2170. doi: https://doi.org/10.1108/BFJ-01-2016-0035.

Murphy, J. (1998) *What is branding? In "Brands, the new wealth creator"*. MacMillian press.

Murtagh, A. (2015). *The defining characteristics of alternative food initiatives in Ireland: a social movement battling for an alternative food future?* [PhD Thesis, University College Cork].

Natcher, D., Yang, Y., Hobbs, J., Hansen, K., Govaerts, F., Elde, S. … Valsdóttir, Þ (2019). *The Arctic as a Food Producing Region*. Sustainable Development Working Group, Arctic Council.

Nordnorsk Landbruksråd. (2018). Nordnorsk matmerke. *Stolt Arktisk Bonde*. https://arktisklandbruk.no/wp-content/uploads/2019/03/Rapport_2018_ORG.pdf

Norsk mat. (2018). Salg av lokal mat og drikke i dagligvarehandelen. *Norsk mat.* https://norskmat.no/no/norsk-mat-i-tall.

SSB. (2016). 08868: *Fangst, etter fiskefartøyet sin landingskommune og hovedgruppe av fangstarter* [Metadata]. https://www.ssb.no/statbank/table/08868?rxid=9f2fb50e-fd24-48b0-908f-76f1c88e4915

SSB. (2019). Nyetablerte foretaks overlevelse og vekst [Data set]. https://www.ssb.no/fordem

SMB Norge. (2018, August 8). Hvorfor mislykkes så mange små bedrifter?. *SMB Norge.* https://www.dinbedrift.no/slik-lykkes-du/

Trognon, L. (1998). The influences of territorial identity on consumer preferences: A contribution based on the RIPPLE programme. *Consumer preferences for Products of Own Region/Country and Consequences for Food Marketing. AIR-CAT Workshop Proceedings, 4* (3).

van Ittersum, K., Candel, M. J., & Meulenberg, M. T. (2003). The influence of the image of a product's region of origin on product evaluation. *Journal of Business Research, 56*(3), 215–226.

12 Marine fisheries and aquaculture in the Arctic

Catherine Chambers, Theresa Henke,
Brad Barr, David Cook, Barry Costa Pierce,
Níels Einarsson, Brooks Kaiser,
Ögmundur Knútsson, Matthias Kokorsch,
& Trent Sutton

Introduction

Arctic peoples and economies have long been linked with reliance upon living marine resources. Fisheries livelihoods center on both capture fisheries and its related industries (fish processing, gear manufacturing, harbor operations, etc.), and subsistence fisheries that contribute to local mixed economies (Vilhjálmsson et al., 2005; SADA Report, 2014; Zeller, Booth, Pakhomov, Swartz, & Pauly, 2011). The rich, productive waters of the Arctic supply over 10% of the world's marine fish catches (CAFF, 2013) and delete for culturally important traditional fisheries to the over 40 ethnic groups of the Arctic (Fondahl, Filippova, & Mack, 2015). Further, although aquaculture production for all of the Arctic hasn't changed much since 2014, it remains an important economic sector and has opportunities for growth (FAO, 2020). The Arctic marine socio-ecological ecosystem is experiencing continuous, rapid change, including shifts in the range of fisheries, decreasing sea ice coverage, increased risk of pollution, and varying forms of economic development and governance changes that can have both positive and negative impacts (Lam, Cheung, & Sumaila, 2014; Burgass et al., 2019; Kaiser, Pahl, & Horbel, 2018). As interest in the Arctic marine space increases, it is ever more important for research and policy to address specific challenges at local, regional, national, Arctic, and global levels. For example, as areas of the Arctic are becoming more and more accessible due to melting sea ice, there is increasing interest in developing new fisheries in the northern seas (Lam et al., 2014). Even though these fisheries would generate a positive economic benefit for the Arctic, developing the area bears a potential ecological risk (Lam et al., 2014). Another example is that of access to local fisheries, whether commercial or subsistence, where access rights to specific resources are governed without the interests of small-scale actors in mind (Briant, 2018; Chambers, Einarsson, & Karlsdóttir, 2020). This creates systems of inequity that can impact the well-being and self-actualisation goals of sustainable development. While fisheries and aquaculture can provide continued and perhaps expanded opportunities for

DOI: 10.4324/9781003172406-12

renewable Arctic economies, barriers and uncertainties related to climate change, socio-economic change, environmental impacts, and institutional and political obstacles can create challenges.

Information reviewed in this chapter will provide background information and identify opportunities and constraints related to fisheries and aquaculture to increase our understanding of renewable economies in the Arctic. The specific objective of this chapter is to use the best available data to contribute to scholarship, policy and future development by identifying opportunities and threats for current and fisheries and aquaculture activities in the Arctic. The chapter is organised as follows. Section 2 reviews the best available knowledge related to fisheries and aquaculture in the Arctic from an interdisciplinary perspective, including the impacts of climate change. To narrow the scope, this chapter does not cover other living marine resources such as marine mammals or seabirds. Next, Section 3 identifies opportunities and threats related to renewable economies based on fisheries and aquaculture in the Arctic related to governance challenges. Section 4 then identifies knowledge gaps relevant for the research community by highlighting areas for future research under the framework of the United Nation Sustainable Development Goals (UN SDGs). The chapter concludes by offering key considerations for Arctic communities and decision-makers interested in renewable economies that include fisheries and aquaculture.

Background

There is a wealth of scholarship on Arctic fisheries and aquaculture (see, for example, Zeller et al., 2011; Kourantidou et al., 2021). Rather than extensively reviewing the literature, this section highlights the current state of knowledge of Arctic marine socio-ecological systems, which will then be used to highlight specific focus areas under the SDGs in Section 3. The PAME (Protection of the Arctic Marine Ecosystems) Working Group of the Arctic Council identifies 18 LME's (Large Marine Ecosystems) of the Arctic. Not all waters of Arctic nations are in the 18 Arctic LME's, and non-Arctic nations may fish in Arctic waters. Therefore, the information presented below denotes the best available data based on either Arctic LME's, sub-Arctic waters, or Arctic nations.

Fisheries

Subsistence fisheries

The Arctic has a long history of Indigenous fisheries, heavily relying on marine resources such as fish and mammals (AHDR, 2004; CAFF, 2013). Various traditionally harvested food sources are an important part of the diet in Arctic Indigenous communities such as in the Canadian and Alaskan Arctic with whitefishes (*Coregonis* sp.), salmon (*Onchorchynchus* sp.), char (*S. alpinus*) and trout (*Salvelinus* sp.) representing the most frequently consumed fish species (Kuhnlein & Receveur, 2007). In Greenland, the population largely consists of Indigenous inhabitants, making up approximately 90% in 2019 (Statistics Greenland, 2019). Up until today, small-scale commercial fisheries and hunting are an important economic aspect in the eastern, southern, and northern parts of Greenland (Pálsson, n.d.).

Commercial fisheries and processing

According to the United Nations Food and Agriculture Organization (FAO), the Arctic can be divided into the Fisheries Statistical Area 18 and 27 (Zeller et al., 2011). Area 18 can be subdivided into the 8 Large Marine Ecosystems (LME) of the Kara Sea, Laptev Sea, East Siberian Sea, Chukchi Sea, Beaufort Sea, Arctic Archipelago, Baffin Bay, and Arctic Ocean (Zeller et al., 2011). Using the PAME designation of the 18 Arctic LME's, the commercial fishery catch in 2014 is outlined below in Table 12.1, using Sea Around Us data (seaaroundus.org).

Table 12.1 Commercial fishery catch in Arctic large marine ecosystems (LME) in 2014

LME	2014 catch (1000 tonnes)
1-Faroe Plateau	413.27
2-Iceland Shelf and Sea	914.71
3-Greenland Sea-East Greenland	172.97
4-Norwegian Sea	1108.23
5-Barents Sea	1355.5
6-Kara Sea	1.64
7-Laptev Sea	2.17
8-East Siberian Sea	1.18
9-East Bering Sea	1001.59
10-Aleutian Islands	725.73
11-West Bering Sea	804.75
12-Northern Bering-Chukchi Sea	377.32
13-Central Arctic Ocean	0
14-Beaufort Sea	.24
15-Canadian High Arctic-North Greenland	0
16-Canadian Eastern Arctic-West Greenland	194.99
17-Hudson Bay Complex	1.47
18-Labrador-Newfoundland	487.15

While some of the most productive fishing grounds are located in the sub-Arctic region, such as the Barents and Bering Sea, with 633 fish species occurring of which 58 are considered commercial, the Arctic basin is comparably deserted with only 3 commercial species out of 63 (Meltofte, 2013; SADA report, 2014). These productive fisheries are usually characterised by only one or two species, with the Bering Sea being dominated by Pollock and the Barents Sea by a cod-capelin system (SADA report, 2014).

The Russian area of the Arctic consists of the Kara Sea, the Laptev Sea and the East Siberian Sea as well as the Chukchi Sea, which is also bordered by the United States (Zeller et al., 2011). In accordance to the overall low fishery production of the Arctic waters, the Russian Arctic areas sustain a low number of fish species (Pauly and Swartz, 2007; Zeller et al., 2011). There were significant increases in the population of Siberia during the Soviet regime, and the current expanding industrialisation of the Russian Arctic between the 1960s and today has further expanded the population of the region. The implications of potentially increasing capture fisheries targeting such limited fish populations in adjacent Arctic waters is uncertain, but reported estimates of current fisheries production from this region may not accurately reflect its potential in this regard (Zeller et al., 2011).

Commercial fishing in the Kara Sea is dominated by six species of white fish, making up about 80% of the landings (Zeller et al., 2011). In the Laptev and East Siberian Seas, there are no commercial fisheries being operated and instead just small-scale fisheries (Newell, 2004). The East Siberian Sea does have populations of pink salmon (*Oncorhynchus gorbuscha*) and Dolly Varden (*Salvelinus malma*) that could sustain commercial harvesting on the grounds of size, but these populations are facing the risk of overfishing (Newell, 2004). In the area of the Chukchi Sea, the human population has been steadily decreasing with about 1,000 people remaining in coastal areas where they rely heavily on marine food sources (Newell, 2004; Zeller et al., 2011).

In 2009, under the United States Magnuson-Stevens Fishery Conservation and Management Act, an Arctic management area was implemented north of Alaska, reaching from the Seward Peninsula to the Canadian border and therefore covering parts of the Chukchi and Beaufort Sea (NPFMC, n.d.). In this area, all commercial fishing for finfish, crustaceans, and other animals except for marine mammals and birds, is prohibited to protect the sensitive ecosystem from potentially destructive fishing activities (NPFMC, n.d.). However, fisheries in the sub-Arctic waters of Alaska accounted for 2.67 M tons of catch or $1.872 billion 2010 USD in 2014 (seaaroundus.org) and are a vital economic activity in coastal communities.

Aquaculture

In the Arctic, harvesting fish from aquaculture is limited, however, there is room for growth in the sector for mainly finfish aquaculture (SADA report, 2014; Barbier & Burgess, 2017). There is also a small but growing niche production of seaweeds and shellfish (Allison, Badjeck, & Meinhold, 2011). Norway, by far, dominates aquaculture production in the Arctic, mainly through salmon farms, followed by the USA and Canada (AP, 2012; SADA report, 2014). Norway's production of Atlantic salmon (*Salmo salar*) was valued at 2.2 billion EUR in 2010 (Norwegian Directorate of Fisheries, 2020). In Alaska, finfish aquaculture is prohibited (ADF&G, n.d.) but the aquaculture sector produces shellfish, such as Pacific oyster (*Magallana gigas*) and blue mussels (*Mytilus trossulus*), and kelp, which generated 1.53 million USD in 2017 (Pring-Ham, 2018). Meanwhile in Canada, as in Alaska, no measurable aquaculture currently takes place in Arctic waters, but in the sub-Arctic Pacific waters of British Columbia, 88,834 tons of finfish were produced in 2018, generating a value of 791 million CAD (Fisheries and Oceans Canada, 2020). In the Canadian Atlantic, Newfoundland, New Brunswick, and Newfoundland combine to produce 51,634 tons in 2018 for a value of 480 million CAD (Fisheries and Oceans Canada, 2020). Salmon pen aquaculture is rapidly expanding in Iceland, with backing from Norwegian companies. In 2018, 19,000 tons were produced, consisting mostly of salmon and Arctic char for an export value of 13.1 billion ISK in 2018 (Statistics Iceland, 2020). In Russia, there are cultivation rights for 30 salmon and trout sites in the Murmansk and Karelia regions, with the strategic goal of increasing by 16–18 farms in 2025 to produce up to 25–30,000 tons of salmonids. This growth coincides with the steady demand for and consumption of salmonids in Russia during the last decade along with the identification of aquaculture as a top priority in the agri-food section by the country's authorities.

Impacts of climate change

Based on model predictions, the effects of climate change will be most acute in the Arctic, with sea surface temperatures expected to rise more than in temperate latitudes (Team, Pachauri, & Meyer, 2014). The rate of climate change is considered to be two times faster in the Arctic than in other regions, and it is projected that the rate of ocean warming might be up to seven times faster than projected changes in terrestrial landscapes (Burrows et al., 2011). As an example, with 1.5°C of ocean warming, mobile species, such as plankton and fishes, might be driven to relocate at higher latitudes. Those species that cannot move, such as kelps and corals, would undergo high mortality, causing damage to dependent ecosystems (Hoegh-Guldberg et al., 2018). With either the mortality of sessile specie or the range expansion of mobile species, the marine ecosystem complexes may experience great change. In the Arctic, change has already shown some impacts on the environment, both the abiotic and biotic components (Wassmann et al., 2011; Hoegh-Guldberg et al., 2018, AMAP, 2018). Other chemical changes occur due to the absorption of anthropogenic carbon dioxide have already been observed in the Beaufort Sea and in Alaska's coastal waters (Sumaila, 2015), resulting in ocean acidification that can affect the growth and development of organisms that build shells. This in turn impacts a wide range of marine species further down the food web that depend on shelled organisms for food (AMAP, 2018). However, documenting particular changes is not easy due to a lack of reliable baseline data, which is needed for a comparison to the current situation (Wassmann et al., 2011).

Expected changes in the Arctic include more than just warming air and water temperatures; for example, the observed mean annual spatial extent of sea ice has been decreasing 3.5–4.1% per decade and precipitation is expected to increase between 30 and 50% (Team, Pachauri, & Meyer, 2014). The reduction in the duration of shore-fast sea ice will likely yield greater coastal erosion due to more impactful winter storms, with coastlines in many locations of the Beaufort Sea expected to undergo high rates of erosion (Gibbs & Richmond, 2015; Jones et al., 2009). Increased coastal erosion in the central Beaufort Sea has amplified the suspension of sediment load, thereby reducing benthic and water column primary production (Bonsell & Dunton, 2018). Such large-scale changes in environmental conditions in the nearshore region of the Arctic have significant implications for the ecological responses of local fishes as well as other marine organisms.

Physical impacts on marine ecosystems

All fishes are impacted by variability in their surrounding environment (Fechhelm, Fitzgerald, Bryan, & Gallaway, 1993). Water temperature and salinity are perhaps most influential, and deviations from "normal" conditions can lead to significant energetic costs through increased temperature regulation, metabolism, and osmoregulation (Bœuf & Payan, 2001; Edsall, 1999). Fish species that are forced to inhabit environments with conditions outside their optimal requirements can experience reductions in growth and recruitment and/or higher mortality (Arnesen, Jørgensen, & Jobling, 1993; Dutil, Lambert, & Boucher, 1997). To optimise growth and survival, fish may seek environments which provide conditions to optimise growth and survival (Monaghan, 2008); to locate such conditions may include shifts in geographic distribution (Hansen & Closs, 2009; Křivan, 2003). As a result, the magnitude, duration, and variability of environmental factors all play a significant role in determining the fish species composition, abundance, and rate-dynamic parameters, especially in dynamic environments such as the Arctic.

Despite the variable environment, nearshore Arctic waters support a variety of fishes. Whitefishes, such as Arctic Cisco *Coregonus autumnalis*, Broad Whitefish *C. nasus*, Least Cisco *C. sardinella*, and Humpback Whitefish *C. pidschian*, are amphidromous species that are tolerant of moderate levels of salinity and often undergo long-range migrations (de March, 1989; Fechhelm et al., 1993). For example, Arctic Cisco in Alaskan waters of the Beaufort Sea originate in the Mackenzie River, Northwest Territories, Canada, and are transported as juveniles >500 km along the shore via easterly wind-driven surface currents (Fechhelm & Fissel, 1988; Fechhelm & Griffiths, 1990). However, other coregonids (for example, Broad Whitefish, Humpback Whitefish, and Least Cisco) spawn in rivers throughout northern Alaska and Canada (Craig, 1984, Craig, 1989; Fechhelm, Bryan, & Griffiths, 1994). Regardless of species, juvenile whitefishes spend the short growing season (late June into September) feeding in coastal estuaries and deltas of the Beaufort Sea and overwinter in freshwater pools or upwelling areas in coastal tributaries (Craig, Griffiths, Haldorson, & McElderry, 1985; Fechhelm, Martin, & Gallaway, 1999; Seigle & Gottschalk, 2013). The gadids Arctic Cod (*Boreogadus saida*) and Saffron Cod (*Eliginis gracilis*) are also ecologically important and support the Arctic marine ecosystem with their high abundance and energetic content (Elliott & Gaston, 2008; Harter, Elliott, Divoky, & Davoren, 2013; Thorsteinson & Love, 2016). These species serve as a key link between lower trophic levels (for example, calanoid copepods and amphipods) and higher trophic organisms (for example, seabirds and marine mammals; Harter et al., 2013). Pacific salmon *Oncorhynchus* spp. have been documented in the Arctic at increasing frequency in recent years, with a natal population in the Mackenzie River in Canada and potentially other rivers of the central Beaufort Sea in Alaska (Gatt, Hamman, Priest, & Sutton, 2019; Irvine, Macdonald, & Brown, 2009).

Impacts to coastal communities: Food security,
economic opportunity, and local infrastructure

In addition to their key role in the ecology and food-web structure of coastal marine ecosystems, nearshore fishes also provide important subsistence food resource for local Indigenous communities (Fechhelm, Streever, & Gallaway, 2007; Thorsteinson & Love, 2016). As a consequence, the effects of climate change could significantly impact fishing activities of Indigenous peoples (Galappaththi, Ford, Bennett, & Berkes, 2019). As higher levels of acidity are already documented for the Beaufort Sea (Zhang, Yamamoto-Kawai, & Williams, 2020) and Alaska's waters (Monacci, Cross, Hurst, Long, & Rossin, 2019), Indigenous communities might eventually not have enough nearshore marine resources available to harvest as fish stocks will be decreasing due to lower prey availability (Sumaila, 2015). As a result, climate change has the potential to disrupt food security for humans. While there are no commercial fisheries in federal waters of Arctic Alaska, subsistence fisheries in nearshore areas are important socio-cultural and nutritional contributions to local Inupiaq communities (Fechhelm et al., 2007; NPFMC, 2009). Communities along the Chukchi and Beaufort Sea coastline, such as Utqiaġvik, Nuiqsut and Kaktovik, are dependent on the subsistence harvest of marine organisms, including abundant whitefishes found in the river deltas and nearshore coastal waters (Craig, 1987; Braund et al., 2012). Although the economic impact of these fisheries is limited, ecological changes in the Arctic would have substantial ramifications for both food-web structure and dynamics as well as human communities (Moerlein & Carothers, 2012). As a result, the importance of Arctic fishes to coastal communities necessitates the examination of potential effects of climate change prior to understanding how ecosystems and food security may be impacted in the future (Reist et al., 2006).

There is also uncertainty involved in projecting fisheries catches for commercial stocks under climate change (Cheung et al., 2016) and how the fishing industry might change. While it can be assumed that fisheries-dependent communities could adapt to fluctuating fish stocks or might even profit in the short term, their overall adaptive capacity is limited due to the threatened infrastructure and other socio-economic factors (Alvarez, Yumashev, & Whiteman, 2020; Ford, McDowell, & Pearce, 2015). The adaptive capacity of Arctic fishing communities can be defined as the ability to cope with external stresses, such as climate change, and the ability to modify or change (Kokorsch & Benediksson, 2018). It is an integral part of community resilience which describes "[...] the ability of a community to cope and adjust to stresses caused by social, political, and environmental change and to engage community resources to overcome adversity and take advantage of opportunities in response to change" (Amundsen, 2012, p. 1). However, the adaptive capacity and community resources are somewhat limited and dependent on economic factors, decision making power and resource access that are usually not part of the local agency (Landauer & Juhola, 2019; Ford et al., 2015). Relevant locational factors include infrastructures for the communities in general and the fishing industry in particular. Especially fisheries dependent communities with a monotonous local or regional economy are predicted to face difficulties due to climate change. Community infrastructure, local agency and a diversified economy are, however, key parameters for resilience building and the preparedness for transformative shocks (drastic shifts or sudden changes) or structural changes (slow and gradual processes) (Kokorsch, 2018).

Arctic coastal communities are not only dependent on the infrastructure at the shoreline, but also the hinterland and transportation systems onshore are of relevance. Thus, climate change related threats to renewable fisheries economies are not limited to melting sea ice and rising sea levels, but include coastal erosion, thawing permafrost, extensive precipitation, flooding and severe weather (Landauer & Juhola, 2019; Moon et al., 2019; Melvin et al., 2017). Hence, some communities, for example in coastal Nunavut or Alaska, will face multiple threats and socio-economic burdens (Alvarez et al., 2020; Melvin et al., 2017; Walker & Peirce, 2015). It can be expected that climate change will lead to significantly higher costs for maintenance, restructuring or new construction of critical infrastructure (Larsen et al., 2008; Pahl & Kaiser, 2018; Streletskiy, Suter, Shiklomanov, Porfiriev, & Eliseev, 2019; Suter, Streletskiy, & Shiklomanov, 2019). While rising costs for infrastructure are not a problem per se, in the Arctic it hits communities and industries that are commonly investment-intensive and face small profit margins (Stephen, 2018; Suter et al., 2019). Recommended proactive investments in adaptation and infrastructure preparedness are subject to debate regarding their financing and responsibility (Landauer & Juhola, 2019; Melvin et al., 2017). From this perspective, it is debatable to what extent renewable economies can be sustained, in a setting in which critical infrastructure is vulnerable. In other terms, the economic fundamentals (literally and figuratively) might not be renewable but need to be constantly renewed.

Apart from the impact on infrastructure and local economies, several socio-cultural and demographic consequences for coastal communities due to climate change can be identified: cultural heritage loss, health disparities and worsening food and water security (Alvarez et al., 2020; Fritz, Vonk, &, Lantuit, 2017; Irrgang, Lantuit, Gordon, Piskor, & Manson, 2019; Stephen, 2018). The combination of these factors might lead to intensified out-migration or the relocation of some communities, thereby impacting the workforce for economic activities. The attractiveness of the fishing sector (capture fisheries and subsistence fisheries) on and offshore will be negatively affected by the threats outlined above; fisheries are a business sector that can be characterised as unstable and climate change is an additional hindrance for new entrants to the industry in the long run. Aquaculture might be an alternative business strategy for Arctic coastal communities and an opportunity to respond to the changes and threats.

Impacts to aquaculture

Aquaculture is likely not as affected by climate driven change in the environment as the commercial fisheries, since location, population density and resource availability are controlled factors (SADA report, 2014). Significant environmental changes might change the distribution of the aquaculture industry throughout the Arctic (SADA report, 2014), but overall aquaculture in the Arctic will likely see positive effects from warming water temperatures (Hermansen & Troell, 2012). For example, the climate-induced temperature rise on the Norwegian coasts is likely to range between 0.5 and 2.5°C and will play out differently during different seasons. Despite large uncertainties, and just a few detailed studies that specifically target climate change impacts on Arctic aquaculture, the direct effects of temperature change on the aquaculture industry can be modelled with fairly good accuracy, including effects on fish growth and impacts on the whole industry. Present optimal conditions for open sea cage salmon farming in Norway lie between 62 and 64° North latitude. Further south, summer temperatures are higher than optimum, and further north, temperatures are too low throughout the year. Increased sea temperatures will generally move this optimal zone further north. For fish farms in colder locations than optimal, production can increase with 11–15% per degree increase in temperature (Lorentzen, 2008). For farms at optimum or higher temperatures, production will decrease. Salmon farms in the Arctic generally experience lower than optimum temperatures and will therefore likely experience improved productivity. Species like cod and halibut have narrower temperature ranges and should respond in a similar way (Troell et al 2017). As a buffer against uncertain climatic impacts of sea-pen aquaculture, employing recirculating aquaculture system (RAS) technologies is seen as a means of reducing exposure to climatic extremes. The RAS are expensive in terms of both capital and operational costs and require high levels of technical expertise (Murray, Bostock, & Fletcher, 2014). The long-term reliability of RAS still needs to be demonstrated. Aquaponics, the production of fish and plants in an integrated system, is proposed as a means of producing food in areas where freshwater is limited (Somerville, Cohen, Pantanella, Stankus, & Lovatelli, 2014).

Opportunities and threats related to governance systems

The management of the Arctic areas of those nations adjacent to the Arctic Circle lies within each country's national governance (Burgass et al., 2019). Under the 1982 United Nations Convention on the Law of the Sea (UNCLOS), each country is obliged to prevent overfishing by national fisheries within the EEZ (Pinsky et al., 2017. However, some issues reach beyond just one jurisdiction and therefore require nations to work together and develop joint management strategies (Pinsky et al, 2017; Van Pelt, Huntington, Romanenko, & Mueter, 2017). Therefore creating new, more legally binding cooperative fisheries governance mechanisms and strengthening existing agreements, has been identified as a pressing need (Molenaar, 2012; Molenaar, 2013).

The Arctic Council was created to provide a forum for enhancing and expanding international collaboration in the Arctic, but any focus on fostering such collaboration with regard to sustainably managing fisheries in this region has largely been "off the table" (Molenaar, 2013). As reported by Schatz et al. (2020), "the Arctic Council decided in 2007 that fisheries issues should be considered 'within the context of existing mechanisms'." However, Molenaar (2012) has observed that "there is no juridical obstacle" for the Arctic Council to take a more active role, and it has engaged in some broader international fishery management-related discussions through its standing committees. While the Arctic Council has not "explicitly reversed" its formal position with regard to governance and international coordination and management of fisheries in the region, it could, and perhaps should, take a more active role (Molenaar, 2013).

As extensively reviewed and summarised by Molenaar (2012, 2013), the existing governance framework for fisheries in the Arctic encompasses global, regional and bilateral fora and instruments, as well as limited Indigenous co-management arrangements (Ayles, Porta, & McV Clarke, 2016). However, all of these arrangements, instruments and fora are constrained by lack of sufficient data, inadequate domestic regulation, gaps in important Arctic Coastal State fora and instruments, and gaps in high seas coverage with Regional Fishery Management Organizations (RFMOs) and other arrangements (Molenaar, 2013). This last identified shortcoming was more recently addressed in 2018 by the signing and ratification of the "Agreement to Prevent Unregulated High Seas Fisheries in the Central Arctic Ocean" (2020). This agreement was signed in Ilulissat, Greenland, by the five Arctic states of Norway, Denmark (for the Faroe Islands and Greenland), Canada, the United States, and Russia as well as by Iceland, Japan, Korea, and China (European Commission, 2018; Harvey, 2018; NOAA Fisheries, 2018). Such broader participation in creating new, and enhancing existing, Arctic fisheries management instruments and agreements continues to be a significant challenge and needs to be actively pursued (Molenaar, 2012) as it was for this agreement.

This CAOF agreement has been represented as the first preventive, international agreement of such magnitude (Harvey, 2018; Rayfuse, 2018; NOAA Fisheries, 2018). Currently there is no fishing activity actively occurring in the CAO, but exploratory fisheries have been conducted in that area in the past (Harvey, 2018). The agreement provides for some limited continuation of this exploratory fishing activity under requirements to be developed within three years of the agreement's ratification. For a period of 16 years following ratification, commercial fishing is limited, "pursuant to conservation and management measures for the sustainable management of fish stocks adopted by one or more regional or sub-regional fisheries management organizations or arrangements" (Article 3,1a), and any "conservation and management measured that may be established" by the parties (Article 3,1b). Encompassing an area of approximately 2.8 million km^2, commercial fishing will only be allowed after this 16-year period if international management strategies are developed and implemented to manage such fisheries (NOAA Fisheries, 2018). Furthermore, the agreement also requires (Article 4,2) the establishment of a "Joint Program of Scientific Research and Monitoring" to better understand the ecosystem and potential impacts of fisheries as well as create a baseline needed to document future development (Harvey, 2018; NOAA Fisheries, 2018). This joint program also explicitly includes the integration of "Indigenous and local knowledge" (Article 4,4). By largely prohibiting commercial fishing activity for the period the agreement is in force and effect, not only are fish populations present in the CAO protected from overexploitation, but potential pollution and damage caused by the vessels will be very limited (Harvey, 2018). However, before the agreement comes into force, all nations that have signed the agreement must ratify their commitment which, as of August 2019, has only been done by Canada, the EU, Russia, and the United States (US Dept of State, 2020). Some analyses of the agreement have predicted that closing the high seas could eventually benefit coastal fisheries as the high seas would become recovery areas from which a spill-over effect could replenish harvested fish stocks in coastal areas (Sumaila et al., 2015). However, the agreement has also been the subject of some criticism. Schatz et al. (2019) provides a comprehensive and exhaustive legal analysis of the agreement that identifies potential conflicts and ambiguities, concluding that "the success of the newly concluded CAOF Agreement cannot be assessed from a purely legal standpoint, but largely depends on the political will of its parties to implement the rights and obligations codified therein in an effective, lawful and legitimate manner."

Arctic Coastal States and non-Arctic countries have participated in both unilateral National actions to address fishery management in their Arctic waters and have engaged in international collaborations. The United States was the first nation to set an example on protecting their Arctic waters by adopting, in 2009, a Fishery Management Plan for their Arctic management area that bans commercial fishing activity (NPFMC, n.d.). China has not only shown interest in the development of the Arctic but has been involved in Arctic politics over the past decades, such as its involvement as an observer in the Arctic Council (Pan and Huntington, 2016; Østhagen, 2019, Kuo, 2019). Particularly on Svalbard, China has established considerable infrastructure to conduct research on natural science as well as on opportunities to utilise Arctic resources to strengthen further the relationship to Arctic states (Pan & Huntington, 2016; Østhagen, 2019; Kuo, 2019). Besides their interest in possible Arctic sea routes, there are incentives for involvement in potential Arctic fisheries to ensure enough resources for China's growing population, recognising that local fish stocks are in decline (Østhagen, 2019, Kuo, 2019). The European Union lacks direct access to the Arctic Ocean, but the EU states of Finland, Sweden, and Denmark are geographically positioned close to the Arctic (Østhagen, 2019). Interest in being involved in the policy of developing the Arctic has been expressed by the EU (Østhagen, 2019). The EU's Arctic Policy as well as their Ocean Governance policy highlights various goals, including sound stewardship of the high seas, developing a responsible strategy to access the Arctic's resources, and at the same time respecting the rights of Indigenous communities in the Arctic (European Commission, 2018).

While it is clear that progress is being made towards creating and implementing a governance framework for fisheries in the Arctic, the challenges are significant. Many countries are involved that often have conflicting interests. It could be argued that the power, and much of the responsibility, lies with the Arctic Coastal States in achieving this goal, but many other countries have interests in the future of Arctic fisheries, and their perspectives must be heard. Most importantly, the Indigenous peoples of the North are very much affected by the outcome of these deliberations over fisheries resource allocations and conservation, and need to be at the table and share equally in the decision making. Many hard choices and trade-offs need to be made as this new and hopefully integrated and comprehensive fisheries governance framework is crafted. As Jensen and Rottem (2010) have observed, "The Arctic region does not suffer under a state of virtual anarchy, despite outward appearances."

The future of aquaculture in the Arctic

Recent international policy directives from the UN recommend replacing meat with seafood to limit global temperature rise to 1.5 degrees C, and that increasing food production in the ocean through aquaculture is a key strategy in reaching the climate goal set in the Paris Agreement. Other directives for growth, such as the EU Blue Economy Report, identify worldwide opportunities for aquaculture production, and Arctic nations are among those with strong potential due to the environmental changes discussed above. However, governance of aquaculture in the Arctic is in constant flux and this presents a challenge for communities planning for growth.

In the United States, the Advancing the Quality and Understanding of American Aquaculture Act was introduced in the US Senate in 2018 (S 3138 – 115th Congress) and the United States House of Representatives (HR6966 – 115th Congress) to encourage aquaculture to meet the global demand for seafood (Resneck, 2018; Bittenbender, 2019). As this new bill could impact the ban of finfish aquaculture in Alaska, the United States Rep. Don Young responded by filing the Keep Fin Fish Free Act to ensure the health of wild fish stocks (Bittenbender, 2019). While the 2018 bills did not progress, a 2020 version was introduced to the United States House of Representatives in March and is progressing through Committee reviews (HR 6191 – 116th Congress). More recently, the United States president has signed an executive order intended to expand United States aquaculture production rapidly due to pressures from the COVID-19 pandemic to increase domestic food production and reduce foreign trade (EO 13921, 85 Fed. Reg. 28,417 (May 12, 2020)).

In Russia, there is a clear example of how the growth of aquaculture is linked with political decisions that are not easily reversed. Sanctions introduced in August 2014 that prohibited imports of trout and salmon to Russia from Norway, the EU, United States, Canada, and Australia meant that Russian companies expanded to replace the near 200,000 tons of those products that previously came from outside Russia (Adamowski, 2017). While the growth in Russia is expected to continue, it is unclear how future governance decisions might impact the growth of aquaculture.

In Norway, aquaculture licensing and permitting processes can be a hindrance in an uncertain future. Legal requirements limit production sites because a license is granted for one region and cannot be transferred to any of the other four regions in place for aquaculture management. Model predictions show a significant improvement in productivity for the northern farms and vice versa for the farms furthest south and a corresponding northward shift in production if the restrictions are lifted (Hermansen & Heen, 2012).

Discussion: Renewable economies, SDGs, and knowledge gaps

Because renewable economies are so tightly linked with sustainable development, the authors chose to organise the discussion around the UN SDGs (UN 2020). The UN SDGs are a list of 17 priority goals that serve as a call to action for countries around the world, and provide a basis of comparison for tracking the progress of sustainable development (UN, 2020). All SDGs are relevant for Arctic nations; however, it has been noted that the SDGs were not created with the Arctic in mind and certain goals are difficult to scale down (Sköld, Baer, Scheepstra, Latola, & Biebow, 2018). Specifically, goals related to poverty, food, education, water, and sanitation are not often thought of as important topics in Arctic nations due to a dichotomy in the UN between developing and developed nations, but these issues can have high variance in the Arctic depending on country and region (Kroll et al 2019; Menezes & Chater, 2018). It has been suggested that the SDGs can be addressed through the development of indicators that are Arctic-specific (Sköld, Baer, Scheepstra, Latola, & Biebow, 2018). Following this line of thought, the authors of this chapter selected seven SDGs that have the most relevance in identifying knowledge gaps for Arctic fisheries and aquaculture. Based on the information provided in Sections 2 and 3, we identify below 2–3 priority points for each SDG that identify areas of future focus for researchers and decision-makers that will help track indicators and therefore the fulfillment of SDG targets. For ease of dissemination, the following section presents a series of graphics. With each SDG, bullet points are listed to provide recommendations by the authors for priority areas of research questions that would create relevant data to assess the goal under the lens of Arctic fisheries and aquaculture.

- Model the contribution of Arctic fisheries and aquaculture to Target 1.2 on reducing poverty by at least half among men, women and children by 2030 according to national definitions.

- Given the sensitivity of the Arctic marine environment and resource dependency of remote communities, effort should be made to gather data on how fisheries and aquaculture can help to build the resilience of the poor and those in vulnerable situations, and reduce their exposure and vulnerability to climate-related extreme events and other economic, social and environmental shocks and disasters.

- Explore potential data gaps in poverty indicators that are not visible at national or regional levels.

Figure 12.1 SDG 1 "No Poverty" and related research topics. (SDG image reproduced from https://sdgs.un.org/goals)

- Quantify the extent to which Arctic fisheries and aquaculture can contribute to ending hunger among Arctic communities, in accordance with the objectives of Target 2.1.
- Assess the extent to which national investment strategies in Arctic nations are targeting the fulfilment of Target 2.A through focus on fisheries and aquaculture.
- Gather data on the extent to which developed aquaculture markets in the Arctic, e.g. in Norway (increasingly Iceland), are contributing to the meeting of Target 2.C on price stability in commodity markets.

Figure 12.2 SDG 2 "Zero Hunger" and related research topics. (SDG image reproduced from https://sdgs.un.org/goals)

- With regards to Target 3.D, data could be gathered on the extent to which Arctic fisheries and aquaculture production is accounted for in early warning, risk reduction and management strategies and policies for national and global health risks.
- Collect data on job satisfaction in fisheries and aquaculture, related to occupational health and safety, right to work, and self-determination in employment.
- Explore how local food systems products from fisheries and aquaculture contribute to good health and well-being.

Figure 12.3 SDG 3 "Good Health and Well-being" and related research topics. (SDG image reproduced from https://sdgs.un.org/goals)

- Explore how Target 5.5 is fulfilled in ensuring women's full and effective participation and equal opportunities for leadership at all levels of decision-making in political, economic and public life.

- Quantify Target 5.A: "Undertake reforms to give women equal rights to economic resources, as well as access to ownership and control over land and other forms of property, financial services, inheritance and natural resources, in accordance with national laws".

- Target 5.C could be considered in the context of the increasing role of technology in fisheries and aquaculture – how to use information and communications technology to promote the empowerment of women.

Figure 12.4 SDG 5 "Gender Equality" and related research topics. (SDG image reproduced from https://sdgs.un.org/goals)

- Explore the contribution of fisheries and aquaculture sectors in Arctic nations to sustaining economic growth, in accordance with Target 8.1. Satellite accounts could be used or developed for the sectors, akin to tourism satellite accounts for GDP used in Iceland.

- Specific to Target 8.3 on job creation, this could be assessed with a particular focus on local community job creation and potential economic multiplier effects.

- Related to Target 8.4, gather more data on the efficiency of production and consumption in Arctic fisheries and aquaculture, particularly the potential decoupling of the growth of the industry from environmental degradation e.g. economic growth versus GHG emissions or energy consumption. Various economic / resource intensity measures that already exist could be applied in this context.

Figure 12.5 SDG 8 "Decent Work and Economic Growth" and related research topics. (SDG image reproduced from https://sdgs.un.org/goals)

- Quantification of inclusiveness in Arctic fisheries and aquaculture, especially to fulfil Target 10.2 on the social, economic and political inclusion of all, irrespective of age, sex, disability, race, ethnicity, origin, religion or economic or other status.
- Overlapping with Targets 8.5 and 8.6: Assess full employment (including for persons with disabilities) and equal pay, and youth employment, respectively.
- Explore regional inequalities in fisheries and aquaculture in Arctic nations in terms of power and access to fishing rights, marine space, and decision-making power.

Figure 12.6 SDG 10 "Reduced Inequalities" and related research topics. (SDG image reproduced from https://sdgs.un.org/goals)

- Given the issue of ocean acidification that already affects Arctic coastal communities, data gathering should focus on Target 8.3, looking at the extent to which Arctic fisheries and aquaculture is endeavouring to minimise and address its impacts.
- Research could focus on the extent to which there is overfishing and illegal fishing in Arctic waters, as per the requirement of Target 14.4.
- Develop better regional models for temperature increases and storm events that may impact the growth of aquaculture.
- Explore how migration of fish stocks could benefit Arctic fishing communities in the short run (like the mackerel in Greenland during the past five years).

Figure 12.7 SDG 14 "Life Below Water" and related research topics. (SDG image reproduced from https://sdgs.un.org/goals)

Conclusion

As reviewed in this chapter, many reports produced for the Arctic Council, among others, have detailed the best available knowledge of Arctic fisheries and aquaculture. The goal of this chapter was to put that knowledge in the context of renewable economies and sustainable development to provide key considerations for Arctic communities and decision makers. The UN SDGs are not a perfect framework to guide policy concerning renewable economies in the Arctic, but instead provide a starting point to build effective future policy and research for renewable, sustainable Arctic fisheries and aquaculture. This chapter has reviewed the prevailing social, economic, and climate-related challenges to Arctic fisheries and aquaculture, and provided a path forward by creating recommendations for future focus related to seven SDGs that are most closely related to fisheries and aquaculture: No Poverty (1), Zero Hunger (2), Good Health and Well-being (3), Gender Equality (5), Decent Work and Economic Growth (8), Reduced Inequalities (10), and Life Below Water (14). Through the focus on SDG's, this chapter highlights the need for Arctic-specific data collection and modeling that is not currently clearly visible through larger indicator data sets in fisheries and aquaculture. For example, poverty indicators, hunger elimination national objectives, job satisfaction and well-being data sets, and gender and technology aspects are not often considered with specific focus on fisheries and aquaculture. However, these industries and livelihoods, when practiced renewably and sustainably, may be key factors in fulfilling Arctic-specific SDGs. The path forward to renewable economies in the Arctic related to fisheries and aquaculture must start with better and more detailed Arctic-specific focus related to commercial and subsistence fisheries and aquaculture.

Acknowledgments

The authors would like to acknowledge funding support from the Arctic Council SDWG for the development of this chapter and the editors for their support. This chapter was also supported in part by financial support from the Nordic Centre of Excellence ARCPATH (Arctic Climate Predictions – Pathways to Resilient, Sustainable Communities) (grant number 76654).

References

Adamowski, J. (2017). Russia to increase salmon aquaculture. *The Fish Site*. https://thefishsite.com/articles/russia-to-increase-salmon-aquaculture

ADF&G. (n.d.) Aquatic farming. *Alaska Department of Fish and Game*. https://www.adfg.alaska.gov/index.cfm?adfg=fishingaquaticfarming.main

Agreement to prevent unregulated high seas fisheries in the central Arctic Ocean. (2020). *Ministry of Foreign Affairs of Japan*. https://www.mofa.go.jp/mofaj/files/000449233.pdf.

Allison, E. H., Badjeck, M. C., & Meinhold, K. (2011). The implications of global climate change for Molluscan aquaculture. In S.E. Shumway (Ed.), *Shellfish aquaculture and the environment* (461–483). Wiley.

Alvarez, J., Yumashev, D., & Whiteman, G. (2020). A framework for assessing the economic impacts of Arctic change. *Ambio, 49*(2), 407–418.

AMAP. 2018. AMAP Assessment 2018: Arctic Ocean Acidification. *Arctic Monitoring and Assessment Programme (AMAP)*.

Amundsen, H. 2012. Illusions of resilience? An analysis of community responses to change in Northern Norway. *Ecology and Society 14*(4): 46.

AP. (2012). Aquaculture in the Arctic. *Arctic Portal*. https://arcticportal.org/yar-features/625-aquaculture-in-the-arctic

Arnesen, A. M., Jørgensen, E. H., & Jobling, M. (1993). Feed intake, growth and osmoregulation in Arctic charr, *Salvelinus alpinus* (l.), following abrupt transfer from freshwater to more saline water. *Aquaculture, 114,* 327–338.

Ayles, B., Porta, L., & McV Clarke, R. (2016). Development of an integrated fisheries co-management framework for new and emerging commercial fisheries in the Canadian Beaufort Sea. *Marine Policy, 72,* 246–254.

Barbier, E. B., & Burgess, J. C. (2017). The sustainable development goals and the systems approach to sustainability. *Economics: The Open-Access, Open-Assessment E-Journal, 11*(2017-28), 1–23.

Bittenbender, S. (2019) With environmentalist support, Alaska Rep. Don Young files anti-aquaculture bill. *SeafoodSource.* https://www.seafoodsource.com/news/aquaculture/with-environmentalist-support-alaska-rep-don-young-files-anti-aquaculture-bill

Braund, S. R. (2012). Quantification of subsistence and cultural need for bowhead whales by Alaska Eskimos. Prepared for the Alaska Eskimo Whaling Commission, Barrow Alaska. IWC/64/AWS 3 Agenda item 6.1.2. https://iwc.int/private/downloads/6BOXTFAz3CvHjbIMkcyDMQ/64-ASW%203.pdf

Bœuf, G., & Payan, P. (2001). How should salinity influence fish growth? *Comparative Biochemistry and Physiology – Part C: Toxicology & Pharmacology, 130,* 411–423.

Bonsell, C., & Dunton, K. H. (2018). Long-term patterns of benthic irradiance and kelp production in the central Beaufort Sea reveal implications of warming for Arctic inner shelves. *Progress in Oceanography, 162,* 160–170.

Burrows, M. T., Schoeman, D. S., Buckley, L. B., Moore, P., Poloczanska, E. S., Brander, K. M. … Holding, J. (2011). The pace of shifting climate in marine and terrestrial ecosystems. *Science, 334*(6056), 652–655.

CAFF. (2013). Arctic Biodiversity Assessment: Status and trends in Arctic biodiversity. *Conservation of Arctic Flora and Fauna, Akureyri.*

Chambers, C., Einarsson, N., & Karlsdóttir, A. (2020). Small-scale fisheries in Iceland: Local voices and global complexities. In C. Pita, J. Pascuel and M. Bavinck (Eds.), *Small scale fisheries in Europe.* Springer.

Cheung, W. W., Jones, M. C., Reygondeau, G., Stock, C. A., Lam, V. W., & Frölicher, T. L. (2016). Structural uncertainty in projecting global fisheries catches under climate change. *Ecological Modelling, 325,* 57–66.

CIESM. (2018). *Engaging marine scientists and fishers to share knowledge and perceptions – Early lessons* (F. Briant, Ed.). CIESM Publishers. http://www.ciesm.org/online/monographs/download.php?file=exec50.pdf

Craig, P. C. (1984). Fish use of coastal waters of the Alaska Beaufort Sea: A review. *Transactions f the American Fisheries Society, 113,* 265–282.

Craig, P. C. (1987). Subsistence fisheries at coastal villages in the Alaskan Arctic, 1970-1986. *Minerals Management Service.*

Craig, P. C. (1989). An introduction to anadromous fishes in the Alaskan Arctic. *Biological Paper of the University of Alaska, 24,* 27–54.

Craig, P. C., Griffiths, W. B., Haldorson, L., & McElderry, H. (1985). Distributional patterns of fishes in an Alaskan Arctic lagoon. *Polar Biology, 4,* 9–18.

Dutil, J. D., Lambert, Y., & Boucher, E. (1997). Does higher growth rate in Atlantic cod (*Gadus morhua*) at low salinity result from lower standard metabolic rate or increased protein digestibility? *Canadian Journal of Fisheries and Aquatic Science, 54,* 99–103.

Edsall, T. A. (1999). The growth-temperature relation of juvenile lake whitefish. *Transactions of the American Fisheries Society, 128*, 962–964.

Elliott, K. H., & Gaston, A. J. (2008). Mass-length relationships and energy content of fishes and invertebrates delivered to nestling thick-billed Murres *Uria lomvia* in the Canadian Arctic, 1981–2007. *Marine Ornithology, 36*, 25–34.

European Commission. (2018). EU and Arctic partners enter historic agreement to prevent unregulated fishing in high seas. *European Commission of the European Union*. https://ec.europa.eu/fisheries/eu-and-arctic-partners-enter-historic-agreement-prevent-unregulated-fishing-high-seas_en

Executive Order 13921, 85 Federal Register 28, 471. (2020). https://www.federalregister.gov/documents/2020/05/12/2020-10315/promoting-american-seafood-competitiveness-and-economic-growth

Fechhelm, R. G., & Fissel, D. B. (1988). Recruitment of Canadian Arctic cisco (Coregonus autumnalis) into Alaskan waters. *Canadian Journal of Fisheries and Aquatic Science, 45*, 906–910.

Fechhelm, R. G., & Griffiths, W. B. (1990). Effect of wind on the recruitment of Canadian Arctic Cisco (*Coregonus autumnalis*) into the central Alaskan Beaufort Sea. *Canadian Journal of Fisheries and Aquatic Science, 47*, 2164–2171.

Fechhelm, R. G., Fitzgerald, P. S., Bryan, J. D., & Gallaway, B. J. (1993). Effect of salinity and temperature on the growth of yearling Arctic Cisco (*Coregonus autumnalis*) of the Alaskan Beaufort Sea. *Journal of Fish Biology, 43*, 463–474.

Fechhelm, R. G., Bryan, J. D., & Griffiths, W. B. et al. (1994). Effect of coastal winds on the summer dispersal of young Least Cisco (*Coregonus sardinella*) from the Colville River to Prudhoe Bay, Alaska: A simulation model. *Canadian Journal of Fisheries and Aquatic Science, 51*, 890–899.

Fechhelm, R. G., Martin, L. R., & Gallaway, B. J. et al. (1999). Prudhoe Bay causeways and the summer coastal movements of Arctic Cisco and Least Cisco. *Arctic, 52*, 139–151.

Fechhelm, R. G., Streever, B., & Gallaway, B. J. (2007). The Arctic Cisco (*Coregonus autumnalis*) subsistence and commercial fisheries, Colville River, Alaska: A conceptual model. *Arctic, 60*, 421–429.

Fisheries and Oceans Canada 2020. Aquaculture. *Government of Canada*. https://www.dfo-mpo.gc.ca/stats/aquaculture-eng.htm. Accessed 7 July 2020.

Ford, J. D., McDowell, G., & Pearce, T. (2015). The adaptation challenge in the Arctic. *Nature Climate Change, 5*(12), 1046–1053.

Fondahl, G., Filippova, V., & Mack, L. (2015). Indigenous peoples in the new Arctic. In *The new Arctic (7–22)*. Springer.

Fritz, M., Vonk, J. E., & Lantuit, H. (2017). Collapsing Arctic coastlines. *Nature Climate Change, 7*(1), 6–7.

Galappaththi, E. K., Ford, J. D., Bennett, E. M., & Berkes, F. (2019). Climate change and community fisheries in the Arctic: A case study from Pangnirtung, Canada. *Journal of Environmental Management, 250*, 109534.

Gatt, K. P., Hamman, C. R., Priest, J. T., & Sutton, T. M. (2019). *Beaufort Sea Nearshore Fish Monitoring Study: 2019 Annual Report*. Report for Hilcorp Alaska, LLC by the University of Alaska Fairbanks, College of Fisheries and Ocean Sciences, Department of Fisheries, Fairbanks, Alaska

Gibbs, A. E., & Richmond, B. M. (2015). *National assessment of shoreline change — historical shoreline change along the North coast of Alaska, U.S.-Canadian border to Icy Cape*. U.S. Geological Survey Open-File Report 2015–1048.

Hansen, E. A., & Closs, G. P. (2009). Long-term growth and movement in relation to food supply and social status in a stream fish. *Behavioral Ecology, 20,*
616–623.

Harter, B. B., Elliott, K. H., Divoky, G. J., & Davoren, G. K. (2013). Arctic cod
(*Boreogadus saida*) as prey: Fish length-energetics relationships in the Beaufort
Sea and Hudson Bay. *Arctic, 66,* 191–196.

Harvey, F. (2018). Commercial fishing banned across much of the Arctic. *The
Guardian.* https://www.theguardian.com/environment/2018/oct/03/commercial-
fishing-banned-across-much-of-the-arctic

Hermansen, Ø., & M. Troell (2012). Aquaculture in the Arctic. A review. Report no. 36,
Nofima, Tromsø.

Hermansen, Ø, & Heen, K. (2012). Norwegian Salmonid farming and global
warming: Socioeconomic impacts. *Aquaculture Economics and Management, 16,*
202–221.

Hoegh-Guldberg, O., Jacob, D., Taylor, M., Bindi, M., Brown, S., Camilloni, I. …
Zhou, G. (2018). Impacts of 1.5°C global warming on natural and human systems. In V. Masson-Delmotte, P. Zhai, H.-O. Pörtner, D. Roberts, J. Skea, P.R.
Shukla, A. Pirani, W. Moufouma-Okia, C. Péan, R. Pidcock, S. Connors, J.B.R.
Matthews, Y. Chen, X. Zhou, M.I. Gomis, E. Lonnoy, T. Maycock, M. Tignor,
and T. Waterfield (Eds.), *IPCC special report on the impacts of global warming of
1.5°C above pre-industrial levels and related global greenhouse gas emission pathways, in the context of strengthening the global response to the threat of climate
change, sustainable development, and efforts to eradicate poverty.* IPCC.

IPCC. (2014). Climate change 2014 synthesis report. In Team CW, R.K. Pachauri,
L. Meyer, and Working groups I, II and III (Eds.), *Fifth assessment report of the
intergovernmental panel on climate change* (1–151). IPCC.

Irrgang, A. M., Lantuit, H., Gordon, R. R., Piskor, A., & Manson, G. K. (2019).
Impacts of past and future coastal changes on the Yukon coast—threats for cultural sites, infrastructure, and travel routes. *Arctic Science, 5*(2), 107–126.

Irvine, J. R., Macdonald, R. W., & Brown, R. J., et al. (2009)/ Salmon in the Arctic
and how they avoid lethal low temperatures. *North Pacific Anadromous Fish
Commission Bulletin, 5,* 39–50

Jensen, Ø, & Rottem, S. V. (2010). The politics of security and international law in
Norway's Arctic waters. *Polar Record, 46*(236), 75–83.

Jones, B. M., Arp, C. D., Jorgenson, M. T., Hinkel, K. M., Schmutz, J. A., & Flint,
P. L. (2009). Increase in the rate and uniformity of coastline erosion in Arctic
Alaska. *Geophysical Research Letters, 36,* 1–5.

Kaiser, B. A., Pahl, J., & Horbel, C. (2018). Arctic ports: Local community development issues. In N. Vestergaard, B. Kaiser, L. Fernandez, J. Nymand Larsen (Eds.),
Arctic marine resource governance and development. Springer Polar Sciences.

Kaiser, B. A., Kourantidou, M., & Fernandez, L. (2018). A case for The com-
mons: The snow crab in the Barents. *Journal of Environmental Management, 210,*
338–348.

Kokorsch, M., & Benediktsson, K. (2018). Where have all the people gone? The limits of resilience in coastal communities. *Norsk Geografisk Tidsskrift-Norwegian
Journal of Geography, 72*(2), 97–114.

Kokorsch, M. (2018). *Mapping resilience – coastal communities in Iceland*
(Unpublished doctoral dissertation or master's thesis). [Ph.D. dissertation,
University of Iceland]. University of Iceland, Reykjavík, Iceland.

Kourantidou, M., Kaiser, B., & Vestergaard, N. (2021). International governance and Arctic fisheries. In A. D. Nuttall & M. Nuttall (Eds.), *Handbook of Arctic politics*. Routledge.

Křivan, V. (2003). Ideal free distributions when resources undergo population dynamics. *Theoretical Population Biology, 64*, 25–38.

Kroll, C., Warchold, A., & Pradhan, P. (2019). Sustainable development goals (SDGs): Are we successful in turning trade-offs into synergies? *Palgrave Communications, 5*(1), 1–11.

Kuhnlein, H. V., & Receveur, O. (2007). Local cultural animal food contributes high levels of nutrients for Arctic Canadian indigenous adults and children. *The Journal of Nutrition*, 1110–1114

Kuo, M. A. (2019) The US and China's Arctic ambitions. *The Diplomat.* https://thediplomat.com/2019/06/the-us-and-chinas-arctic-ambitions/

Lam, V. W. Y., Cheung, W. W. L., & Sumaila, U. R. (2014). Marine capture fisheries in the Arctic: Winners or losers under climate change and ocean acidification? *Fish and Fisheries, 17*(2), 335–357.

Landauer, M., & Juhola, S. (2019). Loss and damage in the rapidly changing Arctic. *Loss and damage from climate change* (pp. 425–447). Springer.

Larsen, P. H., Goldsmith, S., Smith, O., Wilson, M. L., Strzepek, K., Chinowsky, P., & Saylor, B. (2008). Estimating future costs for Alaska public infrastructure at risk from climate change. *Global Environmental Change, 18*(3), 442–457.

de March, B. G. E. (1989). Salinity tolerance of larval and juvenile broad Whitefish (Coregonus nasus). *Canadian Journal of Zoology, 67*, 2392–2397.

Meltofte, H. (Ed.). (2013). Arctic Biodiversity Assessment. Status and trends in Arctic biodiversity. *Conservation of Arctic Flora and Fauna (CAFF)*.

Melvin, A. M., Larsen, P., Boehlert, B., Neumann, J. E., Chinowsky, P., Espinet, X., & Nicolsky, D. J., et al. (2017). Climate change damages to Alaska public infrastructure and the economics of proactive adaptation. *Proceedings of the National Academy of Sciences, 114*(2), E122–E131.

Menezes, D. R., & Chater, A. (Eds.). (2018). *Proceedings of the Sustainable Development Goals (SDGs) in the Arctic High-Level Dialogue Series*, No.1 (Session 1: Arctic Circle Assembly 2017, Reykjavik, Iceland, 14 October 2017). London: Polar Research and Policy Initiative, February 2018.

Moerlein, K. J., & Carothers, C. (2012). Total environment of change: Impacts of climate change and social transitions on subsistence fisheries in Northwest Alaska. *Ecology and Society, 17*, 10.

Molenaar, E. J. (2012). Arctic fisheries and international law: Gaps and options to address them. *Carbon & Climate Law Review*, 63–77.

Molenaar, E. J. (2013). Arctic Fisheries Management. *The Law of the Sea and the Polar Regions* (pp. 243–266). Brill Nijhoff.

Monacci, N. M., Cross, J. N., Hurst, T. P., Long, W. C., & Rossin, A. (2019). Ocean Acidification in Alaska: Chemistry, Clams, Cod, and Crabs. *AGUFM, 2019*, OS11C-1491.

Monaghan, P. (2008). Early growth conditions, phenotypic development and environmental change. *Philosophical Transactions of the Royal Society B: Biological Sciences, 363*, 1635–1645.

Moon, K., Blackman, D. A., Adams, V. M., (2019). Expanding the role of social science in conservation through an engagement with philosophy, methodology, and methods. *Methods in Ecology and Evolution*; 10, 294–302. doi: https://doi.org/10.1111/2041-210X.13126

Murray, F., Bostock, J., & Fletcher, D. (2014). Review of recirculation aquaculture system technologies and their commercial application. *Report prepared for the Highlands and Islands Enterprise*. University of Stirling. http://www.hie.co.uk/common/handlers/download-document.ashx?id=236008c4-f52a-48d9-9084-54e89e965573

Newell, J. (2004). *The Russian far east: A reference guide for conservation and development*. Daniel & Daniel Publishers. http://urbansustainability.snre.umich.edu/wp-content/uploads/2012/10/Review_Geographical-Journal_Oldfield.pdf

NOAA Fisheries. (2018). U.S. signs agreement to prevent unregulated commercial fishing on the high seas of the Central Arctic Ocean. *NOAA*. https://www.fisheries.noaa.gov/feature-story/us-signs-agreement-prevent-unregulated-commercial-fishing-high-seas-central-arctic

NPFMC. (n.d.). Arctic fishery management. *North Pacific Fishery Management Council*. https://www.npfmc.org/arctic-fishery-management/

Østhagen, A. (2019). The new geopolitics of the Arctic: Russia, China and the EU. *Wilfried Martens Centre for European Studies*. https://www.martenscentre.eu/sites/default/files/publication-files/geopolitics-arctic-russia-china-eu.pdf

Pahl, J., & Kaiser, B. A. (2018). Arctic Port development. In N. Vestergaard, B. Kaiser, L. Fernandez, J. Nymand Larsen (Eds.), *Arctic marine Resource governance and development*. Springer Polar Sciences.

Pauly, D., & Swartz, W. (2007). Marine fish catches in north siberia (Russia, FAO area 18). In D. Zeller and D. Pauly (Eds.), *Reconstruction of marine fisheries catches for key countries and regions (1950-2005)*. Fisheries Centre Research Reports, 15(2).

Pálsson, S. K. (n.d.). Fishernet – DRAFT country report for Greenland. *The Stefansson Arctic Institute*. https://www.fishernet.is/images/stories/Country_report_Greenland.pdf

Pan, M., & Huntington, H. P. (2016). A precautionary approach to fisheries in the central Arctic Ocean: Policy, science, and China. *Marine Policy, 63*, 153–157.

Pinsky, M. L., Reygondeau, G., Caddell, R., Palacios-Abrantes, J., Spijkers, J., & Cheung, W. W. L. (2017). Preparing ocean governance for species on the move. *Science, 360*(80), 1189–1191.

Pring-Ham, C. (2018). Aquatic farming industry status [conference presentation]. *Alaska department of fish and game*. https://www.adfg.alaska.gov/static/fishing/PDFs/aquaticfarming/2018_asga_presentation.pdf

Rayfuse, R. (2018). Regulating fisheries in the Central Arctic Ocean: Much ado about nothing?. In N. Vestergaard, B. Kaiser, L. Fernandez, & J. Nymand Larsen (Eds.), *Arctic marine Resource governance and development*. Springer Polar Sciences.

Reist, J. D., Wrona, F. J., Prowse, T. D., Power, M., Dempson, J. B., King, J. R., & Beamish, R. J. (2006). An overview of effects of climate change on selected Arctic freshwater and anadromous fishes. *AMBIO, 35*, 381–387.

Resneck, J. (2018). Alaska wary of federal push for marine aquaculture. *Alaska Public Media*. https://www.alaskapublic.org/2018/09/06/alaska-wary-of-federal-push-for-marine-aquaculture/

SADA Report. (2014). *Strategic Environmental Impact Assessment of Development of the Arctic*. https://www.arcticinfo.eu/images/pdf/SADA_report.pdf

Seigle, J. C., & Gottschalk, J. M. (2013). Fall 2012 subsistence fishery monitoring on the Colville River. *ABR, Inc. – Environmental Research & Services*.

Schatz, V. J., Proelss, A., & Liu, N. (2019). The 2018 agreement to prevent unregulated high seas fisheries in the Central Arctic Ocean: A critical analysis. *International Journal of Marine and Coastal Law, 34*, 195–244.

Sköld, P., Baer, K., Scheepstra, A., Latola, K., & Biebow, N. (2018). *The SDGs and the Arctic: The need for polar indicators [Paper presentation].* Arctic Observing Summit 2018, Davos, Switzerland.

Somerville, C., Cohen, M., Pantanella, E., Stankus, A., & Lovatelli, A. (2014). Smallscale aquaponic food production - Integrated fish and plant farming (FAO Fisheries and Aquaculture Technical Paper No. 589). *FAO.* www.fao.org/3/a-i4021e.pdf

Statistics Greenland. (2019). Greenland in figures 2019. *Statistics Greenland.* http://www.stat.gl/LinkClick.aspx?link=Intranet%2fGIF_2009_WEB.pdf&tabid=57&mid=473&language=en-US

Statistics Iceland. (2020). Aquaculture in Iceland. *Statistics Iceland.* https://www.statice.is/publications/news-archive/fisheries/aquaculture-in-iceland/#:~:text=There%20has%20been%20a%20significant,has%20been%20increasing%20every%20year.

Stephen, K. (2018). Societal impacts of a rapidly changing Arctic. *Current climate change reports, 4*(3), 223–237.

Streletskiy, D. A., Suter, L. J., Shiklomanov, N. I., Porfiriev, B. N., & Eliseev, D. O. (2019). Assessment of climate change impacts on buildings, structures and infrastructure in the Russian regions on permafrost. *Environmental Research Letters, 14*(2), 025003.

Sumaila, R. (2015). Indigenous fisheries in a changing Arctic. *News Deeply.* https://www.newsdeeply.com/arctic/community/2015/12/08/indigenous-fisheries-in-a-changing-arctic

Sumaila, U. R., Lam, V. W. Y., Miller, D. D., Teh, L., Watson, R. A., Zeller, D. ... Pauly, D. (2015). Winners and losers in a world where the high seas is closed to fishing. *Scientific Reports, 5*, 1–6.

Suter, L., Streletskiy, D., & Shiklomanov, N. (2019). Assessment of the cost of climate change impacts on critical infrastructure in the circumpolar Arctic. *Polar Geography, 42*(4), 267–286.

Thorsteinson, L. K., & Love, M. S. (2016). Alaska Arctic marine fish ecology catalog. *U.S. Geological Survey Scientific Investigations Report 2016-5038* (OCS Study, BOEM 2016-048).

Troell, M., Eide, A., Isaksen, J., Hermansen, Ø, & Crépin, A. S. (2017). Seafood from a changing Arctic. *AMBIO, 46*(3), 368–386.

US Dept of State. (2020). The United States Ratifies Central Arctic Ocean Fisheries Agreement. *US Department of State.* https://translations.state.gov/2019/08/27/the-united-states-ratifies-central-arctic-ocean-fisheries-agreement/

Van Pelt, T. I., Huntington, H. P., Romanenko, O. V., & Mueter, F. J. (2017). The missing middle: Central Arctic Ocean gaps in fishery research and science coordination. *Marine Policy, 85*, 79–86.

Vilhjalmsson, H., Hoel, A. H., Agnarsson, S., Arnason, R., Carscadden, J. E., Eide, A., ... Wilderbuer, T. (2005). *Fisheries and aquaculture. Arctic Climate Impact Report* (pp. 691–780). Cambridge University Press.

Walker, D. A., & Peirce, J. L. (2015). Rapid Arctic Transitions due to Infrastructure and Climate (RATIC): a contribution to ICARP III (Alaska Geobotany Center Publication AGC 15Z 02). *University of Alaska Fairbanks.*

Wassmann, P., Duarte, C. M., Agust, S., & Sejr, M. (2011). Footprints of climate change in the Arctic marine ecosystem. *Global Change Biology, 17*, 1235–1249.

Zeller, D., Booth, S., Pakhomov, E., Swartz, W., & Pauly, D. (2011). Arctic fisheries catches in Russia, USA, and Canada: Baselines for neglected ecosystems. *Polar Biology, 34*, 955–973.

Zhang, Y., Yamamoto-Kawai, M., & Williams, W. J. (2020). Two decades of Ocean acidification in the surface waters of the Beaufort Gyre, Arctic Ocean: Effects of sea ice melt and retreat from 1997–2016. *Geophysical Research Letters, 47*(3), e60119.

13 The Arctic as a food-producing region

David Natcher, Ingrid Kvalvik, Ólafur Reykdal,
Kristin Hansen, Florent Govaerts, Silje Elde,
Bjørg Helen Nøstvold, Rune Rødbotten,
Sigridur Dalmannsdottir, Hilde Halland,
Eivind Uleberg, Jón Árnason,
Páll Gunnar Pálsson, Rakel
Halldórsdóttir, Óli Þór Hilmarsson,
Gunnar Þórðarson, & Þóra Valsdóttir

Introduction

In 2016, the Sustainable Development Working Group (SDWG) endorsed *The Arctic as a Food Producing Region* research project. Involving research teams from Iceland, Norway, Canada, Greenland, and Russia, the objective of the project was to assess the potential for increased production and added value of foods originating from the Arctic, with the overarching aim of improving food security, while enhancing the social and economic conditions of Arctic communities. Although the Arctic was recognised as an important food-producing region, there was a shared sense that the Arctic was not meeting its full potential, either in terms of satisfying local food needs or for maximising its domestic or international export potential. Yet beyond speculation, much of which was informed by individual or anecdotal experience, there was little understanding of the current production capacities of Arctic food sectors or where opportunities may lie for sustainable growth. The aim of the project was, therefore, threefold: (1) complete an inventory of the current levels of Arctic food production in terms of products, volumes, revenues; (2) identify the constraints and opportunities for increased production value-added opportunities; and (3) identify potential pathways and new value chains for expanding Arctic food production and distribution opportunities.

We conceptualised the Arctic food systems in terms of primary, secondary, and tertiary production. Primary food-producing industries include fishing/aquaculture, herding, and agricultural production of raw material, including harvesting, milking, and livestock production before slaughter. In this case, the original character of the product is not changed. Secondary processing includes slaughtering, processing, packing, and transport of a product/raw material from primary production. This includes adding some

DOI: 10.4324/9781003172406-13

conservatives (salt and sugar), drying, freezing, etc., to obtain a more value-added consumer-ready product. Tertiary production involves significant changes being made to the product, by adding other ingredients for consumer readiness.

Each country lead selected a number of products/species and food producers in their respective countries to be considered, highlighting the principal value chain characteristics for their regions. While some latitude was exercised in each study region, the research was guided by a common set of questions: What is the status and potential for various food production opportunities in the Arctic? What are the added values of these products when marketed by their special qualities and unique origin? What conditions are important to the further development of the Arctic as a food-producing region?

In this chapter, we provide a summary of food production in three regions of the Arctic. These include: (1) the entirety of Iceland; (2) Norway's three northernmost counties—Nordland, Troms, and Finnmark; and (3) northern Canada, including Yukon, Northwest Territories, Nunavut, Nunavik, and Labrador. In presenting these findings, it is important to acknowledge that we compiled national-level data on commercial food production only. Food production in and for Indigenous communities was emphasised in the areas/countries where this was relevant but did not include subsistence or food production at the household level. Also not captured are the cultural values that Indigenous peoples assign to many of the foods that have both commercial and subsistence value. For example, while Norway's national statistics capture the economic contribution of the reindeer industry, statistics alone fall short of measuring the cultural significance of reindeer to the Saami. Similarly, export data for seals were compiled for northern Canada, including the volumes and revenues garnered from the export of meat, pelts and oils. These data in no way capture the cultural importance of harvesting, sharing, and consuming seal meat by Inuit in northern Canada. Despite this gap, the commercial results we have compiled clearly demonstrate the significant contribution that commercial food production is making to the Arctic economy, and also shows the enormous untapped potential for Arctic food production to contribute to the economic and social well-being of Arctic communities.

A summary of the Arctic commercial food production

Fish and aquatic food resources

Iceland

The export of fish and fish products are by far the most important export items from Iceland and contribute significantly to the Icelandic economy. Fish products are exported from Iceland to more than 90 countries, with the

EU, United States, and Nigeria markets having the most commercial value. Cod, in particular, is the most important export species. The total cod catch in 2016 was 264 154 tons, of which 70,000 tons (26.5%) were caught by small vessels. In 2016, 669 tons of lumpfish caviar, 731 tons of salted lumpfish roes, and 2,700 tons frozen lumpfish blocks were exported from Iceland, worth an estimated 2,084 million ISK. A number of Arctic char farming operations are situated at different locations around Iceland. All operations are land-based and use water that, according to EU's water framework, is classified as being of unique quality extracted from springs, boreholes, and wells.

Northern Norway

From 2012 to 2016, Norway exported fish products worth 339,207,335,000 Norwegian krone (NOK). Norway's biggest market for fish products was Poland, which imported a value of 32,449,669,000 NOK in the same period. Other notable importing countries include France, Denmark, the United Kingdom, Russia, Sweden, Japan, the Netherlands, Spain, Germany, China, the United States, Portugal, and Italy, who together imported over 10 billion NOK, which accounted for 78% of all Norway fish exports from 2012 to 2016. The fish export value in Norway increased from 50.8 billion NOK in 2012 to 89.2 billion NOK in 2016, demonstrating an annual percentage increase of 8.8% in 2013, 11.5% in 2014, 7.3% in 2015, and 23.9% in 2016. Over this period, the value of fish exports from Norway increased by 75.6%. Only a small amount of the Norwegian seafood goes to domestic consumption. In fact, it is estimated that 95% of the fishery and aquaculture products are exported to 140 other countries. The most important species in terms of export volume include salmon, trout, cod, mackerel, herring, and saithe.

Northern Canada

Marine products accounted for 89% (3,470,745 tons) of northern Canada's total food export. Since 1988, approximately 3,470,745 tons of fish and aquatic products were exported from the Canadian north, adding more than $18 billion, or approximately $600 million per year, to the Canadian economy. The major commercial fisheries in northern Canadian are turbot, shrimp (northern and striped), and Arctic char, that are shipped fresh and frozen as a whole and as fillets. Other commercial products include dried fish, fish meal, molluscs, live fish, and other aquatic invertebrates. Among 133 export destinations, the United States, China, Japan, Denmark, and Russia are the leading importers of northern Canadian seafood. The revenue generated from commercial fisheries has increased slightly in recent years, from $709 million in 2010 to $798 million in 2019. The fish harvest in the Canadian Arctic underwent substantial changes from 1950 to 2018. From 1950 to 1960, the total harvest increased from 1,924.7 tons/year to 3,286.9 tons/year. A decline from 1960 until the early 1990s to 850 tons/year brought the harvest down to almost one third of the peak harvest in the 1960s. From 1994 to 2018, a slight increase led to a fish catch of 993.6 tons in 2018. In addition to fisheries, Canada has exported (2005 and 2014) more than $66.6 million worth of seal products (pelts, meat, and oils) to 48 countries.

Meat production

Iceland

There are over 3,000 sheep farms in Iceland. These farms tend to be small and family owned. The sheep farming is as old as the human settlement of Iceland. Iceland is very well suited for sheep farming with plentiful grasslands and highland pastures. Lambs graze in the highlands until slaughtered in the autumn. The grazing areas are comprised of native vegetation, which enhances the quality and reputation of Iceland's lamb industry.

There are ten slaughterhouses for sheep in Iceland. During the 2017 slaughter season, 560,500 lambs were slaughtered. The average carcass weight was 16.5 kg and total production was 9,200 tons of lamb meat (carcasses). The value was 6,200 million ISK for lamb meat sold at the domestic market. The volume of exported lamb and sheep meat was 4,100 tons. The challenges for sheep farmers are low income and the need for off-farm employment. The slaughterhouse industry has a strong position on the market and needs a considerable share of the market value. The sheep farming industry also faces competition from other more profitable meat sectors. Beyond meat production, sheep farming includes secondary production. For example, skyr production has been a part of the Icelandic cuisine since settlement. The industrialisation of skyr started in the 1930s. Gradually domestic production of skyr decreased and by 2010 only a few farms were reported as skyr producers. In the last few years several farms have started to produce skyr according to the traditional methods that have been accompanied in growth in domestic sales. Skyr has in the past been sold mainly within Iceland but in recent years has been increasing in international exports. About 4,500 tons are produced annually. A total volume of about 1,300 tons are exported for a value of about 500 million ISK. The remaining 3,200 tons are sold on the domestic market.

Northern Norway

The largest agricultural production systems in Northern Norway are based on meat production. Abundant high-quality grazing areas are an important reason for quality production, but also benefits from substantial barn capacity. In the primary production, there are 1,312 producers of cattle (mainly dairy farmers), 1,608 sheep producers, 115 pork producers, and approximately 30 dairy goat farmers. The total primary production was 186,462 sheep, 28,061 cattle, and 103,600 pigs slaughtered in 2016. In that same year, there were approximately 1,000 employees in the secondary and tertiary industry in northern Norway. The meat products are sold locally, regionally, and nationally. The industry is characterised of high level of processing. There are possibilities for increased meat production in the north, both by utilising the total potential in the outfield grazing fields for increased primary production and by increasing the value-added by producing local specialty products. Northern Norway also produces dairy cattle, sheep, and goat. This industry is more less dominated by large cooperatives owned by the farmers, including TINE (dairy) and Nortura (meat), which are supplied by roughly 500 local producers.

The reindeer industry in Norway is closely connected to the Saami-culture and heritage. Only persons of Saami descent can be owners of reindeer, a right protected by law. In 2017, 3,233 persons were registered as reindeer owners in Norway. However, slaughtering and processing of the animal products is open to the qualified general public. The majority of reindeer are found in Finnmark (75%), while Troms and Nordland have approximately 6% and Trønderlag has 14%. The reindeer are semi-domesticated, which means they roam freely, where they consume grass during summer and lichen during winter.

Northern Canada

The value of live animals and animal products exported from northern Canadian has increased steadily over the last 30 years. During the 1990s, the average revenue from sales was roughly $858,814 CAD per year. However, by 2000, annual sales increased to $2.5 million CAD, or an increase 300%. Overseas exports fluctuated from 2000 to 2017, with the highest values being $3.5 million in 2003 and $4 million in 2011.

The western Canadian Arctic area near Sachs Harbor and Ulukhaktok is known for its large-scale commercial harvest of muskoxen. Usable parts from a muskox include meat, hide, wool (Qiviut), and horns. Northern communities export these items to markets in southern Canada. In addition to hunting for subsistence and marketing purposes, guided sport hunting-and-outfitting based tourism connected to muskoxen provides a source of food and income for northern communities. In the Northwest Territories (NWT), the Inuvialuit Regional Corporation helps Sachs Harbour residents hold a community muskox hunt on Banks Island. The muskox population on the island fluctuates between 80,000 and 120,000 animals. At present, the community harvests a few hundred muskoxen annually (roughly 4% of the total estimated population). On average, the hunt brings $70,000 worth of wages to Sachs Harbour and employed more than 20 community members in meat processing and an additional 30–35 hunters involves in the actual harvest (Ryan, 2006).

Although caribou hunting in northern Canada supports household subsistence and inter-settlement trade, the feasibility of commercial hunts is limited. As supplies from wild harvests are intermittent, this influences the commercialisation of caribou meat or any other types of sales of caribou in southern markets. The uncertainty of supply makes buyers from outside Canada hesitant to enter into contracts with northern producers, reducing the viability and commercial value of the caribou harvest.

Agricultural production

Iceland

Geothermal energy for heating and electricity produced by hydroelectric power stations for illumination are the basis of vegetable production in greenhouses in Iceland. Most of the greenhouses are located in clusters where geothermal energy is available. Several greenhouses operate through long winters by using electrical illumination. The vegetables typically grown year-round in greenhouses include tomatoes, cucumbers, bell pepper, and lettuce, but in some cases also includes spinach, cabbage, kale, and herbs. Some greenhouses have successfully grown strawberries but have been challenged financially when competing with the lower imported prices. Agricultural products are marketed primarily for domestic markets. The greenhouse production of cucumbers, tomatoes, and bell pepper was 3,500 tons in 2016. The total value of greenhouse and out-door vegetable production in Iceland was 3,800 million ISK 2016. The value of greenhouse products is a considerable part of this, estimated to be about 1,500 million ISK. The number of employees in vegetable production was roughly 237 and for related services 107 employees were involved (outdoor production included). The greenhouse vegetable production in Iceland meets only a part of the domestic demand. The market share for domestic tomatoes and cucumbers is the highest (70–90%) but lowest for lettuce and bell pepper (about 10%). The import of vegetables is therefore considerable. Vegetable production could be increased considerably, however there is import competition. Barley production in the country is 10,000 to 16,000 tons per year. Only about 2% of the barley is used for food, but this proportion could be increased.

Northern Norway

The agricultural production in northern Norway is the northernmost active agricultural system in the world. It is only possible because of the warm air carried by the North-Atlantic current and because of the latitudinal placement of the region; growing conditions that are characterised by short growing season and 24-hour day daylight in mid-summer. The main horticultural production is potatoes that are grown on about 460 hectares. Vegetables and berries are grown on about 65 hectares with about equal distribution in Troms and Nordland counties. Key strengths for horticultural production in Northern Norway are the natural growing conditions with cool summers and long days. These conditions imply little problems with pests and diseases and require low use of pesticides and herbicides. There are also indications of specific quality attributes of horticultural products produced under these growing conditions. It has been documented in several horticultural products grown in Northern Norway that the low temperature, in combination with 24-hour light and longer photosynthetic activity, causes more crispy and juicy products with sweeter taste compared to the same product produced further south. It is usually not a result of the increased amount of sugars, but rather less production of different kind of bitter-tasting substances. Several food producers in Northern Norway are using these particular biological attributes together with the product origin, for marketing products with 'Arctic quality.' This is used in the successful marketing of a locally produced potato variety gulløye, as "the potato of the midnight sun."

Northern Canada

From 1988 to 2017, 3641 tons of roots, tubers, cucumbers, beans, chickpeas and mushrooms were harvested in the Canadian Arctic and exported (mostly frozen) to countries around the world. The highest volume was in peas, with 492 tons exported to China in 2006, and potatoes, with 275 tons ($107,621) exported to Russia in 2014. During that same period, blueberries, cranberries, bilberries, raspberries, and blackberries accounted for 3,377 tons in export. The Netherlands and other European Union countries were the major markets for fruit and nut products from northern Canada. From 1988 to 2017, the total weight of oil seeds, oleaginous fruits, straw, and fodder exported from northern Canada reached 1,202 tons ($1,689,154) and contributed 13% to the total weight of farm exports. Other farm products including live trees, teas, cereals, malts, starches, resins, and vegetable saps, were also exported with a weight of 964 tons ($1,346,561). Unlike the large commercial farms in the Canadian prairies, agricultural production in the Canadian north is conducted small-scale farms, community gardens, and greenhouses. The 2016 Census of Agriculture indicated that there were 142 farms operating in Yukon, encompassing a total farm area of 10,330 hectares, with 6,801 ha involved in pasture and the production of crops. The number of farms operating in Yukon has fluctuated since the 2001 Census of Agriculture, from a high of 170 (2001) to a low of 136 (2006), though the total area in production has not changed as notably.

Discussion

The objective of the *Arctic as a Food Producing Region* project was to identify new food production opportunities that could lead to sustainable economic development for Arctic communities. Preliminary results show that within the Arctic region there are considerable opportunities for commercial food production, both for export and for meeting local food needs.

Food industries are producing large volumes of food commodities that are culturally compatible with Indigenous/local food preferences and also have high export value.

There are, however, large variations in actual and potential production and harvesting volumes, both between Artic Nations, species, and product groups. The volume variations at a national level can, for instance, be seen in the export statistics. These differences will have a large effect on product development and marketing strategies chosen by producers as a whole. Whereas variations exist in production levels, critical infrastructure, marketing access, lack of available raw material, and skilled workforce, and environmental issues are some of the main challenges shared by all. Arctic food industries also experience production or harvesting limitations, due to national regulations (for example, quotas, health, and safety legislations) or natural conditions of the Arctic (for example, climate, availability of food sources, and resource availability). Second, high transport costs and export tariffs on high-value food products, affect the profitability of food-producing companies, for example, contributing to less processing of seafood within the Norwegian or Canadian borders. Third, there is a general shortage of skilled labor in many rural and remote areas. Since food harvesting and/or production in most instances are located in districts/regions with low population density, acquiring knowledgeable and flexible employees can be a constraint to industry development. For example, the population in northern Iceland has been declining over the last few decades and this could develop to a critical constraint to industry growth. Iceland, except for the capital area, is sparsely populated, with changing demographics impacting regional innovation capacity. Qualified human recourses are important to allow for industry innovations that will be necessary to capitalise on existing and emerging markets. Lacking the necessary labor and human resource capacities may then limit investment interest that will be the key to stabilising the populations of remote regions.

The Arctic climate poses obvious constraints to the development of food industries, particularly in the context of agricultural expansion and diversification. This includes hampering the adoption of new food-producing technologies that are more readily available in southern markets. For example, in southern Canada, the largest algae cultivation systems to date use open pond systems. These autotrophic systems, however, have limited applicability in Canada's northern climate. Given the high capital and operating costs of closed photobioreactor systems, most analysts are skeptical that economically sustainable algae cultivation can take place in northern Canada. Factors limiting algae biomass production in northern Canada include limited solar irradiance, high capital costs, high energy and operating costs, few opportunities for colocation with symbiotic industry partners (Pankratz, Oyedun, Zhang, & Kumar, 2017).

Legislation at various levels of government also prohibits or severely restricts local food production, particularly for export markets. The requirement to meet federal food safety and inspection has also restricted the commercial development of these foods for export, which in turn, has limited entrepreneurial development. Other country-specific legislation can further impede industry development. For example, in Canada a federally supported food program (Nutrition North Canada) subsidises the transportation costs for selected foods shipped from the south to northern communities. The objective of the program is to help make healthy foods more affordable and accessible to northern communities. On average, the program provides an annual subsidy of $65 million, nearly half of which subsidises the transport of meat, fruits, and vegetables. As of 2018, 121 northern communities were eligible for subsidised food rates. While well-intentioned, the subsidy program has in some cases caused economic disincentives for local food producers, as the subsidised costs of imported foods are often lower than actual costs of food produced in the north. Adding further challenges to local food production in northern Canada are the Comprehensive Land Claims (CLC). The settlement of Comprehensive Land Claims (CLC) recognises Indigenous ownership of over 600,000 km² of land, protection of traditional ways of life, confers exclusive and preferential harvesting rights, and provides for Indigenous participation in land and resource management decisions. CLC have the potential to facilitate entitlement and access to food as they contain specific provisions that protect Indigenous hunting, harvesting, and fishing rights. While the settlement of CLC has in many ways empowered Indigenous peoples in Canada, the specific provisions found in these agreements have also created barriers to the development of food-related industries. While the provisions vary from one CLC to another, most contain explicit language that prohibits the commercial sale of traditional foods. The only exception being the Nunavut Land Claims Settlement that was signed in 1993.

Opportunities

Notwithstanding the challenges noted above, there are opportunities for increased food production in the Arctic. While the Arctic climate is a constraint to some areas of food production, new opportunities may also arise from global warming, especially within land-based production. Agriculture in Arctic regions is considered marginal, due to short growing seasons with low growth temperatures. The forecasted increase in temperature due to global warming is predicted to be greater in the Arctic than the global average. As the growing season is being prolonged, it creates opportunities for new and or marginally productive agricultural species, especially annual species and even barley cultivation.

Opportunities for increasing food production and adding value, improving product quality, and increasing food tourism and local markets also exist. This has been the case in northern Canada where the production of prepared or value-added foods has been steadily increasing since 1990s.

Since 2000, the export volume of prepared foods has increased by approximately 18,000 tons per year, or an increase of 384%. These exports consist mainly of value-added fisheries products (for example, farmed char), but also includes agricultural products (berries, teas, and beverages) that are contributing to local economies and food security of the region.

Identifying special chemical attributes within raw materials and using this for marketing or as extractions of valuable compounds used within the industry may also be an opportunity for growth. Other food innovations could include the introduction of full utilisation methods and new production methods, for example using waste products from the seafood to produce medicine or fabrics or seaweed production as both a food source and for CO_2 sequestration. Iceland has a variety of biological resources available for sustainable and responsible utilisation, some of which are underutilised. The major underutilised resources include side streams, such as those found in the marine and agricultural industries. However, data on these resources are sparse, resulting in ineffective decision-making and inefficiencies in technological innovations.

Iceland, Norway, and Canada each have the advantages of adding value to the products by further processing and product development or identifying the local value by historic background or Arctic quality for increasingly growing consumer market. Iceland and Norway are already identifying special chemical attributes within the raw material produced in the Arctic climate and using this for marketing or as extractions of valuable compounds used within the industry. For these reasons, some of main opportunities for increased food production and value-adding revolves around niche products and storytelling, better use of surplus biomasses, improving product quality, and increasing food tourism and local markets.

There are also opportunities for increased food production in the Arctic given the demand for high-quality foods, which are culturally compatible, and sustainably produced. Local niche products are being developed for specialised markets. New opportunities might include domestic food production in farms and villages to meet local needs while reducing carbon footprint. These are proving to be an interest to tourists who are visiting northern regions and who are receptive to local foods and food-related experiences. The continued growth of the tourist industry could become important for the food-related economy of the northern regions. The increasing number of tourists also increases the demand for food. To satisfy this demand, local food production will need to increase production or risk an even greater reliance on food imports. The tourism industry provides an opportunity for food producers to increase production and develop new products. Regional products are of particular interest to tourists and also help to increase sustainability and support local food producers and spin-off industries.

Food originating from the Arctic may also have a marketing advantage. Consumers generally prefer food that is healthy, with good taste, and produced in a sustainable manner, and increasingly they prefer food with a unique story. Food from the Arctic may score high on all these characteristics. There are significant opportunities for developing high-value niche "Arctic products" for tourist markets and other selected regional and national markets. Consumer preference studies conducted in Norway (this volume) and Canada (Yang et al., 2020) indicate that consumers feel the consumption of Arctic foods allow them to experience Indigenous cultures and tradition, while supporting Indigenous communities economically. Compared to other places of origin, Canadian and Norwegian consumers place higher value on the features of Arctic origin, and would choose to purchase wild-caught fish by remote/Indigenous producers over other southern-based alternatives. Consumers have also shown support for new certification standards for Arctic sustainability and authenticity, which could be further used as a marketing advantage.

Summary

Our results indicate that within the Arctic region there are considerable opportunities for commercial food production, both for export and for meeting local food needs. Yet Arctic food industries are also challenged by a plethora of social, economic, climatic, and logistical constraints. Despite the growth in commercial food production in the Arctic, there remain a considerable number of constraints to industry development, including lack of necessary infrastructure, fragmented supply chains, limited access to a skilled workforce, absence of innovation in product development, and limited access to, and knowledge of, domestic markets and consumer interest. While these challenges are experienced unevenly across the Arctic regions, Arctic food industries, as a result, tend to be fragmented with tenuous professional connections and limited communication streams. These conditions have in part led to their general overreliance on raw food exports, imported packaged foods, bottlenecking of distribution points, and limited innovation in primary and secondary product development. These conditions, in turn, have accelerated a nutrition transition among populations characterised by an erosion of diet quality, increased consumption of ultra-processed foods.

Land-based plant production in the Arctic has not received much attention over the years, although it may be of global importance to utilise the production capacities of the northern regions. The effects of global warming have the potential to provide new opportunities for growing new crops and cultivars with a higher yield potential within a prolonged growing season. To identify the capacity for increasing plant production and agriculture in Arctic areas, it is of major importance to understand the mechanisms behind the effects of climate change on plant production and plant persistence. Screening of available genetic resources in the Arctic is important for selecting well-adapted plant material both for crop production and for the preservation of biodiversity. Implementing technological solutions into production systems will further increase the production potential. For the successful development of rural plant production, knowledge transfer is crucial. This can be achieved by strengthening and enlarging the R&D network to support collaboration across national borders in the Arctic. Engaging Indigenous communities and industry stakeholders could enhance innovations in agriculture and create new markets and employment opportunities for northern communities. A better insight into the effective use of available plant genetic resources could also give social and economic advantages for the future rural population in Arctic regions. Future opportunities also lie in new and better storage methods for seasonal raw material and full utilisation of raw materials.

Icelandic Food and Biotech R & D (Matis), has already established several food innovation centers at various locations in the country. Support from regional innovation centers has been a successful strategy for small-scale product development. The intention is to improve regional food production further and the Icelandic government is expected to increase funding for local and regional innovation in food sectors. This is important, as small-scale local and regional food producers and entrepreneurs need financial assistance in the early stages of food innovation. Norway has adopted a similar approach to stimulating innovation in the food sectors. There has been a focus on building various types of network cooperation between the local food companies. These networks are often partly funded through Innovation Norway. Most networks are geographically linked like *Nordlandsmat* or *Vesterålsmat*, but some are also focused on a specific product type like Lofotlam (sheep meat) and Arktisk kje (goat meat). These networks focus on overcoming the challenges of being a small producer in the north by achieving large-scale advantages and capitalising on economies of scale in distribution, sales, and marketing. Evaluations conducted by Innovation Norway found that industry success most often requires business objectives to be clearly defined and offer value to all parties in the value chains. In addition, it was found that success takes time and trust between the companies, especially if they see themselves as competitors, is required. Some of the more successful networks in Norway have evolved for over a decade like *LofotenMat* (founded in 2007) and *Fjellfolket* (funded in 2006). During this time governmental support has been critical for build up the local food sector in Norway and in launching the Centre of Expertise in 2002 for local food production, with a northern hub is located at Nibio in Tromsø. The target industry groups are small to medium-sized food producers whose shared goals include adding increased value to Arctic food products through innovation, skills development, and entrepreneurial training.

We believe the time is right for a pan-Arctic focus on food production and sustainable economic development. Investing in Arctic food systems transformation requires a systems-based approach to policy formulation and investment grounded on a solid foundation of interdisciplinary research that connects capacities and institutions across the Arctic. Likewise, achieving the food security for all Arctic peoples requires a robust research agenda that is developed and implemented in a concerted manner with consistent investment. The international experience and development discourse are much better informed by research in the Global South. The International Food Policy Research Institute (IFPRI), for example, was established in the 1970s alongside substantial investments in agricultural research and the establishment of the global, multi-hub Consultative Group on International Agriculture Research (CGIAR), with the explicit purpose of facilitating independent research that would deal with socioeconomic policies for agricultural development and to improve the understanding of national food policies to promote the adoption of innovations in food producing technologies. With the establishment of national 'hubs', a cluster-based approach to Arctic food innovation could similarly draw together Arctic food producers with governments, Arctic Indigenous communities, universities, research centers, vocational training providers, and industry associations and young people. Such a concerted focus on food production and Arctic sustainable development would prove capable of responding to global challenges and would define the Arctic's role in sustainable development locally and around the world.

References

Animalia statistikk. (n.d.) *Slaktestatistikk per dyreslag.* Animalia. http://statistikk. animalia.no/statistikk/

Dalmannsdottir, S., Jørgensen, M., Rapacz, M., Østrem, L., Larsen, A., Rødven, R., & Rognli, O. A. (2017). Cold acclimation in warmer extended autumns impairs freezing tolerance of perennial ryegrass (Lolium perenne L.) And timothy (Phleum pratense L.). *Physiologia Plantarum, 160*, 266–281.

Johansen, T. J., Hykkerud, A. L., Uleberg, E., & Mølmann, J.. (2018). *Arktisk kvalitet – En beskrivelse av nordlige natur- og klimaforhold og virkning på egenskaper hos nordnorske matprodukter.* NIBIO Rapport. http://hdl.handle.net/11250/2494675.

Landbruksdirektoratet. (n.d.). *Statistikk fra søknader om produksjonstilskudd i jordbruket.* Landbruksdirektoratet. https://www.landbruksdirektoratet.no/filserver/prodrapp.htm

Lusk, J. L., & Briggeman, B. C. (2009). Food values. *American Journal of Agricultural Economics, 91*(1), 184–196.

Matmerk. (2016, October 27). *Lokalmatsalget vokser mest.* Matmerk. https://www. matmerk.no/no/matmerk/aktuelt/lokalmatsalget-vokser-mest

NIBIO. (2020). *Kompetansenettverket for lokalmat i Nord.* NIBIO. https://www. nibio.no/tema/mat/kompetansenettverk-for-lokalmat-i-nord

Pankratz, S., Oyedun, A. O., Zhang, X., & Kumar, A. (2017). Algae production platforms for Canada's northern climate. *Renewable and Sustainable Energy Reviews, 80*, 109–120.

Pettersen, I. (2018). *Mat og Industri.* NIBIO. www.matogindustri.no

Proff. (2020). *Proff-the Business Finder.* Proff. www.proff.no

Ryan, R. (2006). Summary of proceedings: Building a kitikmeot economic development strategy. Proceedings Report KEDC. Regional Planning Meeting.

Statistics Norway. (2018). *External Economy: External Trade.* Statistics Norway. https://www.ssb.no/en/utenriksokonomi?de=External+trade+

Statistics Norway. (2018). *Meat Production.* Statistics Norway. https://www.ssb.no/statbank/list/slakt?rxid=2c9310c4-9a42-4bf9-ae33-94ebde71e54a

Stiberg-Jamt, R., Brastad, B., Flatnes, A., Hauge, E., Tobro, M., & Geschwandtnerova, V. (2018). *Evaluering av Bedriftsnettverks-tjenesten til innovasjon Norge.* Oxford Research.

Yang, Y., Hobbs, J., & Natcher, D. C. (2020). The Arctic as a food producing region: Consumer perceptions and market segments. *Canadian Journal of Agricultural Economics.* doi: https://doi.org/10.1111/cjag.12255.

14 The nexus between water, energy, and food (WEF) systems in Northern Canada

David Natcher & Shawn Ingram

Introduction

In 2015, the United Nations introduced Transforming the World: the 2030 Agenda for Sustainable Development. The 2030 Agenda was endorsed by the world's leaders to serve as an action-oriented road map for safeguarding the welfare of current and future generations (Lim, Søgaard Jørgensen, & Wyborn, 2018). At the core of the Agenda are 17 Sustainable Development Goals (SDGs) that serve as benchmarks for achieving equality, prosperity, and environmental sustainability. Each of these 17 SDGs has specific targets (N=169) and associated indicators (N=232) that are used to measure advancements toward the attainment of each SDG.

While Agenda 2030 has been heralded as a platform for protecting our environment for current and future generations, some (e.g., Nilsson, Griggs, & Visbeck, 2016) warn that simply "ticking off" SDG targets without considering cross-sectoral interactions may result in ill-informed and unintended outcomes. For example, Lim et al. (2018) argue that there is an inherent risk in global goal formation, whether in the case of SDGs or its predecessor, the Millennium Development Goals, when targets are compartmentalised, siloed, and viewed through a reductionist lens (Costanza et al., 2015; Nilsson & Costanza, 2015). In these cases, the complexities of individual SDGs may be obscured, and critical interactions in the global system can go unnoticed (Lim, Søgaard Jørgensen, & Wyborn, 2018). The United Nations acknowledges the risks of treating SDGs as discreet and unrelated and has called for greater attention to the interactions between SDG targets. This includes careful consideration of both the synergies and trade-offs associated with SDG attainment. Synergies include the positive effects of achieving multiple SDG targets through simultaneous interventions, for instance through mutually beneficial infrastructure developments, whereas trade-offs occur when advancements toward one target have a negative impact on the ability to reach other targets, whether due to environmental degradation or intensive use of resources (Fader, Cranmer, Lawford, & Engel-Cox, 2018). For example, Pradhan, Costa, Rybski, Lucht, and Kropp (2017) found that SDG 1 (Ending poverty) has synergetic relationships with most of the

DOI: 10.4324/9781003172406-14

other SDGs, while SDG 12 (Responsible consumption and production) is most commonly associated with trade-offs. Accounting for the positive and negative spillover effects of SDG attainment is therefore essential to formulating sustainable solutions to global challenges (McCollum et al., 2018; Rasul, 2016).

Various methodologies have been developed to systematically map and rank the level of interactions between SDG targets (Nilsson, Griggs, & Visbeck, 2016). These approaches have generally been referred to as nexus research and are used to define, measure, and analyse the connections and interactions between SDGs. There are multiple fields of nexus research but the relationship between water (SDG 6), energy (SDG 7) and food (SDG 2) (WEF) has received considerable research attention (Endo, Tsurita, Burnett, & Orencio, 2017). This focus has been attributed, in large part, to the pervasive interactions that occur between WEF systems (Bhaduri, Ringler, Dombrowski, Mohtar, & Scheumann, 2015; Biggs et al., 2015).

Over the past decade, WEF nexus studies have been conducted at the global (Fader, Cranmer, Lawford, & Engel-Cox, 2018), national (Mainali, Luukkanen, Silveira, & Kaivo-Oja, 2018), and regional levels (Kulat, Mohtar, & Olivera, 2019; Liu, 2016), and all have concluded that WEF nexus research is informative for resource planning and developing effective policies for sustainable development (Pittock, Orr, Stevens, Aheeyar, & Smith, 2015). This was the impetus for the Arctic Council to adopt the United Nations' SDGs to inform its own strategic policy direction; noting the 2030 Agenda for Sustainable Development is global in scope but also applicable to Arctic regions. In particular, the Arctic Council's Sustainable Development Working Group (SDWG) endorsed the principles of Agenda 2030 and made a commitment to use SDG targets as guideposts for advancing the sustainable development of Arctic regions (SDWG, 2017). Yet before those guideposts can be determined, the SDWG acknowledged that a better understanding of the potential synergies and trade-offs between SDG targets was needed before regional implementation could be considered.

Although WEF nexus studies have been conducted in regions around the world, no assessment has been conducted in an Arctic setting. It is in this context that our study for northern Canada was conducted. As defined by the Arctic Council, northern Canada includes Yukon, Northwest Territories, Nunavut, Nunavik, and Labrador. This study serves as a preliminary assessment of the nexus between: SDG 2—ending hunger and achieving food security for all; SDG 6—ensuring the availability and sustainable management of water and sanitation for all; and SDG 7—ensuring access to affordable, reliable, sustainable, and modern energy for all (Table 14.1). By evaluating the current state of WEF security in northern Canada, and making visible the synergies and trade-offs between WEF-SDG targets, policy makers in Canada will be in a more informed position

Table 14.1 United Nations sustainable development goals and targets (SDG 2, 6, and 7)

Goal 2	End hunger, achieve food security and improved nutrition and promote sustainable agriculture.
Target 2.1	End hunger and ensure access to safe, nutritious and sufficient food all year round.
Target 2.2	End all forms of malnutrition.
Target 2.3	Double the agricultural productivity and incomes of small-scale food producers
Target 2.4	Ensure sustainable food production systems and implement resilient agricultural practices
Target 2.5	Maintain the genetic diversity of seeds, cultivated plants and farmed and domesticated animals and their related wild species.
Target 2a	Increase investment, including through enhanced international cooperation, in rural infrastructure, agricultural research.
Target 2b	Correct and prevent trade restrictions and distortions in world agricultural markets.
Target 2c	Adopt measures to ensure the proper functioning of food commodity markets.
Goal 6	Ensure availability and sustainable management of water and sanitation for all.
Target 6.1	Achieve universal and equitable access to safe and affordable drinking water for all.
Target 6.2	Achieve access to adequate and equitable sanitation and hygiene for all.
Target 6.3	Improve water quality by reducing pollution.
Target 6.4	Substantially increase water-use efficiency across all sectors and ensure sustainable withdrawals and supply of freshwater.
Target 6.5	Implement integrated water resources management at all levels, including through transboundary cooperation as appropriate.
Target 6.6	Protect and restore water-related ecosystems.
Target 6a	Expand international cooperation and capacity-building support to developing countries in water- and sanitation-related activities and programmes.
Target 6b	Support and strengthen the participation of local communities in improving water and sanitation management.
Goal 7	Ensure access to affordable, reliable, sustainable and modern energy for all.
Target 7.1	Ensure universal access to affordable, reliable and modern energy services.
Target 7.2	Increase substantially the share of renewable energy in the global energy mix.
Target 7.3	Double the global rate of improvement in energy efficiency.
Target 7a	Enhance international cooperation to facilitate access to clean energy research and technology.
Target 7b	Expand infrastructure and upgrade technology for supplying modern and sustainable energy services for all

to carry out integrative planning. This is particularly necessary given the relatively high rates of WEF insecurities currently experienced in northern Canada compared to the rest of Canada (see CCA, 2014; Egeland, 2009; Natcher, Shirley, Rodon, & Southcott, 2016; Poppel, 2015); insecurities that may be compounded by the social and ecological stresses that are expected to accompany climate change.

Following this introduction, we provide a brief review of the current status of WEF security in northern Canada. This includes current assessments of household food security, access to clean water, and dependence of non-renewable energy sources. We then discuss the data used in this assessment and the methodology employed to calculate interactions between WEF-SDG targets. Our results are then presented, and are followed by our conclusion that highlights our key findings and recommendations for future research.

WEF (in)security in Northern Canada

Communities in northern Canada experience higher rates of water, energy, and food insecurity relative to the national average. These conditions have evoked national (e.g., CCA, 2014) and international attention (e.g., De Schutter, 2012). Below, we offer a general portrait of WEF systems in northern Canada.

Water (in)security

Nearly all residents in northern Canada have access to adequate drinking water and sanitation services. For example, roughly 99% of Inuit in Nunavut have hot and cold running water, a flush toilet, and a septic tank or sewage system (Poppel, 2015). However, northern Canada residents do not have universal access to improved water and sanitation services, with most gaps occurring in remote and rural areas (Bressler & Hennessy, 2018). Approximately 74% of communities in Northwest Territories and Nunavut have trucked water supplies and waste disposal systems, with 16% having below/above ground water distribution systems, and 10% using a combination of water buckets, privies, or trucked services (Environment and Climate Change Canada, 2013). The overreliance on trucked water and waste removal has placed increased pressures on water utilities and constrains the ability for northern communities to deliver reliable water and sanitation services.

The quality of drinking water is also variable and is subject to seasonal fluctuations, often due to environmental or climate-related impacts, which at times affects water safety. In Nunavut, 13% of residents have indicated their water is not safe for drinking in general, while 21% indicated it is not safe for drinking at least some times throughout the year (Poppel, 2015). Nunavut residents have also indicated that environmental and climate-related events have led to decreases in water quality and quantity, damage to water and sanitation infrastructure, and water maintenance and treatment issues (Bressler and Hennessy, 2018). The government of Northwest Territories (GNWT) reported that seven communities within the NWT had boil water advisories at some point in 2018 (GNWT, 2019). However, the GNWT anticipates climate change to have a negative impact on water quality and quantity, with detrimental changes caused by increased temperatures, extreme weather events, variability in precipitation, and impacts to critical infrastructure (GNWT, 2018). Climate change is also contributing to water scarcity, with some northern cities like Iqaluit preparing for water shortages by 2024. In addition to population growth and increasing demand, Iqaluit's water system is challenged by warmer weather and declining levels of rainfall that have led to water shortages (Bakaic et al., 2018).

Territorial governments have developed proactive plans to address water insecurity. For example, in 2014, Yukon released its water strategy and action plan with the overall goal of maintaining the quality, quantity, and health of Yukon water for both people and the environment. Priorities for achieving these goals include maintaining and improving access to safe drinking water, promoting the sustainable use of water, improving water management programs, and planning for water needs now and in the future (Government of Yukon, 2014). In Northwest Territories, the territorial government introduced its updated 2018 water plan, *Northern Voices, Northern Waters: NWT Water Stewardship Strategy.* The plan is a guide to long-term stewardship of water resources in the territory for maintaining the quality, quantity, and rates of flow of territorial waters, ensuring residents have access to safe, clean, and plentiful drinking water at all times, and ensuring that aquatic ecosystems remain healthy and diverse (GNWT, 2018). Notwithstanding these policy commitments, communities in northern Canada continue to be challenged by high rates of water insecurity, which leaves residents at heightened risk of experiencing a multitude of adverse health outcomes (Sarkar, Hanrahan, & Hudson, 2015; Bressler and Hennessy, 2018).

Energy (in)security

Nearly all residents in northern Canada have access to electricity. However, the sources for energy vary, with much of the energy produced through non-renewable and inefficient technologies. Because of the long distances between populated areas in northern Canada, there is limited grid connectivity. Additionally, the northern territories generate a negligible amount of their own electricity, combining to produce less than 1% of the total electricity production in Canada (Government of Canada, 2019). While some areas have isolated power grids for electricity transmission, the remoteness of many communities in northern Canada makes it difficult, either due to high costs or the physical geography, to supply power through conventional distribution systems. Northern communities are therefore overly dependent on imported fuel, mainly diesel, to generate their power and electricity. Global shifts in energy prices are expected to further threaten the energy security of northern communities (Larsen & Fondahl, 2015). There is, however, considerable regional variability in energy generation. For example, in Yukon, 95% of the total electricity is derived from hydroelectric generation, although diesel and natural gas generation are required during periods when hydroelectric generation is insufficient to meet peak power demands. Yukon does have electricity transmission lines that connect the majority of the territory to its hydroelectric grid, although five communities remain off-grid and rely on diesel-fired generation exclusively.

In the Northwest Territories (NWT), roughly 75% of power is generated from hydroelectricity. The NWT has two regional hydro-based electricity grids, the Snare Grid north of Great Slave Lake and the Taltson Grid south of Great Slave Lake. In all, 26 of 33 NWT communities are able to receive electricity through transmission and distribution lines; however, remote communities and industries that are not connected to either grid rely primarily on diesel powered generators as their electricity source. Due to the low population density and relatively expensive generation costs, NWT residents pay among the highest electricity rates in the country, reaching approximately 30 cents per kilowatt hour (Government of Canada, 2019). The availability of hydroelectric power is also subject to annual and seasonal weather variabilities. For example, in 2017 which experienced limited precipitation, only 39% of the NWT's energy was derived from hydroelectric generation, with 57% generated from diesel, 2% from wind, 2% from natural gas, and 1% from solar power (Government of Canada, 2019). In response to energy insecurity, the Government of the NWT has introduced a draft 2030 Energy Strategy that calls for the installation of wind turbines and solar panels in smaller, off-grid communities to reduce their reliance on diesel power and to broaden connectivity to the existing hydroelectric power grids.

In Nunavut, nearly all of the territory's energy is produced from diesel-fueled power generation (Government of Canada, 2019). In the absence of electrical transmission grids, communities in Nunavut rely on standalone diesel generators for their energy needs. In total, approximately 55 million liters of diesel are consumed annually for power generation. Diesel fuel supplies are typically transported to communities during the summer and then stored for year-round use (Government of Canada, 2019). The Government of Nunavut has introduced solar technologies in some communities (for example, Iqaluit, Kugluktuk), and liquefied natural gas and biomass energies are being considered. Electrical transmission lines from Manitoba have also been proposed with construction potentially beginning as early as 2022.

Food (in)security

In the 2014 State of Knowledge of Food Security in Northern Canada assessment, a stark picture was presented on the high rate of food insecurity experienced by Canada's northern communities, particularly among Indigenous populations. While northern food insecurity is experienced differently depending on one's age, gender, and the community and region in which one lives (Natcher et al., 2016), the overall statistics for northern Canada are nonetheless alarming. For example, it is estimated that Inuit living in Nunavut have the highest food insecurity rate of any Indigenous population in a developed country (CCA, 2014). Among Inuit children, 90% experience conditions of hunger on a regular basis, 76% miss meals, and 60% often go an entire day without eating (Egeland, 2010). These conditions are contributing to delayed and declining physical, social, and emotional health among Indigenous youth (CCA, 2014). Households with children also report disproportionately high rates of food insecurity relative to households without children (CCA, 2014). For instance, the IPY Inuit Child Health Survey found that 70% of Inuit preschoolers in Nunavut lived in food insecure households, while 24% of children under five years old lived in homes reporting severe food insecurity (Egeland, 2009). Furthermore, according to the 2007–2008 Nunavut Inuit Child Health Survey, only one third of Inuit children had healthy body weights for their age or height and 72% of children had decayed or extracted teeth (Egeland, 2009). Altogether, the high rates of food insecurity experienced by northern communities have contributed to a general decline in the physical (Compher, 2006), nutritional (Kuhnlein, Receveur, Soueida, & Egeland, 2004) and emotional health (CCA, 2014) of northern residents.

The challenges associated with security in the northern food systems have been further exacerbated by extraordinarily high commercial foods costs in northern communities. For example, the Action Canada Foundation (2014) estimated that the purchase price of food items in Nunavut were 140% higher on average than the purchase price for the same food products in southern Canada. In the community of Old Crow, Yukon, residents pay an average of $496/week for a healthy food basket. This same food basket can be purchased for $206 in the Yukon's capital city of Whitehorse (Natcher et al., 2016). These cost differences can be attributed to the added expense of northern transport (estimated >20%) (Sorobey, 2013), higher electricity rates (roughly 84%) (CCA, 2014), and additional labor, storage, and building maintenance costs (Duhaime & Caron, 2013). When combined, these added costs result in northern residents paying as much as $13 for a head of cauliflower, $9 per kilogram (kg) for tomatoes, and $7/kg for carrots (Nunavut Food Price Survey, 2017).

To offset these costs, the federal government has, since 2011, provided subsidies to northern retailers who are then expected to pass those savings on to consumers. For example, in 2016 Nutrition North Canada (NNC) spent $64.8 million to subsidise the northern transport of 25.5 million kg of perishable goods. This included $21 million (32% of budget) to subsidise the shipment of 7.4 million kg of fruits and vegetables. However, a review of NNC found that the volumes and delivery times for food shipments remained highly variable, resulting in compromised food quality and reduced consumer acceptability. Despite the best intentions of NNC, high costs, coupled with poor retail quality, often removes fruits and vegetables from household food baskets; foods that are then replaced by non-perishable foods that lack equivalent nutritional value. These conditions have added to what many characterise as a public health crisis in northern Canada (CCA, 2014).

Data and methodology

Data

When assessing SDG indicator data, the United Nations (2018) suggests that data should be disaggregated as much as possible, for example by sex, age, race or ethnicity, and geographic location. To assess SDG target interactions in northern Canada, we drew on various public data sources made available by federal and territorial governments and various published resources on Arctic well-being and development. For example, much of the primary energy data was from the Government of Canada's National Energy Board (2019), which provided energy profiles describing the energy production, energy consumption, electricity use, and greenhouse gas (GHG) emissions in each territory. We examined these energy profiles for each territory separately to assess energy security throughout the Canadian Arctic. We also considered the GHG emissions related to energy production in order to evaluate the climate-related impacts of producing power in these regions. In addition to these data sources, territorial governments have developed their own strategic planning documents for WEF sectors. These include, for example, the Yukon's 5 Year Strategic Plan for Energy (2019–2024), the Yukon Water Strategy and Action Plan, and the Local Food Strategy for Yukon (2016–2021).

Additional indicator data were derived from various 'state of knowledge' reports, such as the Arctic Human Development Report (Larsen & Fondahl, 2015), the Survey of Living Conditions in the Arctic (SLiCA) (Poppel, 2015), Aboriginal Food Security in Northern Canada (CCA, 2014), and the Inuit Adult and Children's Health Surveys (Egeland, 2009; Egeland, 2010). These data sources may not align perfectly with U.N. recommended data sources, but they do provide a relatively accurate proxy representation of the indicators for northern Canada.

Methods

Various approaches have been used for analyzing and measuring trade-offs and synergies between WEF systems (Endo et al., 2015). For this research we adopted the approach developed by Fader et al. (2018) who, building on Nilsson et al. (2016), provides a step-wise methodology for calculating and ranking the degree of interaction between WEF-SDG targets. Whereas Fader et al. (2018) conducted their analysis on an international scale, we adapted their methodology for regional application in northern Canada.

In this approach, positive interactions between WEF-SDG targets occur when common infrastructure requirements are required to achieve each target and when the targets have a positive net impact on ecosystem services. Conversely, negative interactions occur when two targets require the same scarce resource inputs and if the target pair imposes a negative net impact on ecosystem services. The magnitude of the synergy or trade-off between any two targets is then represented by the sum of the positive and negative interactions between the two targets, where positive sums indicate synergies and negative sums indicate trade-offs between two targets. This methodology has been widely employed due to its transparency and for illustrating the inherent connections between WEF targets.

This methodology does have limitations, most notably in the method's subjectivity. Because each interaction score depends on the expert knowledge, considerations, and information available to those conducting the analysis, some interpretation is required. Thorough and careful consideration needs to be given to how each target could be met, as it could alter the resulting interaction scores. For example, target 2.1, ending hunger, could be met through conventional food production and transportation methods or through traditional food procurement methods such as hunting and fishing, with each method requiring different input and infrastructure needs that require consideration. This subjectivity could lead researchers to reach different conclusions based on their considerations of how each SDG target could or would be met. Notwithstanding this limitation, this methodology does make apparent the various interactions that occur between WEF-SDG targets in an accessible and transparent format.

In our analysis, we first evaluated the resource input needs, infrastructure requirements, and the risks and benefits toward ecosystem services associated with achieving each target. Three resource inputs were assessed for each target: (1) water, (2) land and soil, and (3) electricity and fuel. A negative interaction occurs between two targets if they both require the same input since they are considered to be in competition for that scarce resource. Therefore, a –1 is attributed to the total interaction score for each input that both targets require.

Similarly, three types of infrastructure requirements were evaluated for each SDG target: (1) health care and hospitals, (2) education, technology, and research, and (3) "gray infrastructure", which includes infrastructure such as streets, pipes, rails, airports, dams, energy production, sewage, and water treatment. Contrary to input needs, a positive interaction occurs between two targets when they require the same infrastructure type since it is assumed that the required infrastructure can be used or developed in a way that helps achieve both targets. Therefore a +1 is attributed to the total interaction score for each infrastructure type that both targets require.

Lastly, each SDG target was evaluated in terms of if the potential risks or benefits it posed toward provisioning and regulating ecosystem services. Supporting ecosystem services were included within regulating services for the purpose of this analysis. A value of –1 is assigned for each ecosystem service the target poses a risk to, +1 is assigned for each ecosystem service the target produces benefits toward, and 0 is assigned if the target has no impact on the ecosystem service. If the net benefits from the two groups of ecosystem services outweigh the risks, +1 is attributed to the total interaction score for that target pair. Conversely, –1 is attributed to the total interaction score if the net ecosystem service risks are greater than the benefits. If the risks and benefits to ecosystem services between two targets are equal, no score is attributed to the total interaction score.

Every pairwise combination of targets in SDGs 2, 6, and 7 were evaluated in this manner. The total interaction score (TIS) between two targets is the sum of the negative input interactions, positive infrastructure interactions, and the net effect on ecosystem services. Written in equation form, this

Table 14.2 Scale of possible total interaction scores between WEF-SDG targets

Interaction	Name	Explanation
-4	Cancelling	Makes it impossible to reach another goal.
-3	Restricting	Obstructs the achievement of another goal.
-2	Counteracting	Clashes with another goal.
-1	Constraining	Limits options on another goal.
0	Consistent	No net positive or negative interactions.
+1	Enabling	Creates conditions that further another goal.
+2	Reinforcing	Aids the achievement of another goal.
+3	Supporting	Strongly facilitates the achievement of another goal.
+4	Indivisible	Inextricable linked to the achievement of another goal.

appears as TIS = RI + INF + ES where RI is the resource input trade-offs impact, INF is the infrastructure synergies impact, and ES is the ecosystem services impact. RI can range in value from −3 to 0, INF can range in value from 0 to +3, and ES can take on a value of −1 if ecosystem services risks are greater than benefits, +1 if ecosystem services benefits are greater than risks, and 0 if the benefits and risks are equal or there are no risks or benefits. Therefore, the total interaction score for any pair of targets can range from −4 to +4, where the greater the absolute value of the total interaction score, the greater the magnitude or strength of the trade off or synergy is between the two targets (Table 14.2).

Results

SDG target interactions

Out of 210 interactions, only 12 were found to be negative, while 15 target pairs had no interacting effect (Table 14.3). Only three interaction scores were lower than −1, whereas 132 are greater than +1. Overall, roughly 87% of all interactions were found to be synergistic of some magnitude. This indicates that achieving or addressing one WEF target would have positive spillover effects on the others. Our findings are consistent with other WEF assessments, which have typically found that synergies between targets outweigh the trade-offs (Fuso Nerini et al., 2018; McCollum et al., 2018). For example, Fader et al.'s (2018) global WEF assessment determined 166 synergistic interactions to only 26 trade-offs, with water-related targets having the most synergistic potential.

One reason for the large number of synergies is the positive impacts on ecosystem services. By design, the SDG targets almost universally promote ecosystem services, and one point is added to the total interaction score of any two targets that result in a net positive environmental impact. This is especially important in northern Canada where WEF targets that benefit ecosystem services can assist in climate change mitigation. The need for

Table 14.3 Interaction scores between WEF-SDGs 2,6, 7 targets for Northern Canada

	2.1	2.2	2.3	2.4	2.5	2.a	2.b	2.c	6.1	6.2	6.3	6.4	6.5	6.6	6.a	6.b	7.1	7.2	7.3	7.a	7.b
2.1		-2	-2	0	2	2	-1	-1	1	0	2	2	2	0	2	2	-1	1	1	2	1
2.2	-2		-2	0	2	2	-1	-1	1	0	2	2	2	0	2	2	-1	1	1	2	1
2.3	-2	-2		0	2	2	-1	-1	1	0	2	2	2	0	2	2	-1	1	1	2	1
2.4	0	0	0		2	3	1	1	1	1	2	2	2	0	2	2	1	2	3	3	2
2.5	2	2	2	2		2	1	1	2	2	2	2	2	2	2	2	2	2	2	2	2
2.a	2	2	2	3	2		1	1	3	3	3	3	2	2	2	3	3	3	3	3	3
2.b	-1	-1	-1	1	1	1		0	1	1	1	1	1	1	1	1	0	1	0	1	1
2.c	-1	-1	-1	1	1	1	0		1	1	1	1	1	1	1	1	0	1	0	1	1
6.1	1	1	1	1	2	3	1	1		1	2	2	1	2	2	1	2	3	3	3	2
6.2	0	0	0	1	2	3	1	1	1		2	2	1	2	2	1	2	3	3	3	2
6.3	2	2	2	2	2	3	1	1	2	2		3	2	2	2	2	3	3	3	3	2
6.4	2	2	2	2	2	3	1	1	2	2	3		2	1	2	2	2	3	3	3	3
6.5	2	2	2	2	2	2	1	1	2	2	2	2		2	2	2	2	2	2	2	2
6.6	0	0	0	0	2	2	1	1	1	2	1	2	2		2	2	1	1	2	2	2
6.a	2	2	2	2	2	2	1	1	2	2	2	2	2	2		2	2	2	2	2	2
6.b	2	2	2	2	2	2	1	1	2	2	2	2	2	2	2		2	2	2	2	2
7.1	-1	-1	-1	1	2	3	0	0	1	1	2	2	1	2	2	2		2	3	3	2
7.2	1	1	1	2	2	3	1	1	2	2	3	2	2	1	2	2	2		3	3	3
7.3	1	1	1	3	2	3	0	0	3	3	3	3	2	2	2	2	2	3		3	3
7.a	2	2	2	3	2	3	1	1	3	3	3	3	2	2	2	2	3	3	3		3
7.b	1	1	1	2	2	3	1	1	2	2	2	3	2	2	2	2	3	3	3	3	

0 = Consistent	
1 = Enabling	-1 = Constraining
2 = Reinforcing	-2 = Counteracting
3 = Supporting	-3 = Restricting
4 = Indivisible	-4 = Cancelling

research, technology, education, and improved infrastructure in northern Canada also contributes to the present synergies. For example, building roads, water lines, or clean power sources in or between communities can improve access to food, water, and clean energy that could help achieve multiple targets simultaneously.

All negative interactions involve at least one target from SDG 2 (zero hunger). This is primarily due to the significant amount of resources needed to achieve targets in SDG 2 such as ending hunger (2.1), ending malnutrition (2.2), and increasing agricultural productivity (2.3). These targets all require intensive resource inputs such as land, water, energy, and fuel in order to be met, and since there is a limited amount of these inputs available, trade-offs occur between the targets that need and consume these scarce resources. For example, in order to eliminate hunger, a strategy may be to increase agricultural production. However, increased agricultural output may require a greater use of inputs (for example, synthetic fertiliser), which necessitates increased energy consumption and may pose adverse effects on water quality and quantity, both resulting in trade-offs.

To illustrate the target pair assessment procedure in practice, consider the interaction between the target pair of 2.2, to end malnutrition, and 2.3, to double agriculture productivity and incomes of small-scale food producers, as an example. Ending malnutrition in northern Canada will require all three input groups as water, land and soil, and electricity and fuel will be required whether the food is produced in northern Canada or elsewhere and then transported to the region. Similarly, all three input groups are required to double agriculture productivity and income as they are all necessary for food production by conventional agriculture or by harvesting wildlife from natural habitats and fisheries. Therefore, the two targets are in competition for all three resources, resulting in a –3 contribution to the total interaction score.

In terms of infrastructure needs, all three infrastructure groups are also deemed necessary to end malnutrition. Health care would be needed to assess the prevalence of malnutrition among northern residents and possibly provide nutritional supplements to address and mitigate malnutrition. Technology and research would be required to increase food production while streets, rails, airports, or other types of gray infrastructure would be required to transport food and health care to northern regions with limited access. Increasing food production and distribution would also require education, technology, and research as well as gray infrastructure to produce, transport, and sell commercial food products. However, target 2.3 does not require health care or hospital infrastructure to be achieved. Therefore, there are two infrastructure synergies between targets 2.2 and 2.3, resulting in a +2 contribution to the total interaction score.

We then examine and assess the impact each target could have on ecosystem services. Ending malnutrition could pose benefits and risks to provisioning services as there would be benefits from greater food production, but achieving this could also reduce the availability of other resources or raw materials such as water and forested land. Achieving target 2.3 could pose similar risks and benefits to provisioning services as ending malnutrition. Therefore, the presence of both risks and benefits for each target results in a zero-sum impact. For regulating services, both targets could pose risks as the actions necessary to achieve each target could negatively impact climate regulation, water quality, or wildlife habitats for example. While efforts could be made to minimise the negative impacts on regulating services, any action taken to achieve either target was deemed unlikely to provide benefits toward regulating services. Therefore, the risks toward regulating services outweighed the benefits for the two targets, resulting in a –1 score.

We then calculate the total interaction score for targets 2.2 and 2.3 as the sum of the results from the input trade-offs, infrastructure synergies, and net ecosystem services impact. Both targets required all three inputs, resulting in a –3 score, while the targets shared two infrastructure requirements resulting in a +2 score, and the net ecosystem service impact was deemed to be negative resulting in a score of –1. Therefore, the total interaction score for targets 2.2 and 2.3 was –2, indicating that the targets are counteracting and are in competition with one another (Table 14.4).

For comparison, we can also consider a case where a positive interaction score occurs using the target pair of 6.3, to improve water quality through a variety of methods, and 7.2, to substantially increase the share of renewable

Table 14.4 Example of a negative interaction score

Target	Inputs: -1 if competing; 0 if no interaction; +1 if complimentary			Infrastructure: -1 if competing; 0 if no interaction; +1 if complimentary			Provisioning Ecosystem Services		Regulating Ecosystem Services	
	Water	Land\Soil	Elec.\Fuel	Health. & Hospitals	Ed. & Research	Gray Infrastructure	Risk	Benefit	Risk	Benefit
2.2	-1	-1	-1	1	1	1	-1	1	-1	0
2.3	-1	-1	-1	0	1	1	-1	1	-1	0
Interaction	-1	-1	-1	0	1	1	0		-1	
Total	-3			2			-1			TIS -2

energy. The resource inputs deemed necessary for target 6.3 were electricity and fuel, which could be used to reduce the amount of untreated wastewater, a component of the target, in order to improve water quality. Target 7.2 could require water as a renewable energy source, while the other input groups were deemed unnecessary as any land, soil, electricity, or fuel requirements would be negligible. Therefore, the two targets do not share any input needs and no negative score is attributed to the total interaction. In terms of infrastructure, health care and hospitals would not be required for either target. Education, research, and technology could help reduce pollutants, treat wastewater, and increase recycling to improve water quality while also helping develop and establish renewable energy sources. Similarly, gray infrastructure such as pipes, sewage, and water treatment facilities could improve water quality while energy production infrastructure could aid the development of renewable energy sources. Therefore a +2 is attributed to the total interaction score for these two shared infrastructure needs. Lastly, neither target poses risks toward any ecosystem service group but they do provide benefits. Improving water quality would benefit both provisioning and regulating services, while implementing renewable energy sources would reduce pollution produced from non-renewable energy sources in the region and therefore benefit regulating services. Therefore, achieving targets 6.3 and 7.2 would have a net positive effect on ecosystem services, resulting in a +1 score attributed to the total interaction score. The sum of the interaction scores is then +3 between targets 6.3 and 7.2, indicating that these targets are supporting and each target strongly facilitate the achievement of the other (Table 14.5).

Table 14.5 Example of a positive interaction score

Target	Inputs: -1 if competing; 0 if no interaction; +1 if complimentary			Infrastructure: -1 if competing; 0 if no interaction; +1 if complimentary			Provisioning Ecosystem Services		Regulating Ecosystem Services	
	Water	Land\Soil	Elec.\Fuel	Health. & Hospitals	Ed. & Research	Gray Infrastructure	Risk	Benefit	Risk	Benefit
6.3	0	0	-1	0	1	1	0	1	0	1
7.2	-1	0	0	0	1	1	0	0	0	1
Interaction	0	0	0	0	1	1	1		1	
Total	0			2			1			TIS 3

Conclusion

This research was conducted as a pilot to demonstrate the potential use-fulness of WEF nexus research in Arctic regions. Although WEF nexus research has been conducted in countries and regions throughout the world, no such studies have been conducted in the Arctic. A focus on northern Canada was particularly warranted given the high rates of WEF insecurity that are being experienced by northern communities, particularly among the Indigenous population. These insecurities are reflected in limited access to clean water, an over-dependence on non-renewable energy sources, and having the highest rates of Indigenous food insecurity among all industrial-ised nations. As unacceptable as these current conditions are, these insecu-rities will likely be compounded by the effects of climate change.

Owing to the high rates of WEF insecurity in northern Canada, our methodology was motivated by the need for integrative thinking that makes visible the interconnectedness of WEF systems. Historically, WEF systems have been treated independently with little policy or institutional coordina-tion occurring between sectors (Nilsson, Griggs, & Visbeck, 2016; Rasul, 2016). The goal of this research was to highlight their inherent connections in order to support decision-makers in identifying sustainable solutions to WEF related challenges. In doing so, we found that the synergies between WEF-SDGs far outweigh the potential trade-offs. In total, 87% of all inter-actions were found to be synergistic of some magnitude. This indicates that achieving or addressing one WEF target would have positive spillover effects for the others.

This assessment ultimately illustrates that interactions and connections exist between almost all WEF targets. Policy and decision makers should consider how each target interacts with others when addressing WEF secu-rity in order to take advantage of positive interactions and minimise neg-ative outcomes. Having synergies significantly outweigh trade-offs signals an opportunity to simultaneously address multiple WEF-SDG targets in northern Canada through mutually beneficial actions that capitalise on and promote synergetic policies. This information can now be used to inform integrated planning efforts that are cognizant of respective resource requirements for achieving WEF security in northern Canada.

As informative as our findings may be, we encourage future WEF analy-ses to be conducted at regional and sub-regional scales. Regional differences exist in population, geography, economy, and access to new technologies. Conducted at finer scales, WEF nexus assessments could promote more nimble policy responses than are not so easily achieved at the national level. Regional WEF assessments may also prove more effective at incorporat-ing the social and cultural values of residents as assessment criteria. Where cultural values are known, the methodology can be expanded to include an evaluation of the potential impacts on cultural ecosystem services, Indigenous livelihoods, and the territorial rights and interests of Indigenous peoples. Such considerations are consistent with the Arctic Council's framework for sustainable development, which calls for social equity, protecting and promoting cultures, and strengthening the capacity of Indigenous peoples (SDWG, 2017). The inclusion of Indigenous participation is also reflected in the Canadian government's commitment to respecting the rights of Indigenous peoples in Canada's national SDG strategy (BCCIC, 2019). Until those regional assessments are conducted, we are hopeful that this research can offer a pathway for untangling the inherent complexities of WEF systems in northern Canada.

References

Action Canada Foundation. (2014). Hunger in Nunavut: local food for healthier communities. *Action Canada.* http://www.actioncanada.ca/wp-content/uploads/2014/04/TF-3-Hunger-in-Nunavut-EN.pdf.

Bakaic, M., Medeiros, A. S., Peters, J. F., & Wolfe, B. B. (2018). Hydrologic monitoring tools for freshwater municipal planning in the Arctic: the case of Iqaluit, Nunavut, Canada. *Environmental Science and Pollution Research, 25,* 32913–32925. doi: https://doi.org/10.1007/s11356-017-9343-4

Bhaduri, A., Ringler, C., Dombrowski, I., Mohtar, R., & Scheumann, W. (2015). Sustainability in the water-energy-food nexus. *Water International: Sustainability in the water-energy-food nexus, 40*(5-6), 723–732.

Biggs, E. M., Bruce, E., Boruff, B., Duncan, J. M., Horsley, J., Pauli, N., & Mcneill, K. et al. (2015). Sustainable development and the water-energy-food nexus: A perspective on livelihoods. *Environmental Science and Policy, 54,* 389–397.

Bressler, J. M., & Hennessy, T. W. (2018). Results of an Arctic Council survey on water and sanitation services in the Arctic. *International Journal of Circumpolar Health, 77*(1). doi: https://doi.org10.1080/22423982.2017.1421368.

British Columbia Council for International Cooperation (BCCIC). (2019). Sustainable development goals: developing national SDG implementation strategies. *BCCIC.* Available at: https://www.bccic.ca/bccic-policy-brief-developing-national-sdg-implementation-strategies/.

Compher, C. (2006). The nutrition transition in American Indians. *Journal of Transcultural Nursing, 17*(3), 217–223.

Costanza, R., Alperovitz, G., Daly, H. E., Farley, J., Franco, C., Jackson, T. … Victor, P. (2015). Ecological economics and sustainable development: Building a sustainable and desirable economy-in-society-in-nature. *Routledge International Handbook of Sustainable Development,* 281–294.

Council of Canadian Academies. (2014). *Aboriginal Food Security in Northern Canada: An Assessment of the State of Knowledge,* Ottawa, ON. The Expert Panel on the State of Knowledge of Food Security in Northern Canada, Council of Canadian Academies. https://cca-reports.ca/wp-content/uploads/2018/10/foodsecurity_fullreporten.pdf

De Schutter, O. (2012). *Report of the special rapporteur on the right to food* (pp. 1–21). United Nations General Assembly.

Duhaime, G., & Caron, A. (2013). Consumer price monitoring in Nunavik, 2011-2013. *Canada Research Chair on Comparative Aboriginal People, University of Laval.* https://www.nunivaat.org/doc/publication/Suivi-des-prix-2011-2013-ang.pdf

Egeland, G. (with Qanuippitali Steering Committee Members). (2009). The International Polar Year Nunavut Inuit Child Health Survey 2007-2008. *Centre for Indigenous Peoples' Nutrition and Environment (CINE), McGill University.* https://www.mcgill.ca/cine/files/cine/child_inuit_health_survey_aug_31.pdf.

Egeland, G. (With Nunavut Steering Committee, CINE staff members, and graduate students). (2010). Inuit Health Survey 2007–2008: Nunavut. *Centre for Indigenous Peoples' Nutrition and Environment (CINE), McGill University.* https://www.mcgill.ca/cine/files/cine/adult_report_nunavut.pdf.

Endo, A., Burnett, K., Orencio, P. M., Kumazawa, T., Wada, C. A., Ishii, A. ... Taniguchi, M. (2015). Methods of the water-energy-food nexus. *Water, 7*(10), 5806–5830. https://doi.org/10.3390/w7105806

Endo, A., Tsurita, I., Burnett, K., & Orencio, P. M. (2017). A review of the current state of research on the water, energy, and food nexus. *Journal of Hydrology, 11,* 20–30. https://doi.org/10.1016/j.ejrh.2015.11.010

Environment and Climate Change Canada. (2013). Water–how we use it. *Government of Canada.* http://www.ec.gc.ca/eau-water/default.asp?lang=en&n=0bbd794b-1#a8.

Fader, M., Cranmer, C., Lawford, R., & Engel-Cox, J. (2018). Toward an understanding of synergies and trade-offs between water, energy, and food SDG targets. *Frontiers in Environmental Science, 6.*

Fuso Nerini, F., Tomei, J., To, L. S., Bisaga, I., Parikh, P., Black, M., & Borrion, A. et al. (2018). Mapping synergies and trade-offs between energy and the sustainable development goals. *Nature Energy, 3*(1), 10–15. 10.1038/s41560-017-0036-5.

Government of Canada: National Energy Board. (2019). Provincial and Territorial Energy Profiles. *Government of Canada.* https://www.neb-one.gc.ca/nrg/ntgrtd/mrkt/nrgsstmprfls/index-eng.html.

Government of Yukon. (2014). Water for nature, water for people: Yukon water strategy and action plan. *Government of Yukon.* http://www.env.gov.yk.ca/publications-maps/documents/Yukon_Water_Strategy_Action_Plan.pdf.

Kuhnlein, H. V., Receveur, O., Soueida, R., & Egeland, G. M. (2004). Arctic indigenous peoples experience the nutrition transition with changing dietary patterns and obesity. *The Journal of Nutrition, 134*(6), 1447–1453.

Kulat, M. I., Mohtar, R. H., & Olivera, F. (2019). Holistic water-energy-food nexus for guiding water resources planning: Matagorda county, Texas case. *Frontiers in Environmental Science, 7,* 3.

Larsen, J. N., & Fondahl, G. (Eds.). (2015). Arctic Human Development Report: Regional Processes and Global Linkages. *Nordisk Ministerråd.* http://norden.diva-portal.org/smash/get/diva2:788965/FULLTEXT03.pdf.

Lim, M. M. L., Søgaard Jørgensen, P., & Wyborn, C. A. (2018). Reframing the sustainable development goals to achieve sustainable development in the anthropocene—a systems approach. *Ecology and Society, 23*(3), 22. doi: https://doi.org/10.5751/ES-10182-230322.

Liu, Q. (2016). Interlinking climate change with water-energy-food nexus and related ecosystem processes in California case studies. *Ecological Processes, 5*(1), 1–14.

Mainali, B., Luukkanen, J., Silveira, S., & Kaivo-Oja, J. (2018). Evaluating synergies and trade-offs among sustainable development goals (SDGs): Explorative analyses of development paths in. *South Asia and Sub-Saharan Africa. Sustainability, 10*(3), 815.

McCollum, D. L., Gomez Echeverri, L., Busch, S., Pachauri, S., Parkinson, S., Rogelj, J. ... Stevance, A. S. (2018). Connecting the sustainable development goals by their energy inter-linkages. *Environmental Research Letters, 13*(3), 033006.

Natcher, D., Shirley, S., Rodon, T., & Southcott, C. (2016). Constraints to wildlife harvesting among aboriginal communities in Alaska and Northern Canada. *Food Security, 8*(6), 1153–1167.

Nilsson, M., Griggs, D., & Visbeck, M. (2016). Map the interactions between sustainable development goals. *Nature, 534*(7607), 320–322.

Nilsson, M., & Costanza, R. (2015). Overall framework for the sustainable development goals. *Review of Targets for the Sustainable Development Goals: the Science Perspective*, 7–12.

Nunavut Food Price Survey. (2017). Price comparisons per kilogram and litre. *Government of Nunavut.* https://www.gov.nu.ca/sites/default/files/2017_nunavut_food_price_survey_-_price_comparisons_per_kilogram_and_litre_report_1.pdf.

Pirkle, C., Lucas, M., Dallaire, R., Ayotte, P., Jacobson, J., Jacobson, S. ... Muckle, G. (2014). Food insecurity and nutritional biomarkers in relation to stature in Inuit children from Nunavik. *Canadian Journal of Public Health, 105*(4), e233–e238. doi: https://doi.org/10.17269/cjph.105.4520.

Pittock, J., Orr, S., Stevens, L., Aheeyar, M., & Smith, M. (2015). Tackling trade-offs in the nexus of water, energy and food. *Aquatic Procedia, 5*, 58–68.

Poppel, B. (Eds.). (2015). *SLiCA: Arctic living conditions: living conditions and quality of life among Inuit, Saami, and indigenous peoples of Chukota and the Kola Peninsula.* Nordisk Ministerråd. http://norden.diva-portal.org/smash/get/diva2:790312/FULLTEXT02.pdf.

Pradhan, P., Costa, L., Rybski, D., Lucht, W., & Kropp, J. P. (2017). A systematic study of sustainable development goal (SDG) interactions. *Earth's Future, 5*(11), 1169–1179.

Rasul, G. (2016). Managing the food, water, and energy nexus for achieving the sustainable development goals in South Asia. *Environmental Development, 18*, 14–25.

Sarkar, A., Hanrahan, M., & Hudson, A. (2015). Water insecurity in Canadian Indigenous communities: Some inconvenient truths. *Rural and Remote Health, 15*, 3354. www.rrh.org.au/journal/article/3354

Sorobey, M. (2013). *Northwest Company* [Paper Presentation]. Northern Exposure 2 Conference: Realities of Remote Logistics, Winnipeg, Manitoba, Canada.

Sustainable Development Working Group (SDWG). (2017). Strategic Framework 2017. *SDWG.* https://www.sdwg.org/wp-content/uploads/2017/04/SDWG-Framework-2017-Final-Print-version.pdf.

United Nations. (2018). Global indicator framework for the Sustainable Development Goals and targets of the 2030 Agenda for Sustainable Development. *UN Statistics.* https://unstats.un.org/sdgs/indicators/Global%20Indicator%20Framework%20after%20refinement_Eng.pdf

Index

Note: *Italic* page numbers refer to figures

Printed in the United States
by Baker & Taylor Publisher Services